Professional Patisserie

for Levels 2, 3 and professional chefs

Mick Burke, Chris Barker, Neil Rippington

HODDER
EDUCATION
AN HACHETTE UK COMPANY

Orders: please contact Hachette UK Distribution, Hely Hutchinson Centre, Milton Road, Didcot, Oxfordshire, OX11 7HH. Telephone: +44 (0)1235 827827. Email education@hachette.co.uk Lines are open from 9 a.m. to 5 p.m., Monday to Friday. You can also order through our website: www.hoddereducation.co.uk

If you have any comments to make about this, or any of our other titles, please send them to educationenquiries@hodder.co.uk

British Library Cataloguing in Publication Data

A catalogue record for this title is available from the British Library

ISBN: 978 1 4441 9644 3

This edition published 2013.

Impression number   10 9

Year   2022

Typeset by DC Graphic Design Limited, Swanley Village, Kent.

Printed and bound by CPI Group (UK) Ltd, Croydon, CR0 4YY

Contents

Introduction

Welcome to *Professional Patisserie for Levels 2 and 3 and Professional Chefs.* This book covers the skills and knowledge you'll need when completing a range of patisserie and confectionery courses, taking you from the basics at Level 2, through to more advanced and specialist skills at Level 3, and beyond into your career as a professional patisserie chef.

Professional Patisserie will be invaluable to those working towards the Level 2 Certificate in Professional Patisserie and Confectionery and Level 3 Diploma in Patisserie and Confectionery,

or the Level 3 NVQ Diploma in Professional Cookery (Patisserie and Confectionery). The book is also relevant to other non-specialist courses with patisserie and confectionery elements, including units of the Level 2 NVQ Diploma in Professional Cookery, the Level 3 NVQ Diploma in Professional Cookery and the Level 2 and Level 3 Diplomas in Professional Cookery (VRQ).

The following table outlines where you can find content that is relevant to the units you are studying. You'll also find a helpful summary of units covered at the start of each chapter.

Chapter	1	2	3	4	5	6	7	8	9	10
Level 2 units										
Investigate the catering and hospitality industry	✓									
Healthier foods and special diets	✓									
Catering operations, costs and menu planning	✓									
Applying workplace skills	✓									
Prepare, cook and finish basic pastry products			✓	✓						
Produce paste products			✓	✓						
Prepare, cook and finish basic bread and dough products					✓					
Produce fermented dough products					✓					
Prepare, cook and finish basic hot and cold desserts						✓	✓			
Produce hot and cold desserts and puddings						✓	✓			
Prepare, cook and finish basic cakes, sponges and scones								✓		
Produce biscuit, cake and sponge products								✓		
Level 3 units										
Supervisory skills		✓								
Practical gastronomy		✓								
Maintain health, hygiene, safety and security of the working environment		✓								
Develop productive working relationships with colleagues		✓								
Employment rights and responsibilities in the hospitality, leisure, travel and tourism sector		✓								
Contribute to the control of resources		✓								
Contribute to the development of recipes and menus		✓								
Contribute to the control of resources		✓								
Prepare, cook and finish complex pastry products			✓	✓						
Produce paste products			✓	✓						
Prepare, cook and finish complex bread and dough products					✓					
Produce fermented dough and batter products					✓					
Prepare, cook and finish complex hot desserts						✓				
Produce hot, cold and frozen desserts						✓	✓			
Prepare, cook and finish complex cold desserts							✓			

Prepare, cook and finish complex cakes, sponges, biscuits and scones				✓		
Prepare sauces, fillings and coatings for complex desserts	✓					
Produce biscuits, cakes and sponges				✓		
Produce petits fours					✓	
Prepare, process and finish complex chocolate products					✓	
Produce display pieces and decorative items						✓
Prepare, process and finish marzipan, pastillage and sugar products						✓

How to use this book

Chapters 1 and 2 provide an overview of the key underpinning knowledge required for the theoretical units of your course (Chapter 1 for Level 2 and Chapter 2 for Level 3 courses).

Chapters 3 to 10 guide you through each key area of patisserie and confectionery, from basic preparations and fillings to complex decorative pieces and display items. Each chapter contains the following useful features:

- Carefully selected recipes covering all aspects of patisserie and confectionery. A reference table at the start of each chapter shows you whether the recipe is for Level 2, Level 3 or relevant to both levels.
- Clear photos of almost every finished product, so you can see what you should be aiming to achieve.
- Step-by-step sequences guide you through each stage of key skills and techniques.
- Test yourself questions at the end of each chapter help you to prepare for assessment, and are differentiated for Level 2 and Level 3.

Using the QR codes

There are free videos online to accompany this book. Look out for the QR codes throughout the book. They look like this.

To use the QR codes to view the videos you will need a QR code reader for your smartphone/ tablet. There are many free readers available, depending on the smartphone/tablet you are using. We have supplied some suggestions below, but this is not an exhaustive list and you should only download software compatible with your device and operating system. We do not endorse any of the third-party products listed below and downloading them is at your own risk.

- for iPhone/iPad, Qrafter – http://itunes.apple. com/app/qrafter-qr-code-reader-generator/ id416098700
- for Android, QR Droid – https://market.android. com/details?id=la.droid.qr&hl=en
- for Blackberry, QR Scanner Pro – http://appworld.blackberry.com/webstore/ content/13962
- for Windows/Symbian, Upcode – http://www.upc.fi/en/upcode/download/

Once you have downloaded a QR code reader, simply open the reader app and use it to take a photo of the code. The video will then load on your smartphone/tablet.

If you cannot read the QR code or you are using a computer, the web link next to the code will take you directly to the same video.

The terms and conditions which govern these free online resources may be seen at http://bit.ly/yfVC0P

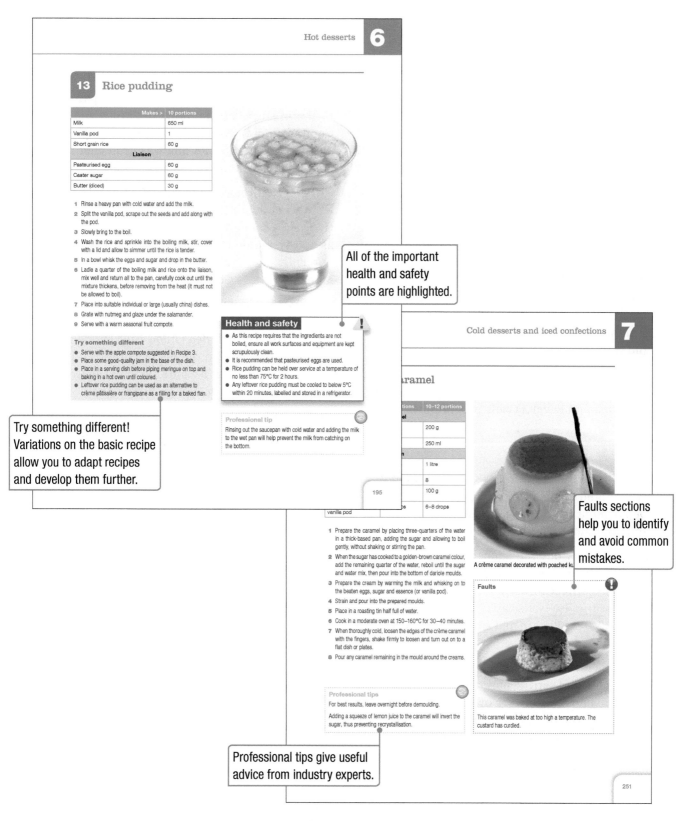

Hot desserts **6**

13 Rice pudding

Makes >	10 portions
Milk	650 ml
Vanilla pod	1
Short grain rice	60 g
Liaison	
Pasteurised egg	60 g
Caster sugar	60 g
Butter (diced)	30 g

1 Rinse a heavy pan with cold water and add the milk.
2 Split the vanilla pod, scrape out the seeds and add along with the pod.
3 Slowly bring to the boil.
4 Wash the rice and sprinkle into the boiling milk, stir, cover with a lid and allow to simmer until the rice is tender.
5 In a bowl whisk the eggs and sugar and drop in the butter.
6 Ladle a quarter of the boiling milk and rice onto the liaison, mix well and return all to the pan, carefully cook out until the mixture thickens, before removing from the heat (it must not be allowed to boil).
7 Place into suitable individual or large (usually china) dishes.
8 Grate with nutmeg and glaze under the salamander.
9 Serve with a warm seasonal fruit compote.

Try something different
- Serve with the apple compote suggested in Recipe 3.
- Place some good-quality jam in the base of the dish.
- Place in a serving dish before piping meringue on top and baking in a hot oven until coloured.
- Leftover rice pudding can be used as an alternative to crème pâtissière or frangipane as a filling for a baked flan.

Health and safety
- As this recipe requires that the ingredients are not boiled, ensure all work surfaces and equipment are kept scrupulously clean.
- It is recommended that pasteurised eggs are used.
- Rice pudding can be held over service at a temperature of no less than 75°C for 2 hours.
- Any leftover rice pudding must be cooled to below 5°C within 20 minutes, labelled and stored in a refrigerator.

Professional tip
Rinsing out the saucepan with cold water and adding the milk to the wet pan will help prevent the milk from catching on the bottom.

195

Cold desserts and iced confections **7**

...aramel

...tions	10–12 portions
...el	
	200 g
	250 ml
...n	
	1 litre
	8
	100 g
...es	6–8 drops
vanilla pod	

1 Prepare the caramel by placing three-quarters of the water in a thick-based pan, adding the sugar and allowing to boil gently, without shaking or stirring the pan.
2 When the sugar has cooked to a golden-brown caramel colour, add the remaining quarter of the water, reboil until the sugar and water mix, then pour into the bottom of dariole moulds.
3 Prepare the cream by warming the milk and whisking on to the beaten eggs, sugar and essence (or vanilla pod).
4 Strain and pour into the prepared moulds.
5 Place in a roasting tin half full of water.
6 Cook in a moderate oven at 150–160°C for 30–40 minutes.
7 When thoroughly cold, loosen the edges of the crème caramel with the fingers, shake firmly to loosen and turn out on to a flat dish or plates.
8 Pour any caramel remaining in the mould around the creams.

A crème caramel decorated with poached ku...

Faults

This caramel was baked at too high a temperature. The custard has curdled.

Professional tips
For best results, leave overnight before demoulding.

Adding a squeeze of lemon juice to the caramel will invert the sugar, thus preventing recrystallisation.

251

All of the important health and safety points are highlighted.

Try something different! Variations on the basic recipe allow you to adapt recipes and develop them further.

Faults sections help you to identify and avoid common mistakes.

Professional tips give useful advice from industry experts.

We hope that *Professional Patisserie* will provide you with a valuable foundation in all of the skills and knowledge you need to begin your career as a successful pastry chef.

About the authors

Chris Barker

Chris Barker completed his apprenticeship with Trust House Forte before opting to specialise in patisserie. He moved to London in 1976 to gain experience, first working at The Intercontinental, Hyde Park Corner under Michael Nadell, then in 1979 moving to the Ritz as Chef Patissier.

While teaching Patisserie at Colchester Institute, Chris was invited to become a City & Guilds examiner for the 706/3 Advanced Pastry, a position he held for 12 years until the series was phased out in 1996. He has been awarded the Palmes Culinaires Lacam (1986) and the Maitrise Lacam (1997) by the Conseil Culinaire de Grande Bretagne in recognition of services to education and training.

Chris is currently a Curriculum Manager with responsibility for a range of courses including all part-time and Level 3 Chefs programmes.

Mick Burke

Mick Burke trained in the UK, France and Switzerland and is currently Deputy Head of Department in Catering, Hospitality, Aviation and Tourism at The Sheffield City College.

Mick was awarded the Education Award by the Craft Guild of Chefs and the Palmes Lacam by the Association Culinaire Francaise for services to patisserie. He is also a Fellow of The Master Chefs of Great Britain and serves on the executive committee.

Mick has worked with different awarding bodies developing and setting national occupational standards for the delivery of patisserie in colleges across the UK. He has also won national competitions for patisserie work and given demonstrations on numerous occasions at chefs' conferences.

Neil Rippington

Neil Rippington's career as a chef has spanned Michelin-starred restaurants in France and the Capital Hotel in Knightsbridge, London. He has also worked in the USA and was head chef in a country house hotel in the New Forest.

In 1994, Neil returned to education as a chef lecturer at South East Essex College. After five years, he moved on to Redbridge College as a Programme Manager for Hospitality and Catering, later taking up the post as Quality Manager for the College.

In 2003, Neil joined Colchester Institute as Head of Centre for Hospitality and Food Studies, where the Centre was awarded Grade 1 for outstanding provision, the first in the history of the college. In August 2010, Neil became Dean of the College of Food at University College Birmingham.

Neil has a Bachelor of Arts Degree in Education and Training and a Master's Degree in Culinary Arts.

Acknowledgements

We are most grateful to Keylink Limited and KeyChoc Limited for their support in the development of this book, including the provision of ingredients and equipment for the photographs.

We would also like to thank Vicky Anderson, Development Chef at MSK Ingredients.

Chris Barker: I would like to thank Stephanie Conway and Paula Summerell for their help and advice, and not least my wife Alex for her patience over many weekends while I locked myself away working on the book.

Photography

Most of the photos in this book are by Andrew Callaghan of Callaghan Studios. The photography work could not have been done without the generous help of the authors, their colleagues and students at The Sheffield City College, in particular:

- Gregg Rodgers, who created the marzipan model
- Sally Griffiths
- Hussein Kamkar-Loghmani
- Josie Marstem
- Emma Turton.

The authors and publishers are grateful to everyone involved for their hard work.

Picture credits

Every effort has been made to trace the copyright holders of material reproduced here. The authors and publishers would like to thank the following for permission to reproduce copyright illustrations.

p.15 Crown copyright; pp.34, 37 ©Bananastock/ Photolibrary Group Ltd/Getty Images; p.40 ©chris32m – Fotolia; pp.126, 220, 221, 315, 403 (bottom left), 405, 406 (right) ©Sam Bailey/ Hodder Education.

Except where stated above, photographs are by Andrew Callaghan and illustrations by DC Graphic Design Limited. Crown copyright material is licensed under the Open Government Licence v1.0.

Keylink

The one-stop shop for everyone working with chocolate

The Chocolate Specialists

Looking for the perfect grade of chocolate or that hard-to-find ingredient?

Keylink has been serving the specialised needs of chocolatiers and pastry chefs for over 40 years and is the UK's only 'one-stop-shop' for everyone working with chocolate. Keylink's vast product range includes chocolate, ingredients, décor, packaging, equipment and moulds – everything a professional could want for their craft.

Supporting Chocolate Excellence

Created in 2011, the prestigious Keylink Junior Chocolatier of the Year competition is the only competition in the UK aimed exclusively at full-time students. Students from colleges throughout the UK, mentored by their tutors, train hard to complete four tasks in four and a half hours; these include hand dipped/rolled truffles, moulded chocolate pralines, a chocolate centrepiece and a plated dessert.

If you're interested in competing, you can take a look at the KJCY website where you can view a full gallery of pictures showing the competitors hard at work as well as their finished creations.

Supporting Learning

If you're looking for information or guidance on working with chocolate, the Knowledge Bank on Keylink's website is a really great resource. You'll find useful articles on many specialist products as well as a series of fourteen modular product and technique videos produced by Keylink with the help of Mick Burke and Len Unwin from Sheffield College.

These training guides are very easy to follow and will give you some really useful and practical tips. Best of all, they're free to access, so take full advantage of them.

The Chocolate Machinery Specialists

Choosing the right machinery for your business is never a decision to make lightly!

KeyChoc is the only specialist manufacturer in the UK of machinery for chocolate work with a range of standard products covering all the items most commonly used by professionals for chocolate work. But if you ever find you need something a little special, they regularly also design and manufacture bespoke machinery to suit the specific needs of individual customers.

- Melting and holding tanks
- Moulding machines
- Vibrating tables
- Automatic tempering machines
- Continuous tempering machines
- Cooling tunnels.

CHOCOLATE
ACADEMY
Keychoc is the official machinery partner of the Barry Callebaut UK Academy

With both a showroom and demonstration area, KeyChoc's engineers are always on hand to help you decide what is right for your business; and as the Official Machinery Partner of the Barry Callebaut (UK) Academy, you can be confident that you're working with a tried and tested local partner.

KeyChoc Services

Repairs & Maintenance – with a fully equipped workshop based in St. Helens, Merseyside, KeyChoc will help you with whatever you need, including preventative maintenance, emergency repairs, reconditioning and moving premises.

Machinery Hire – if you only have a short term need, such as for a big one-off order, hiring a machine can be the most cost-effective solution and a great option.

Marketplace – if you want to buy or sell a used machine, just visit the KeyChoc Marketplace, a free website which puts you in touch with the people you need.

KeyChoc
Market
Place

KeyChoc Limited, Unit 2a Delphwood Drive, Sherdley Road Industrial Estate, St. Helens, Merseyside, WA9 5JE UK
Tel: +44 (0)1744 730086   Fax: +44 (0)1744 731226
Email: sales@keychoc.com
www.keychoc.com

1 Underpinning knowledge for level 2

This chapter covers:
→ **VRQ level 2 Investigate the catering and hospitality industry**
→ **VRQ level 2 Healthier foods and special diets**
→ **VRQ level 2 Catering operations, costs and menu planning**
→ **VRQ level 2 Applying workplace skills**

This chapter provides an overview of the curriculum content required for study towards completion of the theoretical units to complete the Diploma in Professional Patisserie and Confectionery at level 2, with the exception of the units in Food Safety and Health and Safety, which are offered across a range of qualifications and are often delivered as stand-alone, one-day programmes. The unit in Food Safety, for example, is standardised as a stand-alone qualification across all providers offering it, certifying the unit independently. The Royal Society for Public Health (www.rsph.org.uk) and the Chartered Institute for Environmental Health (www.cieh.org) are two leading examples of organisations offering this qualification in this way.

Furthermore, legislation in these two particular areas is subject to review on a frequent basis and therefore the most reliable sources of information are directly available from the Food Standards Agency (www.food.gov.uk) and the Health and Safety Executive (www.hse.gov.uk).

In order to enable a broader coverage of the practical skills and the products associated with the practical components of the level 2 diploma, the intention here is to provide a generic coverage of the key aspects of the theoretical content for students.

Investigate the catering and hospitality industry

Types of establishment in the hospitality industry

Hotels

Hotels provide accommodation in private bedrooms. Many offer other services such as restaurants, bars and room service, reception, porters and housekeepers. What a hotel offers will depend on the type of hotel it is and its star rating.

Hotels are rated from five-star down to one-star. A luxury hotel will have five stars while a more basic hotel will have one star. There are many international hotel chains, such as the Radisson group, Mandarin Oriental, Intercontinental and Dorchester Collection in the five-star hotel market. There are also budget hotels, guesthouses and bed and breakfast accommodation (see below).

In the UK there are more hotel bedrooms in the mid-market three-star hotels than in any other category. Most hotels in this market are independent and privately owned, in other words they are not part of a chain of hotels.

Some hotels have speciality restaurants, run by a high-profile or 'celebrity' chef, for example, or specialising in a particular area, such as steaks, sushi or seafood. To attract as many guests as possible, many hotels now offer even more services. These may include office and IT services (such as internet access, fax machines, and a quiet area to work in), gym and sports facilities, swimming pool, spa, therapy treatments, hair and beauty treatments, and so on.

Country house hotels

Country house hotels are mostly in attractive old buildings, such as stately homes or manor houses, in tourist and rural areas. They normally have a reputation for good food and wine and a high standard of service. They may also offer the additional services mentioned above.

Budget hotels

Budget hotels like motels and Travelodges are built near motorways, railway stations and airports. They are aimed at business people and tourists who need somewhere inexpensive to stay overnight. The rooms are reasonably priced and have tea- and coffee-making facilities. No other food or drink is included in the price. Staff members are kept to a minimum and there is often no restaurant. However, there may be shops, cafés, restaurants close by, which are often run by the same company as the motel. The growth and success of the budget hotel sector has been one of the biggest changes in the hospitality industry in recent years.

There are guesthouses and bed-and-breakfast establishments all over the UK. They are small, privately owned businesses. The owners usually live on the premises and let bedrooms to paying customers. Many have guests who return regularly, especially if they are in a popular tourist area.

Some guesthouses offer lunch and an evening meal as well as breakfast.

Clubs and casinos

People pay to become members of private clubs. Private clubs are usually run by managers (sometimes known as club secretaries) who are appointed by club members. What most members want from a club in the UK, particularly in the fashionable areas of London, is good food and drink and informal service.

Most nightclubs and casinos are open to the public rather than to members only. As well as selling drinks to their customers, many now also provide food services, such as restaurants.

Restaurants

The restaurant sector has become the largest in the UK hospitality industry. It includes exclusive restaurants and fine-dining establishments, as well as a wide variety of mainstream restaurants, fast-food outlets, coffee shops and cafés.

Many restaurants specialise in regional or ethnic food styles, such as Asian and Oriental, Mexican and Caribbean, as well as a wide range of European-style restaurants. New restaurants and cooking styles are appearing and becoming more popular all the time.

There are grading systems for restaurants, such as AA rosettes and Michelin stars. Michelin inspectors grade restaurants every year; even one star is a great achievement and the maximum of three stars is much rarer and indicates a restaurant at the very top of the ladder.

Speciality restaurants

Moderately priced speciality restaurants continue to increase in popularity. In order for them to succeed, it is essential that they understand what customers want and plan a menu that will attract enough customers to make a good profit. A successful caterer is one who gives customers what they want; they will be aware of changing trends and adapt to them. The most successful catering establishments are those that maintain the required level of sales over long periods and throughout the year.

Fast food

Many customers now want the option of popular foods at a reasonable price, with little or no waiting time. Fast-food establishments offer a limited menu that can be consumed on the premises or taken away. Menu items are quick to cook and have often been partly or fully prepared beforehand at a central production point.

Drive-ins (or drive-throughs)

The concept of drive-ins came from America, and there are now many of them across the UK. The most well known are the 'drive-thrus' at McDonald's fast-food restaurants. Customers stay in their vehicles and drive up to a microphone where they place their order. As the car moves forward in a queue, the order is prepared and is ready for them to pick up at the service window when they get there.

Delicatessens and salad bars

These offer a wide selection of salads and sandwich fillings to go in a variety of bread and rolls at a 'made-to-order' sandwich counter. The choice of breads might include panini, focaccia, pitta, baguette and tortilla wraps. Fresh salads, homemade soups, chilled foods and a hot 'chef's dish of the day' may also be available, along with ever-popular baked jacket potatoes with a good variety of fillings.

With such a wide variety of choices these establishments can stay busy all day long, often serving breakfast as well.

Chain catering organisations

There are many branded restaurant chains, coffee shops, and shops with in-store restaurants. Many of these chains are spread widely throughout the UK and, in some cases, overseas. These are usually well-known companies that advertise widely. They often serve morning coffee, lunches and teas, or may be in the style of snack bars and cafeterias.

Coffee shops

The branded coffee shop has been a particularly fast-growing area, providing a wide variety of good-quality coffee and other drinks, along with a limited selection of food items. They provide for either a fast 'takeaway' or more leisurely café style consumption.

Licensed-house (pub) catering

Almost all of the tens of thousands of licensed public houses (pubs) in the UK offer food of some sort or another. The type of food they serve is ideal for many people. It is usually quite simple, inexpensive and quickly served in a comfortable atmosphere. In recent years there have been a number of pub closures due to the availability of cheaper alcohol in supermarkets, and the smoking ban, which has had a major effect on business. For this reason, many pubs have moved into selling food, revisiting their product offer (what they have to offer the customer) and the total pub experience for their customers in order to stay in business, for example adding restaurants, offering more bar food and putting on live entertainment.

There is a great variety of food available in pubs, from those that serve simple sandwiches and rolls to those that have exclusive à la carte restaurants. Pub catering can be divided into five categories:

1 luxury-type restaurants (some have earned Michelin stars)
2 gastro-pubs that have well-qualified chefs who develop the menu according to their own specialities, making good use of local produce
3 speciality restaurants like steak bars, fish restaurants, carveries and theme restaurants
4 bar-food pubs where dishes are served from the bar counter and the food is eaten in the normal drinking areas rather than in a separate restaurant
4 bar-food pubs that just serve simple items such as rolls and sandwiches.

The provision of hospitality in the leisure industry

Museums

In order to diversify and to extend their everyday activities, some museums now provide hospitality services. For example, many museums have one or more cafés and restaurants for visitors. Some run events such as lunchtime lectures, family events and children's discovery days where food is provided as part of the event. Museums can even be used as an interesting venue for private events and banqueting during the hours they are closed to the general public. Sometimes outside caterers are employed for the occasion, but many museums employ their own catering team to provide a wide range of food.

Theme parks

Theme parks are now extremely popular venues for a family day out or even a full holiday. The larger theme parks include several different eating options ranging from fast food to fine dining. Some include branded restaurants (such as McDonald's and Burger King), which the visitor will already know. Theme parks are also used for corporate hospitality, in other words, they are used by companies for conferences or other events. Several have conference and banqueting suites for this purpose, and larger theme parks may even have their own hotels.

Historical buildings

Numerous historical buildings and places of interest have food outlets such as cafés and restaurants. Many in the UK specialise in light lunches and afternoon tea for the general public. Some are also used as venues to host large private or corporate events.

Visitor attractions

Places like Hampton Court, Kew Gardens and Poole Pottery can be categorised as visitor attractions. They will usually have refreshment outlets serving a variety of food and drinks. Some, like Kew Gardens, are also used to stage large theatrical events or concerts in the summer.

Event management

Event management is when a person or company plans and organises events, such as parties, dinners and conferences, for other people or companies. This will include such tasks as hiring the venue, organising the staff, the food and drink, music, entertainment and any other requests the host may have.

Further examples which demonstrate the provision of hospitality in the leisure industry include:
- health clubs and spas
- farms
- youth hostels
- holiday parks
- catering at sports events.

The provision of hospitality in the travel industry

Catering at sea

There is a variety of types of sea-going vessels on which catering is required for both passengers and crew. Catering for passengers on ferries and cruise liners is becoming increasingly important in today's competitive markets.

Sea ferries

There are several ferry ports in the UK. Ferries leave from these every day, making a variety of sea crossings to Ireland and mainland Europe.

As well as carrying passengers, many ferries also carry the passengers' cars and freight lorries. In addition to competing against each other, ferry companies also compete against airlines and, in the case of English Channel crossings, Eurostar and Le Shuttle. In order to win customers, they have invested in (spent money on) improving their passenger services, with most ferries having several shops, bars, cafés and lounges on board. Some also have very good restaurant and leisure facilities, fast-food restaurants and branded food outlets. These are often run by contract caterers (sometimes known as 'contract food service providers' – see page 7) on behalf of the ferry operator. More recently, well-known chefs have become involved in providing top-quality restaurants on popular ferry routes.

Cruise liners

Cruise ships are floating luxury hotels, and more and more people are becoming interested in cruising as a lifestyle. The food provision on a large cruise liner is of a similar standard to the food provided in a five-star hotel and can be described as excellent quality, banquet-style cuisine. Many shipping companies are known for the excellence of their cuisine.

As cruising becomes more popular, cruise companies are acquiring more, larger cruise liners. This means that there are excellent hospitality career opportunities. Many companies provide good training and promotion prospects, and the opportunity to travel all over the world. The caterers produce food and serve customers to a very high standard in extremely hygienic conditions. All of these things mean that working on cruise liners can be an interesting and rewarding career. As an example of working conditions, staff may work for three months and then have, say, two months off. On-board hours of work can be long, perhaps 10 hours a day for 7 days a week, but this appeals to many people who want to produce good food in excellent conditions and a have chance to travel the world.

On cruises where the quality of the food is of paramount importance, other factors such as the dining room's ambience of refinement and elegance are also of great significance. Ship designers generally want to avoid Las Vegas-type glittery dining rooms, but also those that are too austere. Ships must also be designed with easy access to the galley (ship's kitchen) so waiters are able to get food quickly and with as little traffic as possible.

Airline services

Airline catering is a specialist service. The catering companies are usually located at or near airports in this country and around the world. The meals provided vary from snacks and basic meals to luxury meals for first-class passengers. Menus are chosen carefully to make sure that the food can safely be chilled and then reheated on board the aircraft.

The price of some airline tickets includes a meal served at your seat. The budget airlines usually have an at-seat trolley service from which passengers can buy snacks and drinks.

Airports also offer a range of hospitality services catering for millions of people every year. They operate 24 hours a day, 365 days a year. Services include a wide variety of shops along with bars, themed restaurants, speciality restaurants, coffee bars and food courts.

All of these outlets need to have the ability to respond rapidly to fluctuations in demand caused by delayed and cancelled flights and high-volume periods such as bank holidays.

Roadside services

Roadside motoring services (often referred to as 'service stations') provide a variety of services for motorists. These include fuel, car washing and maintenance facilities and a variety of shops. Many are becoming more sophisticated, with baby changing, infant and pet feeding facilities, bathrooms and showers, a variety of branded food and drinks outlets (such as Burger King, Costa Coffee and M&S Simply Food) and often accommodation. The catering usually consists of food courts offering travellers a wide range of meals 24 hours a day, 7 days a week. MOTO is an example of a company that provides these sorts of services nationwide.

Rail travel

Snacks can be bought in the buffet car on a train, and some train operators also offer a trolley service so passengers can buy snacks without leaving their seats. Main meals are often served in a restaurant car. However, there is not much space in a restaurant car kitchen, and with a lot of movement of the train, it can be quite difficult to provide anything other than simple meals.

Two train services run by separate companies run through the Channel Tunnel. One is Euro Tunnel's Le Shuttle train, which transports drivers and their vehicles between Folkestone and Calais in 35 minutes. Passengers have to buy any food and drink for their journey before they board the train. The other company, Eurostar, operates between London St Pancras and Paris or Brussels. This carries passengers only.

Eurostar is in direct competition with the airlines, so it provides catering to airline standards for first- and premier-class passengers. Meals are served by uniformed stewards and stewardesses in a similar service to an airline's club class. This food is included in the ticket price. Economy travellers usually buy their food separately from buffet cars or trolley services. This is another area where catering is often provided by contract food service providers.

Public-sector catering (secondary service sector)

Public-sector organisations that need catering services include hospitals, universities, colleges, schools, prisons, the armed forces, police and ambulance services, local authorities and many more.

The aim of catering in hotels, restaurants and other areas of the leisure and travel industry (known as the private sector) is to make a profit. The aim of public sector catering is to keep costs down by working efficiently. However, these days the business of catering for public-sector organisations is tendered for. This means that different companies (like Compass and Sodexo) will compete to win a contract to provide the catering for these organisations.

Many public-sector catering tenders have been won by contract caterers (contract food service providers), which have introduced new ideas and more commercialism (promoting business for profit) into the public sector. Because much of the public sector is now operated by profit-making contractors, it is sometimes referred to as the secondary service sector.

For a variety of reasons, the types of menu in the public sector may be different from those in the private sector. For example, school children, hospital patients and soldiers have particular nutritional needs (they may need more energy from their food, or more of certain vitamins and minerals), so their menus should match their needs. Menus may also reflect the need to keep costs down. However, the standards of cooking in the public sector should be just as good as they are in the private sector.

Prisons

Catering in prisons may be carried out by contract caterers or by the Prison Service itself. The food is usually prepared by prison officers and inmates. The kitchens are also used to train inmates in food production. They can gain a recognised qualification to encourage them to find work when they are released.

Prisons used to have something called Crown immunity, which meant that they could not be prosecuted (taken to court) for poor hygiene and negligence in the kitchen. However, they no longer have this and they must operate to the same standards as other kitchens in the public sector. In addition to catering facilities for the inmates, there are also staff catering facilities for all the personnel (staff) who work in a prison, such as administrative staff and prison officers.

The armed forces

Catering in the armed forces includes providing meals for staff in barracks, in the field and on ships. Catering for the armed forces is specialised, especially when they are in the field, and they have their own well-established cookery training programmes. However, like every other part of the public sector, the forces need to keep costs down and increase efficiency. Consequently, they

also have competitive tendering for their catering services. The Ministry of Defence contracts food service providers to cater for many of their service operations.

The National Health Service

The scale of catering services in the NHS is enormous. Over 300 million meals are served each year in approximately 1,200 hospitals. NHS Trusts must ensure that they get the best value for money within their catering budget. Hospital caterers need to provide well-cooked, nutritious, appetising meals for hospital patients and must maintain strict hygiene standards.

As well as providing nutritious meals for patients in hospital (many of whom need special diets), provision must also be made for out-patient (people who come into hospital for treatment and leave again the same day) visitors and staff. This service may be provided by the hospital catering team, but is sometimes allocated to commercial food outlets, or there may be a combination of in-house hospital catering and commercial catering.

The education sector school meals service

School meals play an important part in the lives of many children, often providing them with their only hot meal of the day. In 1944 the Education Act stated that all schools should provide a meal to any child who wanted one. The meal had to meet strict nutritional and price guidelines. The school meals service continued to work in this way until the 1980 Education Act. This stated that schools no longer had to provide school meals for everyone, and also removed the minimum nutritional standards and the fixed charge.

In April 2001, for the first time in over 20 years, minimum nutritional standards were re-introduced by the government. These are designed to bring all schools up to a measurable standard set down in law. Since this date, Local Education Authorities have been responsible for seeing that the minimum nutritional standards for school lunches are met. Schools also have to provide a paid-for meal, where parents request one, except where children are under five years old and only go to school part-time. This does not affect the LEA's or the school's duty to provide a free meal to those children who qualify for one.

In 2006 the government announced new standards for school food. These were phased in by September 2009. Together they cover all food sold or served in schools: breakfast, lunch and after-school meals; tuck shops, vending machines, mid-morning break snacks and anything sold or served at after-school clubs. The standards are based on both foods and nutrients.

Catering for business and industry

The provision of staff dining rooms and restaurants in industrial and business settings has provided employment for many catering workers outside traditional hotel and restaurant catering. Working conditions in these settings are often very good. Apart from the main task of providing meals, these services may also include retail shops, franchise outlets (see page 8) and vending machines. It will also include catering for meetings and conferences as well as for larger special functions.

In some cases a 24-hour, 7-days-a-week service is necessary, but usually the hours are more streamlined than in other areas of the hospitality industry. Food and drink is provided for all employees, often in high-quality restaurants and dining rooms. The catering departments in these organisations are keen to keep and develop their staff, so there is good potential for training and career development in this sector.

Many industries have realised that satisfied employees work more efficiently and produce better work, so have spent a great deal of money on providing first-class kitchens and dining-rooms. In some cases companies will subsidise (pay part of) the cost of the meals so that employees can buy the food at a price lower than it costs.

The contract food service sector

The contract food service sector, which has already been mentioned in relation to other sectors of the hospitality industry, consists of companies that provide catering services for other organisations. This sector has developed significantly over recent years.

Contract food service management provides food for a wide variety of people, such as those at work in business and industry, those in schools, colleges and universities, private and public healthcare establishments, public and local authorities, and other non-profit making outlets such as the armed forces, police or ambulance services. It also includes more commercial areas, such as corporate hospitality events and the executive dining rooms of many corporations, special events, sporting fixtures and places of entertainment, and outlets such as leisure centres, galleries, museums, department stores and specific retail stores, supermarket restaurants and cafés, airports and railway stations. Some contractors also provide other support services such as housekeeping and maintenance, reception, security, laundry, bars and retail shops.

Outside catering

When events are held at venues where there is no catering available, or where the level of catering required is more than the normal caterers can manage, then a catering company may take over the management of the event. This type of function will include garden parties, agricultural and horticultural shows, the opening of new buildings, banquets, parties in private houses, military pageants and tattoos, sporting fixtures such as horse racing, motor racing, football or rugby, and so on.

There is a wide variety in this sort of outside catering work, but the standards can be very high and people employed in this area need to be adaptable and creative. Sometimes specialist equipment will be required, especially for outdoor jobs, and employees need to be flexible as the work often involves travel to remote locations and outdoor venues.

Corporate hospitality

Corporate hospitality is hospitality provided by businesses, usually for its clients or potential clients. The purpose of corporate hospitality is to build business relationships and to raise awareness of the company. Corporate entertaining is also used as a way to thank or reward loyal customers. Companies these days understand the importance of marketing through building relationships with clients and through the company's reputation. They are willing to spend large amounts of money to do this well.

Franchising

A franchise is an agreement where a person or group of people pay a fee and some set-up costs to use an established name or brand which is well known and is therefore likely to attract more customers than an unknown or start-up brand.

An example of this is where a contract caterer, for example, Compass Group, buys a franchise in the Burger King brand from Burger King's owner. It pays a fee and a proportion of the turnover (the amount of money it makes). The franchisor (the branded-company franchise provider) will normally lay down strict guidelines or 'brand standards' that the franchise user has to meet. In this example these will affect things like which ingredients and raw materials are used and where they come from, as well as portion sizes and the general product packaging and service. The franchisor will check on the brand standards regularly to ensure that the brand reputation is not being put at risk. The franchisor will normally also provide advertising and marketing support, accounting services, help with staff training and development and designs for merchandising and display materials.

The structure of the catering and hospitality industry

Wherever there are groups of people there is likely to be some kind of hospitality provision – in other words, somewhere people can get food, drink and accommodation. The hospitality industry in the UK employs more than 2 million people and is growing all the time. It provides excellent opportunities for training, employment and progression. Even in the current time of economic recession, the hospitality industry is continuing to grow and it is predicted to grow even more towards 2020.

In addition to the employment opportunities within the UK, skilled and qualified staff are in demand throughout the world, presenting a huge range of opportunities for employment. Such opportunities are rarely so readily available in other industries.

Hotels, restaurants, bars, pubs and clubs are all part of what is known as the commercial sector. Businesses in the commercial sector need to make a profit so that they can survive. Catering provided in places like hospitals, schools, colleges, prisons and the armed services also provides thousands of meals each day (see page 5).

The service sector usually refers to the contract catering market, and is now sometimes referred to as the 'primary service sector'. This includes hospitality in banks, insurance companies, law firms and large corporate businesses.

The three main types of business

The industry can be divided into three main types of business: SMEs, public limited companies and private companies.

Small to medium-sized business enterprises (SMEs)

These have up to 250 employees. In the UK as a whole, SMEs account for over half of all employment (58.7 per cent). These are usually private companies that may become public limited companies if they become very large.

Public limited companies and private companies

The key difference between public and private companies is that a public company can sell its shares to the public, while private companies cannot. A share is a certificate representing one unit of ownership in a company, so the more shares a person has, the more of the company they own.

Before it can start in business or borrow money, a public company must prove to Companies House (the department where all companies in the UK must be registered) that at least £50,000 worth of shares have been issued and that each share has been paid up to at least a quarter of its nominal value (so 25 per cent of £50,000). It will then receive authorisation to start business and borrow money.

Other types of business

The types of business in the catering and hospitality industry can also be divided into sole traders, self-employed, partnership and limited liability companies. These are usually private companies.

Sole trader

A sole trader is the simplest form of setting up and running a business. It is suited to the smallest of businesses. The sole trader owns the business, takes all the risks, is liable for any losses and keeps any profits. The advantage of operating in business as a sole trader is that very little formality is needed. The only official records required are those for HM Revenue and Customs (HMRC), National Insurance and VAT. The accounts are not available to the public.

Self-employed

There is no precise definition of self-employment, although guidance is offered by HMRC.

In order to determine whether an individual is truly self-employed, the whole circumstances of his or her work need to be considered. This may include whether he or she:

- is in control of their own time, the amount of work they take on and the decision making
- has no guarantee of regular work
- receives no pay for periods of holiday or sickness
- is responsible for all the risks of the business
- attends the premises of the person giving him or her the work
- generally uses her or his own equipment and materials
- has the right to send someone else to do the work.

> **Note**
>
> HM Revenue and Customs (HMRC) is the government department responsible for collecting tax.

Partnership

A partnership consists of two or more people working together as the proprietors of a business. Unlike limited liability companies (see below), there are no legal requirements in setting up as a partnership. A partnership can be set up without the partners necessarily being fully aware that they have done so.

The partnership is similar to a sole trader in law, in that the partners own the business, take all the risks, stand any losses and keep any profits. Each partner individually is responsible for all the debts of the partnership. So, if the business fails, each partner's personal assets are fully at risk. It is possible, though not very common, to have partners with limited liability. In this case the partner with limited liability must not play any active part in the management or conduct of the business. In effect, he or she has merely invested a limited sum of money in the partnership.

> **Note**
>
> Personal assets are the possessions or belongings of value that an individual owns and which may not be related to the business (e.g. personal car, house, clothes, personal bank account, stocks and shares).

The advantages of operating a business as a partnership can be very similar to those of the sole trader. Very little formality is needed, although everyone contemplating entering into a partnership should seriously consider taking legal advice and having a partnership agreement drawn up.

The main official records that are required are records for the Inland Revenue, National Insurance and VAT. The accounts are not available to the public. There may be important tax advantages, too, when compared with a limited company. For example, they might be able pay the tax they owe at a later date, or treat deductible expenses more generously.

Limited liability companies

These are companies that are incorporated under the Companies Acts. This means that the liability of their owners (the amount they will have to pay to cover the business's debts if it fails, or if it is sued) is limited to the value of the shares each shareholder (owner) owns.

Limited liability companies are much more complex than sole traders and partnerships. This is because the owners can limit their liability. As a consequence it is vital that people either investing in them or doing business with them need to know the financial standing of the company. Company documents are open to inspection by the public.

The size of the UK hospitality industry

Despite its complexity, catering represents one of the largest sectors of the UK economy and is fifth in size behind retail food, cars, insurance and clothing. It is also an essential support to tourism, another major part of the economy, and one of the largest employers in the country.

The leisure sector, including hospitality, travel and tourism, is large and employs more than 2 million people. Around 7 per cent of all jobs in the UK are in the leisure sector, that is, the sector accounts for about one in every 14 UK jobs, and it is predicted to grow even more.

There are approximately 142,000 hospitality, leisure, travel and tourism businesses in the UK that operate from about 192,100 outlets. The restaurant industry is the largest within the sector (both in terms of the number of outlets and size of the workforce), followed by pubs, bars and nightclubs and the hotel industry. The UK leisure industry (tourism and hospitality) is estimated by the Government to have been worth £85 billion in 2009. Including all business expenditure, the total market is worth much more than this. Expenditure on accommodation is difficult to pin down, but it is estimated that it contributes £10.3 billion (excluding categories such as camp sites and youth hostels) to the total leisure market.

Industry size

- 2.07 million people employed
- £90 billion annual turnover, worth £46 billion to the UK economy
- £7.5 billion on accommodation

Employment in the hospitality industry

Staffing and organisation structure

Hospitality companies need to have a structure for their staff in order for the business to run efficiently and effectively. Different members of staff have different jobs and roles to perform as part of the team so that the business is successful.

In smaller organisations, some employees have to become multi-skilled so that they can carry out a variety of duties. Some managers may have to take on a supervisory role at certain times.

A hospitality team will consist of operational staff, supervisory staff, management staff and, in large organisations, senior management. These roles are explained below.

Operational staff

These are usually practical, hands-on staff. These will include the chefs de partie (section chefs), commis chefs, waiters, apprentices, reception staff and accommodation staff.

Table 1.1 Number of businesses (enterprises) by industry

Industry	Number of businesses	%
Restaurants	63,600	45
Pubs, bars and nightclubs	49,150	35
Hotels	10,050	7
Travel and tourist services	6,750	5
Food and service management (contract catering)	6,350	4
Holiday centres and self-catering accommodation	3,650	3
Gambling	1,850	1
Visitor attractions	450	*
Youth and backpacker hostels	150	*
Hospitality, leisure, travel and tourism total	142,050	100

Supervisory staff

Generally the supervisors work with the operational staff, supervising the work they do. In some establishments, the supervisors will be the managers for some of the operational staff. A sous chef will have supervisory responsibilities, and a chef de partie will have both operational and supervisory responsibilities.

Management staff

Managers have the responsibility of making sure that the operation runs smoothly and within the budget. They are accountable to the owners to make sure that the products and services on offer are what the customer expects and wants and provide value for money. Managers may also be responsible for planning future business.

They will be required to make sure that all the health and safety policies are in place and that health and safety legislation is followed. In smaller establishments they may also act as the human resources manager, employing new staff and dealing with staffing issues.

A hotel will normally have a manager, assistant managers, accommodation manager, restaurant manager and reception manager. So in each section of the hotel there could be a manager with departmental responsibilities. A head chef is a manager, managing kitchen operations, planning purchasing and managing the employees in his/her area.

Employment rights and responsibilities

For those employed in the hospitality industry it is important to understand that there is a considerable amount of legislation that regulates both the industry itself and employment in the industry. Employers who contravene (break) the law or attempt to undermine the statutory (legal) rights of their workers – for example, paying less than the national minimum wage or denying them their right to paid annual holidays – are not only liable to prosecution and fines but could be ordered by tribunals and courts to pay substantial amounts of compensation.

Employers must provide the employee with:
- a detailed job description
- a contract of employment with details of the job itself, working hours, the annual holiday the employee will have and the notice period.

An essential feature of a contract of employment is the 'mutuality of obligation'. This means that the employer will provide the employee with work on specified days of the week for specified hours and, if employed under a limited-term contract, for an agreed number of weeks or months. In return, the employee agrees to carry out the work for an agreed wage or salary. Employers must follow the relevant laws, such as employment law, health and safety law, and food safety law.

Employees must work in the way that has been agreed to in the contract and job description and follow all the organisation's policies and practices.

In the hospitality industry, particularly in front-of-house positions, it is quite common for uniform to be provided as this will help to provide a corporate and professional image. Organisations are also obliged to provide personal protective equipment (PPE) if this is required to undertake a particular task. Some employers will also provide uniforms or kit for staff working in other areas, such as chefs in the kitchen. A laundry service can also be offered to clean uniforms and kit.

Workers and employees

An employee is a person employed directly by a company under a contract of employment or service. A worker or contractor is someone who works for another company (a sub-contractor) that has won a contract to carry out work or provide services (that is, they are not actually an employee of the company itself).

Workers are protected by:
- health and safety legislation
- Working Time Regulations 1998
- Equality Act 2010
- Public Interest Disclosure Act 1998
- National Minimum Wage Act 1998
- Part-time Workers (Prevention of Less Favourable Treatment) Regulations.

Recruitment and selection

When advertising for new staff it is important to be aware of the following legislation:
- Children and Young Persons Act 1933
- Licensing Act 1964
- Rehabilitation of Offenders Act 1974
- Data Protection Act 1988

- Asylum and Immigration Act 1996
- National Minimum Wage Act 1998
- Working Time Regulations 1998
- Equality Act 2010
- Human Rights Act 1998.

Some positions may also be restricted by age restrictions. Selling or serving alcohol, for example, would require the employee to be over 18 years old unless each transaction was authorised by a licence holder. Furthermore, to become a licence holder, in order to apply, the applicant must be aged 18 years or over, and (in almost all cases) hold a licensing qualification (for example, a BII Level II examination certificate or a similar accredited qualification such as the EDI NCPLH level 2 qualification).

Job advertisements

It is unlawful to discriminate against job applicants on grounds of:
- sex, marital status or gender
- colour, race, nationality, or national or ethnic origins
- disability
- sexual orientation
- religion or beliefs
- trades union membership or non-membership.

The following words and phrases should be avoided in a job advertisement:
- young/youthful
- pleasing appearance
- strong personality
- energetic
- articulate
- dynamic
- no family commitments.

These could be construed (understood), or misconstrued, as indicating an intention to discriminate on grounds of sex, race or disability.

Use of job titles with a gender connotation (e.g. 'waiter', 'barmaid', 'manageress') will also be taken to indicate an intention to discriminate on the grounds of a person's sex, unless the advertisement contains an indication or an illustration to the contrary.

Job applications

Job application forms must be designed with care. If 'sensitive personal information' is needed, such as a health record or disability disclosure, the reason for this should be explained, and the candidate reassured that the data will remain confidential and will be used and stored in keeping with the provisions of the Data Protection Act 1998.

Human Rights Act

Candidates must be informed at application stage, and at interview, if they have to wear uniforms or protective clothing on duty. Any surveillance monitoring the company is likely to carry out must also be disclosed to applicants.

Asylum and Immigration Act

It is an offence under the Asylum and Immigration Act 1996 to employ a foreign national subject to immigration control (that is, one who needs a visa or work permit, for example) who does not have the right to enter or remain in the UK, or to take up employment while in the UK. Job application forms should caution future employees that they will be required, if shortlisted, to produce documents confirming their right to be in, and to take up employment in, the UK.

Job interviews

The purpose of the job interview is to assess the suitability of a particular applicant for the vacancy. The interviewer should ask questions designed to test the applicant's suitability for the job, covering qualifications, training and experience, and to find out about the individual's personal qualities, character, development, motivation, strengths and weaknesses.

If a job applicant resigned or was dismissed from previous employment, the interviewer may need to know why. Any health problems, injuries and disabilities the candidate has disclosed may also need to be discussed in order to determine the applicant's suitability for employment – for example, in a high-risk working environment. Employers may lawfully ask an applicant if he or she has been convicted of any criminal offence, but must be aware of the right of applicants, under the

Rehabilitation of Offenders Act 1974, not to disclose details of any criminal convictions that have since become 'spent' (that is, were so long ago that they have been dealt with and no longer count).

The interviewer should not ask questions about sexuality or religion. However, questions on religion may be asked if, for example, aspects of the job may directly affect the beliefs of an individual – an example would be the handling of alcoholic drinks or meat, such as pork.

Job offers

An offer of employment should be made or confirmed in writing, and is often conditional on the receipt of satisfactory references from former employers. Withdrawing an offer of employment once it has been accepted could result in a civil action for damages by the prospective employee.

Employers may also require a health check before confirming an offer of employment.

Statutory Sick Pay

Employers in Great Britain are liable to pay up to 28 weeks' Statutory Sick Pay to any qualified employee who is unable to work because of illness of injury. Employers who operate their own occupational sick pay schemes may opt out of the Statutory Sick Pay scheme, as long as the payments available to their employees under such schemes are equal to or greater than payments they would be entitled to under Statutory Sick Pay, and so long as these employees are not required to contribute towards the cost of funding such a scheme. Payments made under Statutory Sick Pay may be offset against contractual sick pay, and vice versa.

Working Time Regulations

The Working Time Regulations apply not only to employees but also to every worker (part-time, temporary, seasonal or casual) who undertakes to do work or carry out a service for an employer. The 1998 Regulations are policed and enforced by employment tribunals (in relation to a worker's statutory rights to rest breaks, rest periods and paid annual holidays) and by local authority Environmental Health Officers.

Sources of information

Information regarding the hospitality industry and the opportunities available can be found from a wide variety of sources. Those requiring such information could range from a student seeking some part-time employment to an organisation researching trends with a view to expanding their business. This list below provides a range of sources that would be able to assist in providing such information.

Local and regional information
- publicity brochures
- tourist information centres
- libraries
- job centres
- staff recruitment agencies
- newspapers.

Professional associations and publications

Two well-known magazines for the industry are:
- *Caterer and Hotelkeeper*
- *Restaurant.*

Some examples of professional associations are:
- Academy of Culinary Arts
- Craft Guild of Chefs
- Master Chefs of Great Britain
- Association of Pastry Chefs
- PACE (Professional Association of Catering Education)
- Institute of Hospitality
- People 1st (Sector Skills Council for Hospitality)
- Springboard UK
- Savoy Educational Trust
- Association Culinaire Française
- Euro Toques
- World Toques.

These organisations are very useful for networking and sharing information and good practice. They promote the industry and may organise competitions and demonstrations of excellence. Membership is usually subscription-based, as the organisations are self-funding; some groups have different levels of membership.

1 Give an example of a five-star and a four-star hotel that you have heard of. What additional services do they offer?
2 Give an example of a country house hotel. Where is it located?
3 Why are budget hotels often built near motorways, railway stations and airports?
4 Choose a local restaurant which you are familiar with. List the types of food it has on offer and what special features the restaurant has.
5 Find out the name of three major contract catering companies.

Healthier foods and special diets

Government guidelines

Use the eatwell plate to get the nutritional balance right. It shows how much of what you eat should come from each food group. The eatwell plate is a concept produced by the Food Standards Agency, suggesting how much of each food should be consumed from each food group.

Additional information can also be found from the British Nutrition Foundation (BNF), the Department of Health and the Department for Environment, Food and Rural Affairs (DEFRA).

Other sources available to find the nutritional values of different foods include:
● manual of nutrition
● internet
● food labels
● promotional leaflets.

Essential nutrients and their sources

Carbohydrates

We need carbohydrates for energy. They are made by plants and then either used by the plants as energy or eaten by animals and humans for energy or as dietary fibre.

There are three main types of carbohydrate:
● sugar
● starch
● fibre.

Sugars

Sugars are the simplest form of carbohydrates. When carbohydrates are digested they turn into sugars.

Table 1.2 Summing up healthy eating

What it is about	What it is not about!	Immediate benefits!
Eating more fruit and vegetables	Cutting down on food	Better weight control
Supplementing nutrients with bread, pasta, rice or potatoes	Going hungry	Improved self-esteem
Eating a little less of some food items	Depriving yourself of treats	Looking and feeling better
Enjoying good food	Spending more money on food	Feeling fitter, with more energy
Making small, gradual changes	Not enjoying food	Enjoying a wide variety of foods
Knowing more about food	'Brown and boring food'	Not buying expensive 'diet' products
Altering food shopping patterns	Just salads	Knowing that changes made today will have long-term benefits
Feeling satisfied and good about food	Making major changes	
	Going on a 'special diet'	

The eatwell plate

Use the eatwell plate to help you get the balance right. It shows how much of what you eat should come from each food group.

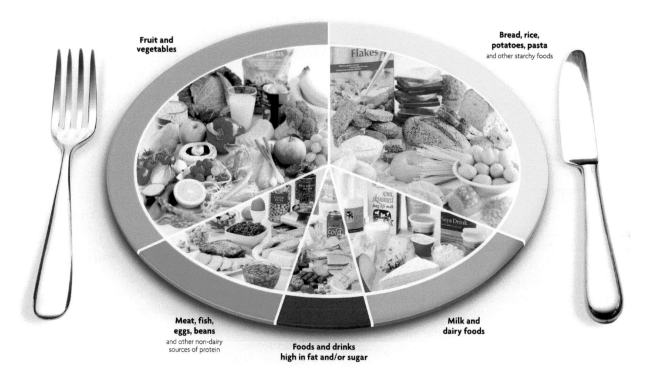

There are several types of sugar:

- glucose – found in the blood of animals and in fruit and honey
- fructose – found in fruit, honey and cane sugar
- sucrose – found in beet and cane sugar
- lactose – found in milk
- maltose – found in cereal grains and used in beer-making.

Starches

Starches break down into sugars. Starches are present in many foods, including:

- pasta, such as macaroni, spaghetti, vermicelli
- cereals, such as cornflakes, shredded wheat
- cakes, biscuits, bread (cooked starch)
- whole grains, such as rice, barley, tapioca
- powdered grains, such as flour, cornflour, ground rice, arrowroot
- vegetables, such as potatoes, parsnips, peas, beans
- unripe fruit, such as bananas, apples, cooking pears.

Fibre

Dietary fibre is a very important form of starch. Unlike other carbohydrates, dietary fibre cannot be digested and does not provide energy to the body.

However, dietary fibre is essential for a balanced diet because it:

- helps to remove waste and toxins from the body and maintain bowel action
- helps to control the digestion and processing of nutrients
- adds bulk to the diet, helping us to stop feeling hungry; it is used in many weight reduction foods.

Fibre is found in:

- fruits and vegetables
- wholemeal and granary bread
- wholegrain cereals
- wholemeal pasta
- wholegrain rice
- pulses (peas and beans) and lentils.

Protein

Protein is an essential part of all living things. Every day our bodies carry out millions of tasks (bodily functions) to stay alive. We need protein so that our bodies can grow and repair themselves.

There are two kinds of protein:
● animal protein
● vegetable protein.

The lifespan of the cells in our bodies varies from a week to a few months. As the cells die they need to be replaced. We need protein for our cells to repair and for new ones to grow. We also use protein for energy. Any protein that is not used up in repairing and growing cells is converted into carbohydrate or fat.

What is protein?

Protein is made up of chemicals known as amino acids. The protein in cheese is different from the protein in meat because the amino acids are different. Some amino acids are essential to the body, so they must be included in a balanced diet. Ideally our bodies need both animal and vegetable protein so that we get all the amino acids we need.
● Animal protein is found in meat, game, poultry, fish, eggs, milk, cheese.
● Vegetable protein is found in vegetable seeds, pulses, peas, beans, nuts, wheat.

Fats

Fats are naturally present in many foods and are an essential part of our diet. The main functions of fat are to protect the body, keep it warm and provide energy. Fats form an insulating layer under the skin and this helps to protect the vital organs and to keep the body warm. Fat is also needed to build cell membranes in the body.

Some fats are solid at room temperature and others are liquid at room temperature. Solid fats are mainly animal fats or hydrogenated fats such as pastry margarine.

Animal fats include butter, dripping (beef), suet, lard (pork), cheese, cream, bacon fat, meat fat, oily fish.

Vegetable fats include margarine, cooking oils, nut oils, soya bean oils.

Too much fat is bad for us. It can lead to:
● being overweight (obesity)
● high levels of cholesterol, which can clog the heart's blood vessels (arteries)
● heart disease
● bad breath (halitosis)
● type 2 diabetes (the other type of diabetes, type 1, is something people are born with and is not from eating too much fat).

There are two types of fats:
● saturated fats
● unsaturated fats.

A diet high in saturated fat is thought to increase the risk of heart disease.

Table 1.3 The proportion of protein in some common foods

Animal foods	Protein (%)	Plant foods	Protein (%)
Cheddar cheese	26	Soya flour, low fat	45
Bacon, lean	20	Soya flour, full fat	37
Beef, lean	20	Peanuts	24
Cod	17	Bread, wholemeal	9
Herring	17	Bread, white	8
Eggs	12	Rice	7
Beef, fat	8	Peas, fresh	6
Milk	3	Potatoes, old	2
Cream cheese	3	Bananas	1
Butter	< 1	Apples	< 1
		Tapioca	< 1

Table 1.4 The percentage of saturated fat in an average diet

Type of food	% saturated fat
Milk, cheese, cream	16.0
Meat and meat products	25.2
This splits down into:	
● Beef	4.1
● Lamb	3.5
● Pork, bacon and ham	5.8
● Sausage	2.7
● Other meat products (e.g. burgers, faggots, paté)	9.1
Other oils and fats (e.g. olive oil, margarine, sunflower oil)	30.0
Other sources, including eggs, fish, poultry	7.4
Biscuits and cakes	11.4

Vitamins

Vitamins are chemicals which are vital for life. They are found in small amounts in many foods. If your diet is deficient in any vitamins, you can become ill or unhealthy.

Vitamins help with many of our bodily functions, such as growth and protection from disease. There are two main groups: fat-soluble and water-soluble.

Vitamin A

Vitamin A (water-soluble) is found in fatty foods. Dark green vegetables are a good source of vitamin A.

Other sources of vitamin A are:
● halibut and cod liver oil
● milk, eggs and cheese
● butter and margarine
● watercress
● herrings
● carrots
● liver and kidneys
● spinach
● tomatoes
● apricots.

Fish liver oils are the best source of vitamin A.

Vitamin A:
● helps children to grow
● helps the body to resist infection
● helps to prevent night blindness.

Vitamin D

An important source of vitamin D is sunlight. Other sources are fish liver oils, oily fish, egg yolk, margarine and dairy produce.

Vitamin D (fat-soluble):
● controls the way our bodies use calcium
● is necessary for healthy bones and teeth.

Vitamin B

There are three main types of vitamin B (water-soluble):
● thiamin, also known as B1 – this helps our bodies to produce energy and is necessary for our brain, heart and nervous system to function properly
● riboflavin, also known as B2 – this helps with growth, and helps us to have healthy skin, nails and hair, among other things
● niacin or nicotinic acid – this is vital for normal brain function and it improves the health of the skin, the circulation and the digestive system.

Vitamin B:
● helps to keep the nervous system in good condition
● enables the body to get energy from carbohydrates
● encourages the body to grow.

When you cook food the vitamin B in it can be lost. It is important to learn how to preserve vitamin B when you cook foods. You will learn about this as you study cookery in more depth.

Table 1.5 Sources of vitamin B

Thiamin (B1)	Riboflavin (B2)	Nicotinic acid
Yeast	Yeast	Meat extract
Bacon	Liver	Brewers' yeast
Oatmeal	Meat extract	Liver
Peas	Cheese	Kidney
Wholemeal bread	Egg	Beef

Vitamin C

Vitamin C (water-soluble) is also known as ascorbic acid. It:
● is needed for children to grow
● helps cuts and broken bones to heal
● prevents gum and mouth infections.

When you cook food the vitamin C in it can be lost. It is important to learn how to preserve vitamin C when you cook foods. One of the most important sources of vitamin C in our diet is potatoes.

Other sources of vitamin C are:
- blackcurrants
- green vegetables
- lemons
- grapefruit
- bananas
- strawberries
- oranges
- tomatoes
- fruit juices.

Minerals

There are 19 minerals in total, most of which our bodies need, in very small quantities, to function properly.
- We need minerals to build our bones and teeth.
- Minerals help us to carry out bodily functions.
- Minerals help to control the levels of fluids in our bodies.

Calcium keeps bones and teeth strong, as well as helping blood to clot and muscles to work. To use calcium effectively, the body also needs vitamin D. Sources of calcium include:
- milk
- green vegetables
- drinking water
- the bones of tinned oily fish, such as sardines.

We need iron to build haemoglobin, which is a substance within blood that transports oxygen and carbon dioxide around the body. Iron is most easily absorbed from lean meat and offal, but other sources include:
- green vegetables
- wholemeal flour
- egg yolks
- fish.

Potassium and magnesium are also important minerals for our bodies.

Fruit and vegetables

Fruit and vegetables contain valuable vitamins, minerals, fibre and folic acid, which help to protect us from illness. They also contain substances called phytochemicals, which are antioxidants. Antioxidants help to protect the body's cells from damage.
- Some vegetables and fruits protect us particularly well against strokes because they have such high levels of antioxidants. These vegetables include cauliflower, cabbage, broccoli and Brussel sprouts. The fruits include blackcurrants, oranges, kiwis, and red and yellow peppers.
- The fibre found in fruit and vegetables plays an important role in preventing a stroke too. It also helps to lower cholesterol and maintain a healthy digestive system. (It is important to remember that fibre cannot be digested. However, we need it to remove waste from our bodies.)
- Folate or folic acid is found in dark green vegetables like broccoli and spinach. This also helps to protect us against strokes.
- Potassium is another essential mineral, like sodium, that we need to balance the fluids in our bodies. It is also important in keeping our heart rate normal and helps our nerves and muscles to function. It is found in bananas, avocados, citrus fruits and green leafy vegetables.
- Ideally you should aim to eat at least five portions of fruit and vegetables a day, for example, one apple, one banana, two plums, a heaped tablespoon of dried fruit like raisins (15g) or three broccoli or cauliflower florets.

Water

Water is vital to life. Without it we cannot survive for very long. We lose water from our bodies through urine and sweat, and we need to replace it regularly to prevent dehydration. It is recommended that we drink water regularly throughout the day.

Our organs require water to function properly.
- Water regulates our body temperature – when we sweat the water evaporates from our skin and cools us down.
- Water helps to remove waste products from our bodies. If these waste products are not removed, they can release poisons, which can damage our organs or make us ill.
- We need water to help our bodies absorb nutrients, vitamins and minerals and to help our digestive system.

- Water acts as a lubricant, helping our eyes and joints to work and stay healthy.

Sources of water:
- drinks of all kinds
- foods such as fruits, vegetables, meat, eggs.

Table 1.6 The impact of having too little or none of some nutrients in your diet

Nutrient	Effect of absence in diet
Carbohydrate	Lack of energy
	Weight loss
	Low immune system
Fat	Weight loss
	Lack of energy
	Low immune system
Protein	Water retention (when the body does not get rid of enough water)
	Muscle wastage (when muscles wither away)
	Hair loss
Dietary fibre	Bowel disorders
	Bowel cancer
	Constipation
Vitamin A	Sight problems
	Hydration problems
Vitamin B1	Nervous disorders
Vitamin B2	Growth disorders
	Skin disorders
Vitamin B6	Anaemia
	Blood disorders
Vitamin B12	Anaemia
	Blood disorders
	Possible mental problems
Vitamin C	Scurvy (a disease that can cause bleeding gums and other symptoms)
	Tiredness
	Blood loss if injured, as blood does not clot properly
	Bruising
Vitamin D	Rickets (a disease of the bones)
Niacin	Possible mental problems and depression
	Diarrhoea
Iron	Tiredness
	Lack of energy and strength
Calcium	Weak bones and teeth
Potassium	High blood pressure
Magnesium	Slower recovery from injury or illness
Folic acid	Blood disorders

Special diets

There are various reasons why people may follow a particular type of diet.

Groups with special dietary requirements

There are certain groups of people who have special nutritional needs. You may have different needs at different points in your life, for example, if you change your lifestyle or occupation and when you grow older.

Pregnant and breastfeeding women

Pregnant and breastfeeding women should avoid soft mould-ripened cheese, paté, raw eggs, undercooked meat, poultry and fish, liver and alcohol.

Children and teenagers

As children grow their nutritional requirements change. Children need a varied and balanced diet rich in protein. Teenagers need to have a good nutritionally balanced diet. Girls need to make sure that they are getting enough iron in their diet to help with the effects of puberty.

Vegetarians and vegans

Vegetarians generally do not eat meat, fish or any food products made from meat or fish. Some vegetarians do not eat eggs. Vegetarians have a lower risk of heart disease, stroke, diabetes, gallstones, kidney stones and colon cancer than people who eat meat. They are also less likely to be overweight or have raised cholesterol levels.

Vegans are vegetarians who also do not eat eggs or milk, or anything containing eggs or milk. They may also not consume animal products such as honey, and may refuse to use products made from leather. A vegetarian or vegan diet may be followed for ethical reasons, certain religions (such as Hindus) usually follow a vegetarian diet, or it may be followed for health reasons.

The elderly

As we get older our bodies start to slow down and our appetite will get smaller. However, elderly people still need a nutritionally balanced diet to stay healthy.

People who are ill

People who are ill, at home or in hospital, need balanced meals with plenty of the nutrients they need to help them recover. Good nutritional food is part of the healing process. In the days of Florence Nightingale, hospital wards were closed for two hours a day while patients ate their nutritious meals. No doctor was allowed in the wards at mealtimes. Florence Nightingale saw food as medicine.

Other reasons for special diets

Many people require special diets for various reasons, such as:

- health (diabetes, obesity, heart disease)
- allergies
- religious or moral beliefs.

Allergies and intolerances

Some people may be intolerant of or allergic to some types of food, so caterers must tell customers what is in the dishes on the menu. It is vital that food is clearly labelled and that staff are fully aware of the content of dishes when communicating with customers. It is also vital that foods that are likely to cause allergic reactions, such as nuts, are not accidentally mixed or placed into direct contact with dishes that do not.

Food allergies are a type of intolerance where the body's immune system sees harmless food as harmful, thus causing an allergic reaction. Some food allergies can cause something called anaphylactic shock, which makes the throat and mouth swell, making it difficult to swallow or breathe. They can also cause skin reactions, nausea, vomiting and unconsciousness. Some allergic reactions can be fatal.

An **allergy** involves the immune system reacting to or rejecting certain foods or ingredients. **Food intolerance** does not involve the immune system, but it does cause a reaction to some foods.

Foods that sometimes cause an allergic reaction include:

- milk
- dairy products
- fish
- shellfish
- eggs
- nuts (particularly peanuts, cashew nuts, pecans, Brazil nuts, walnuts).

Gluten-free diet

A gluten-free diet is essential for people who have coeliac disease, a condition where gluten causes the immune system to produce antibodies that attack the lining of the intestines.

Gluten is a mixture of proteins found in some cereals, particularly wheat. A gluten-free diet is not the same as a wheat-free diet, and some gluten-free foods are not wheat free.

Diabetes

Everything we eat is broken down into sugars in our body, but different foods break down at different speeds. People with diabetes need to pay attention to what they eat and when, to control their blood sugar.

The ideal diabetic meal will be balanced, with a variety of foods, in an appropriate portion size. The details will vary depending on the person, their level of physical activity and the type of diabetes they have. There are some detailed suggestions at www.diabeticdietfordiabetes.com.

Religion-based diets

- Jewish: do not eat pork or pork products, shellfish or eels. Meat and milk are not eaten together. Meat must be kosher (killed according to Jewish custom and rules).
- Muslim: animals for meat must be slaughtered according to custom (halal). No shellfish or alcohol is consumed.
- Hindu: strict Hindus will not eat meat, fish or eggs. Those following a less strict regime will still not eat beef.
- Rastafarian: are often vegetarian. Even if they are not vegetarian, they will not eat any processed food, eels, tea, coffee or alcohol.

Moral beliefs

Some people choose special diets based on their moral beliefs, for example, some people do not eat meat because they do not believe in killing animals. Some who do eat meat will only buy meat from certain sources, which state that the animals have been ethically reared, free to roam, and humanely killed.

The chef's role – good catering practice

Chefs have a vital role in making healthy eating an exciting reality for us all. Customer trends show that many people are looking for healthier options on menus, particularly if they eat away from home every day. Healthy eating is one of the major consumer trends to emerge over the past decade and is an important commercial opportunity for caterers across the UK. This is not a passing fad; healthy eating is here to stay. Some sectors of catering have strict requirements relating to health and nutrition. For example, by law, school caterers have to provide meals that meet a minimum nutritional standard. Often there are health-related specifications for workplace catering contracts because employers feel they have a commitment to the health of their staff.

Chefs can be highly influential in the area of healthy eating. The amounts and proportions of the ingredients used, plus the way they are cooked and served, can make an enormous difference to the nutritional content of a dish or meal. Research has shown that the most effective approach to healthy catering is to make small changes to popular dishes. This may involve the following measures:

- Making small changes in portion size, or adding a bread roll or jacket potato to a meal. Adding bread or potato to a meal means that there is more starch in proportion to fat (effectively diluting the fat).
- Making adjustments to preparation and cookery methods, such as trimming all the fat off meat, 'dry frying' and not adding butter to cooked vegetables.
- Making slight modifications to recipes for composite dishes (dishes made from several different ingredients). For instance, a pizza could be made with a thicker base, adding mushrooms and roasted peppers, and topping it with less mozzarella but adding a sprinkling of Parmesan for flavour. Instead of adding salt, use the Parmesan, black pepper and chopped oregano to add flavour.

Chefs can easily contribute to a healthier diet by making such adjustments to their practices. Small steps help to make gradual improvements and promote healthier choices when people dine away from home.

Further examples include:
- adjusting cooking methods to use less fat, for example, avoiding deep-frying
- selecting cooking methods that retain high levels of nutrients such as steaming
- grilling food rather than shallow frying, where this is suitable
- reducing holding times to retain nutrients.

This is where chefs are vital in developing healthier recipes that work. Their skill is in deciding when and where dishes can be modified without losing quality. Some highly traditional dishes are best left alone, while subtle changes can be made to others without losing their texture, appearance or flavour. The 'healthy eating tips' throughout the recipe sections can help in making some of these changes.

In summary, the key to healthier catering is to:
- make small changes to best-selling items
- increase the amount of starchy foods
- increase the amount of fruit and vegetables
- increase the fibre content of dishes where it is practical and acceptable
- reduce fat in traditional recipes
- change the type of fat used (e.g. olive oil instead of butter)
- select healthier ways to prepare dishes, and be adventurous
- be moderate in the use of sugar and salt.

Following a healthy diet, linked to an active lifestyle, gives the body a much greater chance to function correctly and should provide the person with the energy and general feeling of well-being that they need to enjoy life to the full. This will also reduce the risk of high blood-pressure, high cholesterol as well as the threats of obesity, heart disease and cancers.

The consequences of an unhealthy diet can lead to all of the conditions mentioned above but also increase the threat of strokes, malnutrition, if the body does not get enough of the nutrients it needs, and other visible signs such as tooth decay. This signifies the importance of following a healthy, balanced diet and lifestyle.

1 There are many sources of nutritional guidelines available from a variety of organisations. Research the information currently available from the following organisations and outline their main recommendations.

  a British Nutrition Foundation

  b Department of Health

  c Ministry of Food and Fisheries

2 Complete the following table to identify the need and sources for each of the following nutrients.

Mineral	Need in the body	Sources
Calcium		
Iron		
Phosphorus		
Sodium		
Iodine		

3 Choose two special diets and design a three-course meal to meet the needs of each.

Catering operations, costs and menu planning

The partie system

In the late nineteenth century, when labour was relatively cheap, skilled and plentiful, the public wanted elaborate and extensive menus. In response to this, Auguste Escoffier, one of the most respected chefs of his era, devised what is known as the partie system, in which different sections of the kitchen were delegated to carry out specific jobs, such as preparing or cooking the fish, meat or vegetables. This system is still used in many establishments today. The number of parties (different areas) required, and the number of staff in each, will depend on the size of the establishment.

The organisation of the staffing hierarchy depends on the establishment. A large hotel or restaurant will have a head chef or executive head chef with one or two sous chefs, who deputise for the head chef. They may also run a department – for example, the pastry chef may also be the sectional chef, chef de partie and one of the sous chefs.

The chef de partie is in charge of a section such as sauces or vegetables. There may also be a demi chef de partie, who works on the opposite shift to the chef de partie. There will also be a number of assistant chefs (commis chefs), and there could be apprentices and trainees. The latter will move from section to section to complete their training.

Today, kitchens are organised in many different ways but, in each case, a senior member of staff will be responsible for the smooth operation of the kitchen. This person must have leadership skills, human resource management skills and detailed product knowledge. In order to achieve an efficient and effective system that satisfies customers' needs, it is important to work as a team and develop good working relationships in the kitchen and with the food service staff. This will also contribute to high staff morale and will improve the productivity of staff.

Technology features heavily in the modern kitchen with many labour- and energy-saving forms of machinery and devices available to assist the head chef and their team in their work. Society also changes over time and food, like fashion, follows in line. With this in mind, the kitchen has adapted to meet the needs and expectations of its customers. Food styles change and chefs, particularly in the high-quality/fine-dining segment of the industry, regularly use equipment and food technologies to produce ground-breaking and innovative cuisine.

The staffing hierarchy in professional kitchens

Head chef (Chef de cuisine): has overall responsibility for the organisation and management of the kitchen, including staffing, training, menus, budget control, the implementation of food safety, health and safety and sourcing of food.

Second chef (Sous chef): is the deputy to the head chef and would take overall responsibility in the head chef's absence. The sous chef may also have specific areas of responsibility, such as food safety, health and safety, quality control or training of staff.

Head of section (Chef de partie): is in charge of a specific section within the kitchen, such as meat, vegetables or fish. There may be a demi chef de partie working on an opposite shift and covering days off. Commis chefs and apprentices usually work with these chefs in different sections.

Commis chef: is the junior chef in the kitchen and works under overall supervision of the sous chef. A commis chef will work around the various kitchen sections.

Apprentice: is similar to a commis chef and will complete similar tasks but is usually on a planned programme of learning, often managed by a college or training provider.

Other roles in the restaurant include:
- head waiter
- restaurant manager
- bar manager.

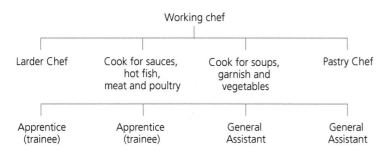

Example of a kitchen brigade for a restaurant serving 40 meals a day

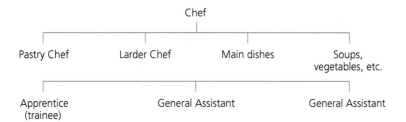

Example of a kitchen brigade for an industrial catering kitchen

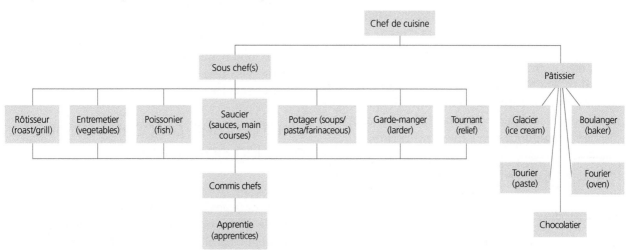

Example of a classical kitchen brigade for the kitchen of a hotel or restaurant

Working relationships

Chefs working in a structure similar to the one described above need to operate as a team in order to succeed as a whole. Working in a structure where relationships between team members are positive is more likely to result in a more efficient operation with staff enjoying a high morale amongst the team. Communication will be clearer and the productivity of the team collectively is going to be better than one where there are problems in these areas. A team working collectively with a positive attitude and approach to their work is much more likely to produce high-quality products and therefore a better service to their customers.

Kitchen layout and workflow

Properly planned layouts with adequate equipment, tools and materials to do the job are essential if practical work is to be carried out efficiently. If equipment is in the right place then work will proceed smoothly and in the proper sequence, without back-tracking or criss-crossing.

Work surfaces, sinks, stores and refrigerators should be within easy reach in order to avoid unnecessary walking. Food deliveries should have a separate entrance because of the risk of contamination. It is also a good idea to have a separate staff entrance to the kitchen and, for food safety reasons, it is essential to have separate changing facilities for employees wherever possible – they should not have to use customer facilities.

The layout of the preparation areas (for vegetables, meat, poultry, dairy products, etc.) is also important. In large catering establishments the preparation areas will be zoned to assist with the workflow. The flow of work through the kitchen and serving areas is essential to the smooth running of any operation. Where possible, the layout of the kitchen should focus on a linear workflow.

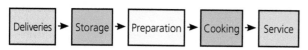

A linear workflow

Effective workflow will:
- help to establish good communication between departments

- improve efficiency
- improve the quality of the finished product
- reduce the risk of accidents
- promote good health and safety and food safety,

all of which will provide a better service to customers.

Future trends in design and decor will be affected by:
- changes in technology
- social changes in eating habits
- lifestyle changes.

When considering kitchen layout, remember:
- health and safety
- food safety
- time and motion (workflow).

It is essential that food businesses comply with legislation surrounding food safety and health and safety. There are experts in commercial kitchen design that can provide advice with regard to layout and safe and efficient workflows. Environmental Health Practitioners are also in a position to offer help and advice in this area.

Different kitchens operate a variety of production systems from cook-to-order (à la carte) to self-service counter operations. Whichever system is deployed, the layout of the kitchen and the work-flows within its operation will play a large part in its overall success. A well-designed layout will help the kitchen operate efficiently and will be more likely to provide:
- established routes of communication
- better quality products
- a reduced risk of accidents
- good health and safety practices
- good food hygiene practices
- better working relationships
- better services to customers.

Planning and preparing menus for catering operations

Meal occasion and types of menu

The main types of menu in use are:
- Table d'hôte or set-price menu: a menu forming a meal, usually of two or three courses at a set price. A choice of dishes may be offered at all courses.

- À la carte: a menu with all the dishes individually priced. Customers can therefore compile their own menu, which may be one, two or more courses. A true à la carte dish should be cooked to order and the customer should be prepared to wait.
- Special party or function menus: menus for banquets or functions of all kinds.
- Ethnic or speciality menus: these can be set-price menus or with dishes individually priced, specialising in the food of a particular country (or religion) or in a specialised food itself, such as ethnic (Chinese, Indian, kosher, African-Caribbean, Greek), or speciality (steak, fish, pasta, vegetarian, pancakes).
- Breakfast menus: these menus can be offered as continental, table d'hôte, à la carte or buffet. For buffet service, customers can self-serve the main items they require with assistance from counter hands. Ideally, eggs should be freshly cooked to order.
- Luncheon and dinner menus: these can range from the following:
  - A set-price one-, two- or three-course menu, ideally with a choice at each course.
  - A list of well-varied dishes, each priced individually so that the customer can make up his or her own menu of whatever number of dishes they require.
  - Buffet, which may be all cold or hot dishes, or a combination of both, either to be served or organised on a self-service basis. Depending on the time of year and location, barbecue dishes can be considered.
  - Special party, which may be either: set menu with no choice (often used when serving a banquet to a large number of people); set menu with a limited choice, such as soup or melon, main course, choice of two sweets; served or self-service buffet.
- Tea menus: these vary considerably, depending on the type of establishment, and could include, for example:
  - assorted sandwiches
  - bread and butter (white, brown, fruit loaf)
  - assorted jams
  - scones with clotted cream, pastries, gâteaux
  - tea (Indian, China, iced, fruit, herb)
  - assorted pastries, gâteaux.

- Dessert menus: as well as the desserts offered as part of table d'hôte and à la carte menus, dessert menus are appearing as speciality menus in some restaurants in their own right. This provides customers with the opportunity to sample a range of desserts as tasting-size portions.
- Tasting menu: several small set courses reflecting the main menu. Often matched to wines.

Factors to consider when planning a menu

Presentation

Ensure that the menu is presented in a sensible and welcoming way so that the customer is put at ease and relaxed. An off-hand, brusque presentation (written or oral) can be off-putting and lower expectations of the meal.

Planning

Consider the following:
- the type and size of establishment – pub, school, hospital, restaurant, etc.
- customer profile – different kinds of people have differing likes and dislikes
- special requirements – kosher, halal, etc.
- time of the year – certain dishes acceptable in summer may not be so in winter
- foods in season – are usually in good supply and reasonable in price
- special days – Christmas, Shrove Tuesday, etc.
- time of day – breakfast, brunch, lunch, tea, high tea, dinner, supper, snack, special function
- price range – charge a fair price and ensure good value for money; customer satisfaction can lead to recommendation and repeat business
- number of courses
- sequence of courses
- use menu language/terminology that customers understand
- sensible nutritional balance
- no unnecessary repetition of ingredients
- no unnecessary repetition of flavours and colours.

Be aware of the Trade Descriptions Act 1968: 'Any person who in the course of a trade or business applies a false trade description to any goods or supplies or offers to supply any goods to which a false trade description is applied shall be guilty of an offence.'

The Food Safety Act 1990 (2009 edition) says that food must be of the nature, substance and quality demanded.

It is also a requirement to inform customers of other charges that may be incurred including items such as VAT and service charges.

It is also important for the caterer to consider resourcing implications such as:
- availability of equipment
- staffing numbers and skills
- space required to deliver effectively.

Understanding basic costs associated with the catering industry

Cost control

It is important to know the exact cost of each process and every item produced, so a system of cost analysis and cost information is essential.

Cost analysis is the process of breaking down the costs of an operation into all its separate parts so that it is possible to look at the exact cost of each process and every item produced, and judge the efficiency and cost effectiveness of each.

The advantages of an efficient costing system are as follows.
- It provides information towards the net profit made by each section of the organisation and shows the cost of each meal produced. It will reveal possible ways to economise and can result in a more effective use of stores, labour, materials, and so on.
- Costing provides the information necessary to develop a sound pricing policy.
- Cost records help to provide speedy quotations for all special functions, such as parties, wedding receptions, and so on.
- It enables the caterer to keep to a budget.

No one costing system will automatically suit every catering business, but the following guidelines may be helpful.
- The cooperation of all departments is essential.
- The costing system should be adapted to the business, not vice versa. If the accepted procedure in an establishment is altered to fit a costing system, then there is a danger of causing resentment among staff and as a result losing their cooperation.
- Clear instructions in writing must be given to staff that are required to keep records.

Element of cost

Costs can come in many different forms, some more obvious than others at first. To simplify cost structures they are broken down into various categories or elements. Some costs are owned directly by a department or cost centre. For example, the cream used to make a dessert can be attributed to the pastry kitchen whereas advertising a hotel is a cost that could be shared by the various departments that make up the hotel. Hopefully, the advertising will benefit all departments so therefore it could be argued that the cost should be shared.

The main elements of cost (types of cost) are split into the following:
1. Food and materials costs: these are known as variable costs because the level will vary according to the volume of business. In an operation that uses part-time or extra staff for special occasions, the money paid to these staff also comes under variable costs; by comparison, salaries and wages paid regularly to permanent staff are fixed costs.
2. Labour costs: regular charges come under the heading of fixed costs, which include labour. Labour costs in the majority of operations fall into two categories:
   - direct labour cost, which is the salaries and wages paid to staff such as chefs, waiters, bar staff, housekeepers, chambermaids, and which can be allocated to income from food, drink and accommodation sales
   - indirect labour cost, which would include salaries and wages paid, for example, to managers, office staff and maintenance staff who work for all departments (so their labour cost should be charged to all departments).

3 Overheads consist of rent, rates, heating, lighting, equipment and other sundry expenses such as cleaning materials. Some overheads are fixed costs, such as rent and rates. Others are variable, going up and down with use, such as lighting, gas use and cleaning materials.

Factors to be considered

Sourcing food items

Selecting suppliers is an important part of the purchasing process. First, think about how a supplier will be able to meet the needs of your operation. Consider:
● price
● delivery
● quality/standards.

Information on suppliers can be obtained from other purchasers. Also, visits to suppliers' establishments are to be encouraged. When interviewing prospective suppliers, you need to question how reliable a supplier will be compared to competitors, and how stable they will be under varying market conditions. Seasonality and availability are also key considerations to ensure a consistent supply at a price that is sustainable to operate.

Principles of purchasing

A menu dictates what an operation needs. Based on this, the buyer searches for a market that can supply these requirements. Once the right market is found, the buyer must investigate the various products available. The right product must be obtained – it must be suitable for the item or dish required and of the quality desired by the establishment. Other factors that might affect production needs include:
● type and image of the establishment
● style of operation and system of service occasion for which the item is needed
● amount of storage available (dry, refrigerated or frozen)
● finance available and supply policies of the organisation
● availability, seasonality, price trends and supply.

The skill of the employees, catering assistants and chefs must also be taken into account, as well as the condition, the processing method and the storage life of the product.

Accurate weighing and measuring

It is important that, when designing standardised recipes, the correct weights and measures are recorded, to achieve consistency, so no matter who prepares the dish the same standard portion size and quality are achieved.

Each recipe should tell you the following:
● the ingredients to be used
● the exact amounts of ingredients required
● how the dish is prepared
● how the dish is cooked
● the number of portions it will produce (yield).

Always read through the recipe carefully and check that you:
● have the right ingredients and equipment
● have the correct weights
● have enough time to prepare the dish.

To facilitate menu planning, purchasing and internal requisitioning, food preparation and production, and portion control, you should:
● know the food cost per portion
● know the nutritional value of a particular dish.

The standard recipe will also help new staff in the preparation and production of standard products, which can be made easier by using photographs or drawings illustrating the finished product.

Other factors to be considered

It is important that waste is minimised in all forms. The issues connected to waste are not only financial (throwing away food items that have been purchased) but ethical (throwing away food that is perfectly safe to eat) and environmental (the food has to be disposed of, perhaps contributing to land-fill waste and also utilising energy to process the waste).

It is vital that food is handled and stored correctly so that it is safe to use and to ultimately serve to customers. Any food that becomes unsafe due to poor handling practices needs to be disposed of, contributing to the factors described above.

Losses during preparation and cooking

Some loss is hard to avoid such as trimming roots from vegetables or removing fat or bones from meat products. However, care should be taken to avoid excess trimming of waste and disposing of perfectly good food. Food items, such as meat products also lose weight during cooking.

Applying calculations in catering operations

Gross profit

Gross profit (or kitchen profit) is the difference between the cost of an item and the price it is sold at. If gross profit is set as a fixed percentage mark-up, the food cost of each dish is calculated and a fixed gross profit (such as 100 per cent) is added. So, for instance, if the food costs £2 it is sold for £4.

It is usual to express each element of cost as a percentage of the selling price. This enables the caterer to control profits.

Net profit

Net profit is the difference between the selling price of the food (sales) and total cost of the product (food, labour and overheads). If the selling price of a dish is expressed as 100 per cent (the total amount received from its sale), it can be broken down into the amount of money spent on food items and the gross profit. This can be expressed in percentages as shown below.

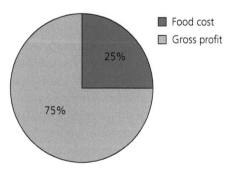

This can also be presented in monetary terms, as shown in the following diagram, if the dish was sold at £10.00, for example.

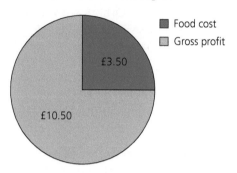

Calculating the selling price

If food costs come to £3.50, to calculate the selling price on the basis of a 65 per cent gross profit, the following calculation can be used.

$$\frac{£3.50 \text{ (food costs)}}{35 \text{ (food cost as a \% of the sale)}} \times 100 = £10.00$$

This calculation brings the food cost to 1 per cent of the selling price before multiplying by 100 to bring the selling price to 100 per cent.

To demonstrate this further, if the gross profit requirement was raised to 75 per cent, this would reduce the food cost as a percentage of the selling price to 25 per cent. Therefore the selling price would have to be higher if the food cost remained at £3.50.

$$\frac{£3.50 \text{ (food costs)}}{25 \text{ (food cost as a \% of the sale)}} \times 100 = £14.00$$

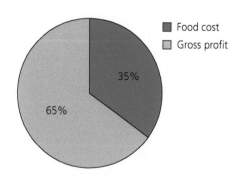

The percentages still add up to 100 per cent, but the proportion spent on food is smaller in terms of the selling price. The diagrams illustrate the breakdown on this basis.

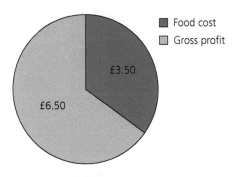

■ Food cost
■ Gross profit

£3.50

£6.50

To check that this is correct, the following calculation can be applied.

$$\frac{£14.00 \text{ (selling price)}}{100 \text{ (brings £14.00 down to 1\%)}} \times 25 = \begin{array}{l} £3.50 \\ \text{(brings up} \\ \text{to 25\%)} \end{array}$$

and

$$\frac{£14.00 \text{ (selling price)}}{100 \text{ (brings £14.00 down to 1\%)}} \times 75 = \begin{array}{l} £10.50 \\ \text{(brings up} \\ \text{to 75\%)} \end{array}$$

£10.50 (75%) + £3.50 (25%) = £14.00 (100%)

Test yourself

If a dish costs £3.00 to produce, what should the selling price be to achieve a 70% gross profit?

Sales and profit

Some basic principles:

● sales – food cost = gross profit (kitchen profit)
● sales – total cost = net profit
● food cost + gross profit = sales

For example:

Food sales for 1 week	= £25,000
Food cost for 1 week	= £12,000
Labour and overheads for 1 week	= £9,000
Total costs for 1 week	= £21,000
Gross profit (kitchen profit)	= £13,000 (£25,000 – £12,000)
Net profit	= £4,000 (£25,000 – £21,000)

To verify:

Food sales– food cost (£25,000– £12,000)	= £13,000 (gross profit)
Food sales– net profit (£25,000– £4,000)	= £21,000 (total costs)
Food cost + gross profit (£12,000 + £13,000)	= £25,000 (food sales)

Profit is commonly expressed as a percentage of the selling price for comparative and monitoring purposes.

Therefore the net profit percentage for the week is:

$$\frac{\text{Net profit (£4,000)}}{\text{Sales (25,000)}} \times 100 = 16\%$$

Financial breakdown

		Percentage of sales
Food cost	£12,000	48%
Labour	£6,000	24%
Overheads	£3,000	12%
Total costs	**£21,000**	**84%**
Sales	£25,000	100%
Net profit	**£4,000**	**16%**

If the restaurant served 1,000 customers then the average amount spent by each customer would be:

$$\frac{£25,000 \text{ (total sales)}}{1,000 \text{ (customers)}} = £25.00$$

As the percentage composition of sales for a month is now known, the average price of a meal for that period can be further analysed:

Average price of a meal = £25.00

$$\frac{£25.00}{100} = £0.25 \ (1\%)$$

From this, it is possible to calculate the contribution of each element towards the sale:

Average price of meal		= £25.00
Food cost	= £0.25 × 48	= £12.00
Labour	= £0.25 × 24	= £6.00
Overheads	= £0.25 × 12	= £3.00
Net profit	= £0.25 × 16	= £4.00

Test yourself

1 How can the layout of the kitchen you work in be improved?
2 Think of a successful team. Say why you think it is successful and what you could learn from this team's success.

Applying workplace skills

Professional presentation

In the hospitality industry it is important to be smart and to wear the appropriate clothing for the job, whether this is chefs' whites or a doorman's uniform. You must present a well-groomed appearance and wear smart, clean clothing that is in good repair.

Employers want people with the right attitude, who are able to show initiative, be punctual, flexible and dependable. They want people who can organise themselves, communicate and manage their time effectively.

Professional presentation includes:
- behaviour
- attitude
- conduct
- standards
- punctuality
- dependability.

These points demonstrate personal pride and develop your confidence. A good attitude helps promote health and safety and food safety. Professional presentation and attitude provide good role models.

Reasons for presenting a professional image

As well as developing your own personal pride and confidence, presenting a professional image is also important to allow identification, to promote you and the job role, adding status and commanding respect. It is also necessary to be clean and tidy to comply with health and safety and food safety legislation. Image can also be used to match branding (for example, uniforms) and to meet job requirements (such as wearing chefs' whites for cooking or wearing a smart suit when working front of house).

The way you carry out the job and are able to use the appropriate skills is also part of your personal presentation. Good presentation, both front and back of house, helps increase customer satisfaction, encourages repeat business, improves business and improves staff morale and staff satisfaction, leading to happy customers and increased profits. These, in turn, enhance the reputation of the establishment.

Professional presentation and clothing

Being smart and wearing the correct clothing is an important part of working in the hospitality industry. All the clothing you wear must be smart, clean and in good repair. Your hair must be short or tied back neatly and you should be clean-shaven or have a neat beard/moustache. You should always wear the correct uniform in the kitchen. You must change your chefs' or cooks' uniform regularly (at least every day) and you should never wear the uniform outside the working premises as this is unhygienic. Bacteria from outside can be carried on the uniform into the kitchen and may cause harm.

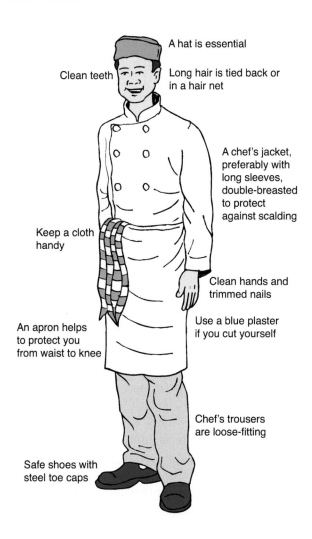

A hat is essential

Clean teeth

Long hair is tied back or in a hair net

A chef's jacket, preferably with long sleeves, double-breasted to protect against scalding

Keep a cloth handy

Clean hands and trimmed nails

Use a blue plaster if you cut yourself

An apron helps to protect you from waist to knee

Chef's trousers are loose-fitting

Safe shoes with steel toe caps

Skills required to maintain the working area

Some of the most important skills required in the workplace are forecasting, planning, organising, commanding, coordinating and controlling, and cleaning and tidying. We will take a brief look at each of these below.

Forecasting

Forecasting is the ability to plan ahead, in order to foresee possible and probable actions and allow for them. For example, if the chef de partie knows that the following day is their assistant's day off, she or he can look ahead and plan accordingly. Forecasting requires good judgement acquired from previous knowledge and experience.

Planning

From forecasting comes planning: how many meals to prepare; how much to have in stock (in case the forecast is not completely accurate); how many staff will be needed, which staff and when; are the staff capable of what is required of them?

Organising

In the hospitality industry, organisational skills are applied to food, equipment and staff. Organising in this context consists of ensuring that what is wanted is where it is wanted, when it is wanted, in the right amount and at the right time. Organisation involves the production of duty rotas and training programmes, as well as cleaning schedules.

Commanding

This means giving instructions to staff on how, what and where. This means that orders have to be given, and a certain degree of order and discipline must be maintained.

Coordinating

Coordinating is the skill that is used to get staff to cooperate and work together – to coordinate the work of each section. Different tasks also need to be coordinated to allow for smooth running of the kitchen and completion of work on time. This coordination is essential to the success of any organisation in the hospitality industry.

Controlling

This involves the skills needed to control the whole operation in order to monitor and improve performance. It would include:

- checking that staff are on time
- checking that standards are maintained
- checking quality
- checking quantity
- ensuring that there is no unnecessary waste
- checking equipment
- monitoring and checking hygiene standards
- monitoring and checking to ensure the area is kept clean and tidy.

Having a professional attitude and developing a professional organisation promotes the organisation: as customers gain confidence in the business and the staff it employs, it develops a good reputation. If the business has a good reputation this boosts staff morale – if staff have confidence in the business, this has a positive effect on profits.

Customer care

You should always:

- put the customer first
- make the customer feel good
- make the customer feel comfortable
- make the customer feel important
- make the customer want to return to your restaurant or establishment.

Dealing with customer problems

If a customer is rude or aggressive to a member of the waiting staff (for instance, blames a waiter for the chef's mistake), the waiter should not be rude or aggressive in return. If the waiter can use his/her skills to remain calm and patient, the customer will often apologise for their anger. Behaviour is a choice; choose behaviour that is appropriate to the customer and the situation.

When dealing with customers, behaviour should be:

- professional
- understanding – customers in a restaurant want a service and are paying for it; learn to understand their needs
- patient – learn to be patient with all customers

- enthusiastic – enthusiasm can be contagious
- confident – confidence can increase a potential customer's trust in you
- welcoming – this can satisfy a customer's basic human desire to feel liked and be approved of
- helpful – customers warm to helpful staff
- polite – good manners are always welcomed
- caring – make each customer feel special.

If customers complain or they are unhappy about anything that is served to them, they should be encouraged to inform the member of staff who served them. This will give the establishment the opportunity to rectify the fault immediately. Ask them about their eating experience; this information will be vital for future planning.

Treat customer complaints seriously. It may be appropriate to offer free drinks or a reduction on the bill. Show them empathy, use the appropriate body language, show concern, sympathise. Always apologise. If you handle the complaint well you will make the customer feel important.

Note

Remember! Customer care = happy customers = profit = jobs.

The term 'empathy' means being able to put yourself in someone else's shoes; understanding and sharing someone else's feelings.

Working with colleagues

It is important to develop good working relationships with your colleagues, work as a team and be supportive of each other. Seek guidance from other team members and your line manager, and identify role models for yourself.

In some cases, it may be necessary to develop a work plan for the day and for the week. Discuss the rotas and work schedule – who will cover which tasks and what needs to be done. Discuss targets and outcomes. Evaluate your performance and the team's performance. This can be done in an informal way, or formally through a team meeting. Identify how the team's and your performance are being measured.

Skills required when working with customers and colleagues

A command of interpersonal skills or what are often referred to as 'soft skills' is required when working with people, whether customers or colleagues. Cooperation is key in aiming for a positive outcome or working relationship.

'Soft skills' is a phrase that relates to an individual's range of personality traits, social skills, communication, language, personal habits, friendliness, and optimism that characterise relationships with other people. Soft skills complement hard skills, which are referred to as the occupational requirements of a job and associated activities.

Soft skills are the personal attributes that enhance an individual's interactions. Unlike hard skills, which are about an individual's skill set and ability to perform a certain type of task or activity, soft skills relate to a person's ability to interact effectively with co-workers and customers and are broadly applicable both in and outside the workplace.

Some of the characteristics referred to as soft skills include:
- a strong work ethic
- a positive attitude
- good communication skills
- good time management
- problem-solving skills
- acting as a team player
- self-confidence
- the ability to accept and learn from criticism
- being flexible and adaptable
- working well under pressure.

The combination of technical ability and soft skills helps to produce a well-rounded employee who possesses the necessary skills-set required to meet targets and plan work in a logical and coordinated way.

Targets are often the result of individuals meeting their collective personal targets. As such, meeting targets in the workplace involves sharing of information (dissemination) and will often require guidance from supervisors and/or managers.

Other important factors when evaluating work roles and performance is the ability to reflect on final outcomes and the recording of levels of performance for future benchmarking (measurement) as well as for future training and development purposes (staff development).

Throughout this process, it is important to evaluate successes and areas requiring improvement. Where elements are not going to plan, it is vital to identify them as quickly as possible in order to take corrective action to get back on track. Consistent review helps to avoid complacency and provides a platform for the consistent strive for excellence.

Applying for jobs

When applying for a job you will usually be asked to supply various documents and letters and then, if the first stage of your application is successful, you will be asked to go for an interview.

Producing a CV

When you apply for a job you will probably be asked to send in a current CV (curriculum vitae). This requires you to list all your educational qualifications and work history, your interests and other activities you participate in. Employers will generally want to know where you have demonstrated certain skills, how you have dealt with certain situations in the workplace and whether you carry out any voluntary work.

When preparing a CV, you should also bear in mind the following points:
- All your work experience and work placements should be included in the CV.
- Always keep your CV up to date and keep track of all your qualifications, experience, jobs, dates of employment, employers, and competitions you have entered and won, membership of associations (that are relevant to the employment you are seeking) and referees.
- Always check spelling, layout and punctuation, and include appropriate vocabulary and information.
- Always update your personal records.
- As you develop your career, personal qualities and skills, write a short profile about yourself.

Producing a covering letter

A covering letter introduces you to the company; it explains why you are suitable for the job on offer and the skills and qualities you can bring to it. In some cases, it may also give you an opportunity to say how you would be able to contribute positively to the establishment and organisation as a whole. The letter will usually accompany a copy of your CV.

- Write down what inspires you and how you use existing skills.
- Specify what your long-term goals are, as well as your immediate goals and targets.
- Identify ways of broadening your outlook, your range of skills, and your ability to deal with a range of different people, personalities and cultural diversity.

Present your work experience and skills in your CV and covering letter

Interview techniques

First impressions are important. Always prepare thoroughly for an interview. Good preparation will help to ensure that you are in control of the interview.

When preparing for an interview, you should also bear in mind the following points:

- Think about communication in all its forms. The way in which you present yourself provides a first impression to your potential characteristics. A professional, smart image will always help to provide a positive first impression.
- Speak clearly, concisely and think carefully about your responses. Try to keep conversation and any comments relevant to the application
- Prepare any questions in advance, and research the role thoroughly.
- Consider how you are going to introduce yourself at the start of the interview, and take appropriate/relevant records and documents with you including a copy of your CV and covering letter.
- Make sure you are well groomed, smart and look professional.
- It is sometimes useful to practise interviews beforehand – this is known as role play.
- Be honest, showing integrity and ambition. Self-analyse your skills and areas for development. Use the opportunity to promote yourself but be open about your intentions to develop yourself further.
- Before the interview, plan the journey and work out the travelling time – allow yourself plenty of time to get there so you do not feel rushed. This is an example of good time management which will impress the interviewer.

At the interview, always maintain eye contact with the interviewer and smile occasionally. Be confident and polite. Think about the questions you are asked before you answer them. Be clear and concise. If you do not understand a question, ask for it to be clarified.

When the interview is over, reflect on your performance. If you are unsuccessful, ask for feedback and learn from the experience. Identify what went well and think about how you might improve in the future.

Following an interview, always try to learn from your experience:

- always ask for feedback from any interview or job application.
- assess your skills.
- how could you improve?
- what did you do well?

Producing a plan to develop skills (personal development plan)

It is necessary to evaluate and check your progress from time to time. Feedback from your peers and managers is a useful way of evaluating your performance. Keeping records (such as personal development plans) as a way of checking your progress is also an important way of referring back to your targets and thinking about the final outcome.

Key stages of monitoring performance to see if it meets targets include:

- work plans (for example, personal development plans) – seeking guidance on them
- targets – evaluating them, taking corrective action if necessary
- outcome – must be measurable in order to know if it has been achieved.

Gathering information on your learning journey to improve your workplace skills is useful so that, once you have achieved a successful outcome, you can use that information to inform and help others, disseminating it as necessary.

Having a personal development plan will help you identify targets and timescales to improve your skills and advance your career for personal and professional success, providing both a record of your skills development, goals and a motivational tool. The next step is to identify which skills you need to develop further, and ways you could do this.

Your personal development plan will help you to evaluate your performance feedback from your mentor, manager or tutor, and will help you to improve your own performance. An important aspect of this process is to evaluate your own performance, using reviews and other forms of performance evaluation to make plans to develop yourself in the future. This will help you form your career path. The plan will help you to achieve your aims and become successful – to be where you want to be and who you want to be.

Test yourself

1 Identify five protective features of the chef's uniform.
2 Give four examples of professional behaviour.
3 What is meant by the term 'soft skills'?
4 State three checks you should make before sending out your CV.

2 Underpinning knowledge for level 3

This chapter covers:
→ **VRQ level 3 Supervisory skills**
→ **VRQ level 3 Practical gastronomy**

The chapter also supports the development of knowledge and understanding in a number of NVQ units across the various routes at level 3.

Supervisory Skills covers content required for the following NVQ units:
→ **Maintain the health, hygiene, safety and security of the working environment**
→ **Develop productive working relationships with colleagues**
→ **Employment rights and responsibilities in the hospitality, leisure, travel and tourism sector**
→ **Contribute to the control of resources.**

Practical Gastronomy covers content required for the following NVQ units:
→ **Contribute to the development of recipes and menus**
→ **Contribute to the control of resources.**

This chapter provides an overview of the curriculum content required for study towards completion of the theoretical units to complete the Advanced Diploma in Professional Patisserie and Confectionery at level 3. This is with the exception of the unit in Supervising Food Safety. This unit is often delivered as a stand-alone programme and is standardised across all providers offering this qualification, certifying the unit independently. The Royal Society for Public Health (www.rsph.org.uk) and the Chartered Institute for Environmental Health (www.cieh.org) are two leading examples of organisations offering this qualification in this way.

Furthermore, legislation in this particular area is subject to review on a frequent basis and therefore the most reliable sources of information are directly available from the two organisations mentioned above as well as the Food Standards Agency (www.food.gov.uk) and the Health and Safety Executive (www.hse.gov.uk).

In order to enable a broader coverage of the practical skills and the products associated in the practical components of the course, the intention here is to provide a generic coverage of the key aspects of the theoretical content for students.

Supervisory skills in the hospitality industry

As a chef acting as a supervisor, you need to be involved in planning health and safety initiatives, training employees and monitoring health and safety performance standards. In order to identify hazards you must carry out safety inspections. This means observing how people carry out their daily work. If you have any concerns you must discuss them with the employees and the manager.

As a working chef you will also need to identify the training needs of the establishment. You need to review each individual training need and assess how safe people are when doing their jobs. Often the chef as a supervisor will be part of a committee

on health and safety – every establishment is required to have one of these to ensure safety inspections are carried out and so that accidents are investigated.

Supervising work tasks

Chefs are responsible for overseeing and carrying out work tasks. These tasks and procedures must adhere to the establishment's health and safety policy. The employee must carry out all procedures and tasks that he or she needs to complete safely and to a high standard. It is the supervisor's role to monitor this work to ensure good health and safety practices are followed throughout the operation.

The supervisor must, at all times, give guidance, demonstrate good practice and ensure that all work activities are carried out in a disciplined manner. Different employees will need varying levels of supervision. Competent people also need supervising to check that they do not fall into bad habits or take dangerous short cuts. Particular attention must be paid to people who are vulnerable to a higher risk of injury.

Team development

In the hospitality industry it is essential that as a supervisor you develop effective working relationships with other supervisors. It is important to discuss health and safety and monitoring issues at team meetings as well as any major issues relating to health and safety. When new equipment has been installed, for example, there will be a need for written safety procedures and for staff to be trained in how to use the equipment.

Effective working relationships are vital in the hospitality industry

What makes a good supervisor?

A good supervisor is someone who:
- is open
- is fair
- is well informed
- is well organised
- is a good communicator
- shows respect for others
- gives support to others to establish policies and procedures.

Health and safety

Maintaining security and health and safety procedures

As a supervisor it is essential that you maintain security and health and safety procedures in your own areas of responsibility. To ensure that legislation regarding safety and security is implemented, it is necessary:
- for the legislation to be known
- that the requirements are carried out
- that a system of checks makes certain that the legislation is complied with.

First, all people involved in an establishment must be made aware of the need for safety and security and their legal responsibilities towards themselves, their colleagues, their employers and members of the public.

It is the responsibility of everyone at the workplace to be conscious of safety and security and to pass on to the appropriate people recommendations for improving the procedures for maintaining safety and security. The types of equipment that need to be inspected to make certain that they are available and ready for use include security equipment, first-aid and fire-fighting equipment. The supervisor or person responsible for these items needs to regularly check and record that they are in working condition and that, if they have been used, they are restored ready for further use. Security systems and fire-fighting equipment are usually checked by the makers. It is the responsibility of the management of the establishment to ensure that this equipment is maintained correctly. First-aid equipment is usually the responsibility of the designated first-aider, whose functions include replenishing first-aid boxes. However, a chef de partie or supervisor will be aware that if fire extinguishers and first-aid equipment are used, he or she has a responsibility to take action to maintain the equipment by reporting to the appropriate person.

It is advisable that all staff are trained in the use of fire extinguishers. Routine checks or inspections need to be carried out in any establishment to see that standards of hygiene, health and safety are maintained for the benefit of workers, customers and other members of the public. Visitors, suppliers and contractors are also entitled to expect the premises to be safe when they enter.

Maintain a healthy and safe working environment

It is necessary to be aware of the policy and procedures of the organisation in relation to health and safety legislation. Every individual at work anywhere on the premises needs to develop an attitude towards possible hazardous situations in order to prevent accidents to themselves and others. Training is also essential to develop good practice and should include information on what hazards to look for, hygienic methods of working and the procedures to follow in the event of an incident. You should keep records of staff training in these areas.

Every organisation will have procedures to follow in the event of a fire, accident, flood or bomb alert; every employee needs to have knowledge of these procedures. Every establishment must have a book to record accidents. It is also desirable to have a book to record items that are in need of maintenance due to wear and tear or damage, so that these faults can be remedied.

Details of incidents, such as power failure, flooding, infestation, contamination, and so on, which do not result in an accident, should be recorded in an incident book. Records should be kept of items lost, damaged or discarded, giving details of why and how it happened and what subsequent steps have been taken.

The kitchen is an area in which products and pieces of equipment are lifted and moved around on a regular basis. Items do not have to be particularly heavy in order to cause an injury. Therefore, organisations should have a policy to protect people from injury in the form of a manual handling policy. This policy will provide advice on how to handle and transport items safely in ways that minimise potential injuries, particularly to the back. It will recommend situations in which loads should be split or shared and cases where it would not be safe for someone to lift items on their own. It will also provide advice on techniques, such as bending legs and keeping the back straight as well as planning routes to make sure they are clear of obstacles.

Health and Safety at Work Act (1974)

Work under the Health and Safety at Work Act 1974 covers work as an employee or as a self-employed person. An employee is a person who works under a contract of employment. A self-employed person can be described as someone who works for reward or gain for his or herself rather than under a contract of employment. A self-employed person may or may not employ others.

The Health and Safety at Work Act makes it clear that employees are at work when carrying out tasks that are in their job description, for which they are paid by their employer; so employees are not 'at work' when they are engaged in activities that are not within the course of their specific employment. Employees cannot be asked to carry out a work task that is not in their job description or that they are not qualified to do. Voluntary workers and those on work experience are also regarded as being employees as far as health and safety at work is concerned.

Legislation and enforcement

Many of the health and safety regulations in the UK are a result of directions from the European Union.

The main piece of legislation dealing with workplace health and safety in England, Wales and Scotland is the Health and Safety at Work Act 1974. Largely similar provision is covered in Northern Ireland under the Health and Safety at Work (Northern Ireland) Order 1978.

The aims of the legislation are:
- to secure the health, safety and welfare of people at work
- to protect people other than those at work against the risks of health and safety that arise out of or in connection with the activities of people at work
- to control the keeping and use of explosive, highly flammable or otherwise dangerous substances at work.

The legislation is there to control and to provide regulations and approved codes of practice which set the standards of health, safety and welfare.

The Health and Safety Commission

The Health and Safety Commission (HSC) is an organisation appointed to regulate health and safety at work.

The HSC is appointed by the government. The Commission is responsible for proposing health and safety law and standards. It consults professional bodies with an interest in health and safety such as trade unions and industry.

The HSC:
- appoints the Health and Safety Executive (HSE) which regulates health and safety law in industry and public areas
- gives local authorities delegated power to regulate health and safety law in premises such as retail shops, offices, catering services, restaurants, hotels, etc.

Health and safety law is enforced by:
- health and safety inspectors from the HSE
- environmental health practitioners (EHPs) and technical officers from local authorities
- fire officers from the fire service
- for factories, farms and hospitals, a health and safety inspector from the HSE
- for shops, restaurants and leisure centres, the local EHP. Fire officers can visit all these premises for the purposes of enforcing the law on fire safety and fire precautions.

Legal responsibilities

The Act makes it clear that everyone at the workplace (employers, managers and workers) has a responsibility for health and safety.

Employers and managers, for example, have the obligation to identify hazards, assess the risks that are present in the workplace and introduce precautions and preventative measures to reduce risks.

Employers' responsibilities to their employees

Every employer has a duty to ensure health and safety and welfare at work of all employees in so far as is reasonably practicable. In hospitality this means:
- providing and maintaining kitchens, restaurants, accommodation and systems of work that are safe and without risk to health
- making sure that storage areas and transporting articles such as food, and so on, are safe and present no risk to health
- there is information and instruction on training and supervision that is necessary to ensure the health and safety of employees at work
- maintaining the premises and building to make sure they are safe and pose no risk to health
- maintaining entrances and exits to the workplace and access to work areas that are safe and without risk to health
- providing a clean and safe environment with good welfare facilities
- where necessary, providing health surveillance of employees.

Employers must provide a written statement about:
- the general policy towards employees' health and safety at work
- the organisation and arrangements for carrying out that policy

An organisation with a board of directors/governors must formally and publicly accept its collective role in providing health and safety leadership in the organisation.

Duties of employees

Every employee has a duty to take reasonable care of their own health and safety and that of other people who may be affected by what they do or do not do in the course of carrying out work.

Employees must cooperate with the employer to enable the employer to comply with the relevant employer's duties.

Regulations

The main current health and safety regulations were originally made in 1992. Since 1992 some of the regulations have been updated:

- Management of Health and Safety at Work Regulations 1999
- Personal Protective Equipment Regulations 1992
- Provision and Use of Work Equipment Regulations 1998 (PUWER)
- Manual Handling Operations Regulations 1992
- Workplace (Health, Safety and Welfare) Regulations 1992
- Health and Safety (Display Screen Equipment) Regulations 1992.

Other regulations which also apply to the workplace include:

- Lifting Operations and Lifting Equipment Regulations 1998 (LOLER)
- Control of Substances Hazardous to Health Regulations 2002 (COSHH)
- Noise at Work Regulations 1989
- The Health and Safety (First Aid) Regulations 1981
- Reporting Injuries, Diseases and Dangerous Occurrences Regulations 1995 (RIDDOR).

RIDDOR puts duties on employers, the self-employed and people in control of work premises (the Responsible Person) to report certain serious workplace accidents, occupational diseases and specified dangerous occurrences (near misses).

Risk assessment

The prevention of both accidents and food poisoning in catering establishments is essential. It is necessary to assess each situation and decide what action should be taken.

Working in the kitchen environment requires the use of kit and equipment in order to work safely and hygienically. Chefs' clothing not only looks professional but also helps to promote a hygienic environment as well as providing protection to the wearer. The material used in the production of chefs' whites is designed to protect the body from scalds from hot liquids – the apron, for example should be worn below knee length to protect the legs.

Everyday kitchen tasks have the potential to cause injuries

Equipment and clothing designed for this purpose is referred to as personal protective equipment (PPE). Other examples include the use of masks, to avoid breathing in fumes from cleaning chemicals, for example, or goggles to protect the eyes.

Injuries also result from slips, trips, falls and knife cuts, and people may suffer scalds and burns as a result of being in contact with hot liquids, hot surfaces and steam. Despite the existence of these hazards, experienced and knowledgeable chefs tend to foresee and avoid them.

It is important in the professional kitchen that potential hazards are identified and risks assessed. From the point of identification, it is important that measures are put into place to eliminate or control the risk and this information should be recorded to support the development of policies and procedures. With such measures in place, the working environment will be a safer place for all concerned.

An awareness of how to work in a kitchen and avoid these hazards develops through experience, but can also be facilitated through induction and training. It is important to understand the meaning of the following three terms, which are in regular use:

1 Hazard: the potential to cause harm.
2 Risk: the likelihood that harm will result from a particular hazard (the catering environment may have many hazards but the aim is to have few risks).
3 Accident: an unplanned or uncontrolled event that leads to or could have led to an injury, damage to equipment or other loss.

Carrying out a risk assessment

A risk assessment can be divided into four levels:

1 Minimal risk: safe conditions with safety measures in place.

2 Some risk: acceptable risk; however, attention must be paid to ensure that safety measures are in operation.

3 Significant risk: where safety measures are not fully in operation (also includes food most likely to cause food poisoning); requires immediate action.

4 Dangerous risk: where processes and operation of equipment should stop immediately; the system or equipment should be completely checked and operation recommenced after clearance.

To carry out a risk assessment:

● assess the risks
● determine preventative measures
● decide who carries out safety inspections
● decide frequency of inspection
● determine methods of reporting back and to whom
● detail how to ensure inspections are effective
● carry out safety training related to the job.

The purpose of the exercise of assessing the possibility of risks and hazards is to prevent accidents. Under the Control of Substances Hazardous to Health Regulations (COSHH) 1999, it is necessary for employers to carry out risk assessments of all hazardous chemicals and substances that employees may be exposed to at work and to survey all areas, in order to ascertain the chemicals and substances in use.

Some examples of chemical substances found in kitchens are:

● cleaning chemicals, alkalis and acids
● detergents, sanitisers, descalers
● chemicals associated with burnishing
● pest control chemicals, insecticides and rodenticides.

Chefs and kitchen workers must also be aware of the correct handling methods required.

Evacuation procedures

There are a number of situations which may require an evacuation of the premises to remove all staff, customers and visitors in order to protect them from any harm. This may be due to fire, a security risk or even due to a false alarm.

Organisations need to have clear evacuation procedures in place with clear signage and instructions as to what people should do in the case of a raised alarm. Staff may have a designated role to marshal and assist others leaving the premises and heading for the designated meeting point. Organisations should regularly test their alarms and have regular staff training and practice evacuations to ensure that plans and systems work effectively.

For further information regarding health and safety legislation, refer to the Health and Safety Executive (HSE) website – www.hse.gov.uk

The HSE will also be able to provide advice and guidance on food safety management systems such as 'Safer Food Better Business'. Other forms of guidance with reference to the safe use of products and equipment are supplied in the literature from the manufacturers concerned.

Applying staff supervisory skills within a small team

Certain leadership qualities are needed to enable the supervisor to carry out their role effectively.

These qualities include the ability to:

● communicate
● initiate
● make decisions
● coordinate
● mediate
● motivate
● inspire
● organise.

Those under supervision should expect their supervisor to show:

● consideration
● understanding
● loyalty
● respect
● cooperation
● consistency.

The good supervisor is able to obtain the best from those for whom he or she has responsibility and can also completely satisfy the management of the establishment that a good job is being done. The job of the supervisor is essentially to oversee the work of the staff, identifying training and support needs and ensuring that staff work within the guidelines as described in the various work policies.

In the catering industry, the name given to the supervisor may vary, including sous-chef, chef de partie, kitchen supervisor or section chef. In some operations such as hospitals, the kitchen supervisor will be responsible to the catering manager, while in hotels and restaurants a chef de partie will be responsible to the head chef. The exact details of the job will vary according to the different areas of the industry and the size of the various units, but generally the supervisory role involves three functions: technical, administrative and social.

Generic function

In the role of a chef and supervisor, the essence of supervision should lead to the production of high-quality products at a consistent level. To achieve this aim, supervision provides the means to assess individual and team behaviour and to evaluate whether behaviour and performance are likely to meet these targets. A good supervisor will ensure that the various functions and outputs of job roles are completed on schedule in a safe and hygienic manner. This will help to build a working environment that is valued by the employees and one in which legislative requirements are met.

Technical function

Culinary skills and the ability to use kitchen equipment are essential for the kitchen supervisor. Most kitchen supervisors will have worked their way up through the section or sections before reaching supervisory responsibility. The supervisor needs to be able 'to do' as well as know 'what to do' and 'how to do it'. It is also necessary to be able to do it well and to be able to impart some of these skills to others.

Administrative function

The supervisor or chef de partie will, in many kitchens, be involved with the menu planning, sometimes with complete responsibility for the whole menu, but more usually for part of the menu (as happens with the larder chef and pastry chef). This includes ordering foodstuffs (which is an important aspect of the supervisor's job in a catering establishment) and, of course, accounting for and recording materials used. The administrative function includes the allocation of duties and, in all instances, basic work-study knowledge is needed to enable the supervisor to operate effectively. The supervisor's job may also include writing reports, particularly in situations where it is necessary to make comparisons and when new developments are being tried.

Social function

The role of the supervisor is perhaps most clearly seen in staff relationships because the supervisor has to motivate the staff under his or her responsibility. To 'motivate' could be described as the initiation of movement and action. Having got the staff moving, the supervisor then needs to exert control. Then, in order to achieve the required result, the staff need to be organised. Therefore, the supervisor has a threefold function regarding the handling of staff; namely to organise, to motivate and to control. This is the essence of staff supervision.

Elements of supervision

The accepted areas of supervision include: forecasting and planning, organising, commanding, coordinating, controlling, delegating, coaching and motivating. Each of these will be considered within the sphere of catering.

Forecasting

Before making plans it is necessary to look ahead, to foresee possible and probable outcomes and to allow for them. For example, if the chef de partie knows that the following day is her/his assistant's day off, she/he looks ahead and plans accordingly. When the catering supervisor in the hospital knows that there is a flu epidemic and two cooks are feeling below par, he/she plans for their possible absence. If there is a spell of fine, hot weather and the cook in charge of the larder foresees a continued demand for cold foods or when an end to the hot spell is anticipated, then the plans are modified.

For the supervisor, forecasting is the good use of judgement acquired from previous knowledge and experience. For example, because many people are on holiday in August fewer meals will be needed in the office restaurant, or perhaps there are no students in residence at a college/university halls, but a conference is being held and 60 meals are required. Events such as a motor show, bank holidays, the effects of a rail strike or a wet day, as well as less predictable situations, such as the number of customers anticipated on the opening day of a new restaurant, all need to be anticipated and planned for.

Planning

From forecasting comes the planning, such as how many meals to prepare; how much to have in stock (should the forecast not have been completely accurate); how many staff will be needed; which staff and when. Are the staff capable of what is required of them? If not, the supervisor needs to plan some training. This, of course, is particularly important if new equipment is installed. Imagine an expensive item, such as a new type of oven, ruined on the day it is installed because the staff have not been instructed in its proper use, or equipment lying idle because the supervisor may not like it, may consider it is poorly sited and does not train staff to use it. As can be seen from these examples, it is necessary for forecasting to precede planning and from planning we now move to organising.

Organising

In the catering industry organisational skills are applied to food, to equipment and to staff. Organising in this context consists of ensuring that what is wanted is where it is wanted, when it is wanted, in the right amount and at the right time. Such organisation involves the supervisor in the production of duty rotas, maybe training programmes and also cleaning schedules. Consider the supervisor's part in organising an outdoor function where a wedding reception is to be held in a church hall: a total of 250 guests require a hot meal to be served at 2 p.m. and in the evening a

dance will be held for the guests, during which a buffet will be provided at 9 p.m. The supervisor would need to organise staff to be available when required, to have their own meals and maybe to see that they have got their transport home. The food would need to be ordered so that it arrived in time to be prepared. If decorated hams were to be used on the buffet then they would need to be ordered in time so that they could be prepared, cooked and decorated over the required period of time. If the staff had never carved hams before, instruction would need to be given – this entails organising training. The correct quantities of food, equipment and cleaning materials would also have to be at the right place when wanted and if all the details of the situation were not organised properly, problems could occur.

Commanding

The supervisor has to give instructions to staff on how, what, when and where; this means that orders have to be given and a certain degree of order and discipline maintained. The successful supervisor is able to do this effectively having made certain decisions and, usually, having established the basic priorities. Explanations of why a food is prepared in a certain manner, why this amount of time is needed to dress up food, say for a buffet, why this decision is taken and not that decision, and how these explanations and orders are given, determine the effectiveness of the supervisor.

Coordinating

Coordinating is a skill required to get staff to cooperate and work together. To achieve this, the supervisor has to be interested in the staff, to deal with their queries, to listen to their problems and to be helpful. Particular attention should be paid to new staff, easing them into the work situation so that they quickly become part of the team or partie. The other area of coordination for which the supervisor has particular responsibility is in maintaining good relations with other departments.

However, the most important people to consider will always be the customers such as, for example, the patients or school children who will receive the service. Good service is dependent on cooperation between waiters and cooks, nurses and catering staff, stores' staff, caretakers, teachers and suppliers. The supervisor has a crucial role to play here.

Controlling

This includes controlling people and products, preventing pilfering, as well as improving performance, checking that staff arrive on time, do not leave before time and do not misuse time in between. Controlling also involves checking that the product, in this case the food, is of the right standard, that is, of the correct quantity and quality, checking to prevent waste, and also to ensure that staff operate the portion control system correctly. This aspect of the supervisor's function involves inspecting and requires tact; controlling may include inspecting the waste bin to observe the amount of waste, checking the disappearance of a quantity of food, supervising the cooking of the meat so that shrinkage is minimised and reprimanding an unpunctual member of the team.

The standards of any catering establishment are dependent on the supervisor doing his or her job efficiently and standards are set and maintained by effective control, which is the function of the supervisor.

Delegating

It is recognised that delegation is the root of successful supervision; in other words, by giving a certain amount of responsibility to others, the supervisor can be more effective. The supervisor needs to be able to judge the person capable of responsibility before any delegation can take place. But then, having recognised the abilities of an employee, the supervisor who wants to develop the potential of those under his or her control must allow the person entrusted with the job to get on with it.

Coaching

Coaching is an important skill that requires a varied approach depending on a number of factors. For example, what level of experience and/or skills set does the member of staff possess? Some staff may be developing their skills from a fairly basic level whereas others possess a range of highly developed skills. As individuals, people naturally have different personalities. This is another factor that can influence the way in which people respond to help and advice. A good supervisor will learn how to approach such a range of staff and how they can positively develop their progress.

Motivating

Since not everyone is capable of, or wants, responsibility, the supervisor still needs to motivate those who are less ambitious. Most people are prepared to work in order to improve their standard of living but there is also another very important motivating factor – most people wish to get satisfaction from the work they do. The supervisor must be aware of why people work and how different people achieve job satisfaction and then be able to act upon this knowledge. A supervisor should have received training that enables them to attempt to understand what motivates people, as there are a number of theories that she or he can use to stimulate ideas.

The importance of communication

A supervisor must be able to communicate effectively. To convey orders, instructions, information and manual skills requires the supervisor to possess the right attitude to those with whom he or she needs to communicate. The ability to convey orders in a manner that is acceptable to the one receiving them is dependent not only on the words but on the emphasis given to the words, the tone of voice, the time selected to give them and on who is present when they are given. This is a skill that supervisors need to develop. Instructions and orders can be given with authority without being authoritative.

The supervisor needs technical knowledge and the ability to direct staff and to carry responsibility so as to achieve the specified targets and standards required by the organisation. He or she is able to do this by organising, coordinating, controlling and planning but, most of all, through effective communication.

Supervisory skills

Supervisors need a wide range of technical, people and conceptual skills in order to carry out their work.

Henry Mintzberg (*The Nature of Managerial Work*, Harper & Row, 1973) suggested that the supervisor has three broad roles that use these skills:

1 interpersonal: people skills
2 informational: people and technical skills
3 decision making: conceptual skills.

Technical skills

These are the skills that chefs, restaurant managers and the like need in order to do the job. The supervisor must be skilled in the area they are supervising because they will be required in most cases to train other staff under them. Supervisors who do not have the required skills will find it hard to gain credibility with the staff.

People skills

Supervisors are team leaders, therefore they must be sensitive to the needs of others. They must be able to communicate effectively and be able to build a team to achieve the agreed goals. Skills such as listening, questioning, communicating clearly, handling conflicts, and providing support and praise when praise is due are important for this aspect of the role.

Conceptual skills

A supervisor must be able to think things through, especially when planning or analysing why things are not going as expected. A supervisor must be able to solve problems and make decisions. For supervisors, conceptual skills are necessary for reasonably short-term planning. Head chefs and hospitality managers require conceptual skills for long-term strategic planning.

Supervisors and ethical issues

Ethical treatment of staff is the fair treatment of staff. A good supervisor will gain respect if they are ethical.

A supervisor must be consistent when handling staff, avoiding favouritism and perceived inequity.

Such inequity can arise from the amount of training or performance counselling given, from the promotion of certain employees and from the way in which shifts are allocated. Supervisors should engage in conversation with all staff, not just a selected few, and should not single out some staff for special attention.

Confidentiality

Confidentiality is often an important issue for a supervisor. Employees or customers may wish to take the supervisor into their confidence and the supervisor must be careful not betray this.

Leadership styles

Leadership style is the way in which the functions of leadership are carried out, the way in which the supervisor typically behaves towards members of the team. There are many dimensions to leadership and many possible ways of describing leadership style.

A leader or supervisor may be described as:

- Dictatorial – a supervisor who is dictatorial is autocratic and often oppressive and overbearing.
- Bureaucratic – a bureaucratic supervisor is one who follows official procedure and is very often office-bound. They stick to the rules and operate within a hierarchical system.
- Benevolent – a benevolent supervisor is kind, passionate, human, kind-hearted, good, unselfish and charitable.
- Charismatic – a supervisor who is charismatic has a special charm that inspires loyalty and enthusiasm from the team.
- Consultative – the consultative supervisor discusses issues with the team through team meetings.
- Participative – a participative supervisor gets involved with the team, taking part in activities and issues and making an active contribution to the success or failure of the team.
- Unitary – the unitary supervisor unites the team, bringing them together as a whole unit.
- Delegative – the delegative supervisor entrusts others in the team to make decisions, assigning responsibility or authority to others.

- Autocratic – this is where the supervisor holds on to power and all the interactions within the team move towards the supervisor. The supervisor makes all the decisions and has all the authority.
- Democratic – the team has a say in decision making. The supervisor shares the decision making with the team. The supervisor is very much part of the team.
- Laissez-faire – a genuine laissez-faire style is where the supervisor observes that members of the group are working well on their own. The supervisor passes power to the team members, allows them freedom of action, does not interfere but is available for help if needed. The word genuine is used because this is contrary to the type of supervisor who does not care, keeps away from trouble and does not want to get involved.

A good leader, regardless of their inherent style of leadership, will develop working relations built upon trust and respect. They will be a good observer and listener and will learn how to approach situations by making more informed judgements.

Reflect on your own practice as a leader and supervisor. Consider what training you think you need to improve your ability to supervise and lead a team.

Continuous training

Supervisors and chefs should encourage continuous training to improve knowledge and skills and change attitudes. This can lead to many benefits for both the organisation and the individual. Training leads to improvements in an individual's skills development and performance but will also help to develop teamwork and an understanding of the way in which an organisation operates, or aspires to operate, in order to achieve its aims and objectives. Training staff to develop communications skills and workplace behaviour, for example, helps to promote a positive working ethos and this will be easily transferable and visible to customers. It is also important that legislative aspects are continually updated in areas such as health and safety and food hygiene. Training is a continuous process for all concerned, including supervisors and managers.

Training and personnel development can help to:
- increase confidence, motivation and commitment of staff
- provide recognition, enhanced responsibility and the possibility of further career development
- give a feeling of personal satisfaction and achievement and broaden wider opportunities
- improve the availability and quality of staff.

Training is therefore a key element of improved organisational performance. Training improves knowledge, skill, confidence and competence.

Main styles of training

The main styles of training are output training, task training, performance training and strategic training. Examples of each style are given in Table 2.1 below.

Training should be viewed as an investment in people. Training requires the cooperation of the managers and supervisors with a genuine commitment from all levels in the organisation. There are different methods of training; as well as the formal methods such as attendance on training courses and working for qualifications, there are more informal methods where people learn by doing, from close observation of 'role models' and from being in challenging situations which

Table 2.1 Training styles

Style of training	Example
Output training	Investing in a new employee or new machine will endeavour to generate output as quickly as possible.
Task training	Involves selected individuals being sent on short training or college-based courses, i.e. hygiene courses, health and safety courses, financial training.
Performance training	Implemented when the organisation has grown substantially and becomes well established. Training is viewed positively, with a person responsible for overseeing training. Plans and budgets are now some of the tools used to manage the training process.
Strategic training	Implemented when the organisation recognises and practises training as an integral part of the management of people and the culture of the organisation.

require initiative and positive leadership. A great deal can be learnt by shadowing a supervisor or work colleague. Having a good mentor also helps personal development. A good supervisor is also able to mentor and coach members of the team to achieve their goals and objectives.

The importance of teamwork

Groups help to shape the work pattern of organisations as well as group members' behaviour and attitudes to their jobs.

Two types of team are often identified within an organisation:

1 The 'formal' team is the department or section created within an organised structure to pursue specified goals.
2 The 'informal' team is created to deal with a particular situation; members within this team have fewer fixed organisational relationships; these teams are disbanded once they have performed their function.

Both formal and informal teams have to be developed and led. Thought has to be given to relationships and the tasks and duties the team has to carry out. Selecting and shaping teams to work within the kitchen is very important. This is the job of the head chef. It requires management skills. Matching each individual's talent to the task or job is an important consideration.

A good, well-developed team will be able to do the following:

- create useful ideas
- analyse problems effectively
- get things done
- communicate with each other
- respond to good leadership
- evaluate logically
- perform skilled operations with technical precision and ability
- understand and manage the control system.

Maintaining the health of the team and developing it further demands constant attention. The individual members of a group will never become a team unless effort is made to ensure that the differing personalities are able to relate to one another, communicate with each other and value the contribution each employee or team member makes.

The chef, as a team leader, has a strong influence on his/her team or brigade. The chef in this position is expected to set examples that have to be followed. She/he has to work with the brigade, often under pressure, and sometimes dealing with conflict, personality clashes, change and stress. The chef has to adopt a range of strategies and styles of working in order to build loyalty, drive, innovation, commitment and trust in team members.

The team needs to identify its strengths and weaknesses, and develop ways to help those team members affected to overcome any weaknesses they may have.

Test yourself

1 Name three jobs in the kitchen that you consider present a higher risk of injury than others.
2 List the qualities you admire most in a good supervisor.
3 List the duties a sous-chef or kitchen supervisor could or should delegate to a chef de partie.
4 Think about yourself and list what motivates you at work. On the contrary, think of things that cause poor levels of motivation.
5 As a supervisor, how would you treat staff fairly?
6 Name three leaders of whom you are aware (managers, executives, politicians) and why you consider them to be good leaders.
7 Explain what kind of leadership style you would adopt as a supervisory manager.
8 Write down what it is you admire in a good team.

Practical gastronomy

Influences on eating and drinking cultures

Gastronomy

Over the years there have been a number of definitions of gastronomy. In very simple terms it is the study of how food influences habits and the influence of history and location in society. Choosing what to eat is a complex development process which we learn from childhood and the way we are socialised into food habits through family and relationships. This is how our taste for certain foods is developed.

Taste

Why do we eat what we eat, select one dish from the menu in preference to another, choose one particular kind of restaurant or use a takeaway? Why are these dishes on the menu in the first place? Is it because the chef likes them, the customer or consumer wants them, or is this the only food available? What dictates what we eat?

Hospitality reflects the eating habits, history, customs and taboos of society, but it also develops and creates them. You have only to compare the variety of eating facilities available on any major high street today with those of a short while ago. Taste affects food choice and is based on biological, social and cultural perspectives. The perception of taste results from the stimulation of the taste cells that make up the taste buds. Taste is not specific to individual foods but to the balance between four main types of chemical compound.

Factors affecting what we eat

There are many factors influencing our choice of what we eat. These include our individual preferences, our relationships and emotional needs. Other factors such as what is acceptable to us as food, images of food, as well as the needs and preferences of people we eat with also affect our choice.

Lifestyle changes have led to increased consumption of ready meals, and more eating out.

The same changes have also created opportunities for people to experience a wider range of foods: improvements in technology, and more travel, make exotic foods more accessible.

The individual

Everyone has needs and wishes, which are met according to their own satisfaction. Tastes and habits in eating are influenced by three main factors:
- upbringing
- peer-group behaviour
- social background.

For example, children's tastes are developed at home according to the eating patterns of their family, as is their expectation of when to eat meals. Teenagers may frequent hamburger or other fast-food outlets and adults may eat out once a week at an ethnic or fine-dining restaurant, steakhouse or gastro pub.

How hungry an individual feels will affect their choice of what, when and how much to eat. However, extremes do exist – some people in the Western world overeat and food shortages cause under-nourishment in poorer countries.

Everyone ought to eat enough to enable body and mind to function efficiently; if you are hungry or thirsty it is difficult to work or study effectively. Health considerations may influence an individual's choice of food, either because they need a special diet for medical reasons or (as the current emphasis on healthy eating shows) because of a belief that everyone needs a nutritionally balanced diet. Many people nowadays feel it is healthier to avoid eating meat or dairy products. Others are vegetarian or vegan for moral or religious reasons.

Relationships

Eating is a necessity, but it is also a means of developing social relationships. You should also consider the needs and preferences of the people you eat with. This applies in the family or at your place of study or work. The provision of suitable foods and dishes for pupils at school mealtimes in an appropriate environment can be a means of developing good eating habits and fostering social relationships. For people at work, it is in canteens, dining rooms and restaurants that relationships may develop.

Often the purpose of eating, either in the home or outside it, is to be sociable and to meet people, or to renew acquaintances, or provide the opportunity for people to meet each other. Frequently there is a reason for the occasion (such as birthday, anniversary, wedding or awards ceremony), requiring a special party or banquet menu, or it may just be for a few friends who choose to have a meal at a restaurant.

Business is often conducted over a meal, usually at lunchtime but also at breakfast and dinner. Eating and drinking help to make work more enjoyable and effective, using food to build relationships!

Emotional needs

Sometimes we eat not because we need food but to meet an emotional requirement:

- For sadness or depression – in eating a meal we may comfort ourselves or give comfort to someone else; after a funeral, people eat together to comfort one another.
- For a reward or treat, or to give encouragement to oneself or to someone else; an invitation to a meal is a good way of showing appreciation.

Beverages that complement different foods

Food and drink are natural partners. These can come in the form of both non-alcoholic and alcoholic drinks. In some cultures, wine is seen almost as a food itself and is enjoyed as part of the dining experience. It is rarely consumed on its own as a beverage. However, in other countries, wine is consumed as a beverage in its own right and enjoyed in a social context, for example.

Wine is probably the first beverage that is thought of when matching food and drink together. The production of wine is historic, going back thousands of years but its consumption as a natural partner to food is as popular in modern times as at any point in its past. The production of wine is huge across the world with many new countries coming to the forefront of production

and providing a true alternative and competition to the historical producers of countries such as France and Italy.

Wine is an excellent example of a commodity that really tastes of its provenance. There are hundreds of grape varieties lending themselves better to particular climates, soil types and geography. This results in a huge variety of wines of different styles and complexity. Wine develops many characteristics and flavours across a spectrum that is almost unique. Strangely, wines develop the characteristics of so many other fruits, for example, it is rare that it actually tastes of grapes.

Certain wines lend themselves to certain types of food. For example, dry white wines are natural accompaniments to white fish dishes, whereas a full-bodied red wine will complement a rich red meat dish. There are also many wines produced that are natural partners to desserts and sweet dishes – Sauternes, from the Sauternais region of Graves in Bordeaux, and wines produced from the Muscat grape, for example.

However, there is a huge range of other drinks that are consumed with food. For example, some restaurants have beer lists in addition to wine lists to promote the way in which beers can complement food. It is also important to note that many people may prefer to avoid the consumption of alcohol. Therefore, alternatives must be considered, but in a similar way that will complement food rather than ways that distract or overpower the natural flavours and characteristics concerned.

The various cultures and regions of the world have their own histories and developments. Although wine is now being produced and consumed in more countries than ever before, this is a recent trend, particularly in areas of Asia. Such regions have traditions dating back thousands of years of drinking beverages such as green tea (China) or sake (Japan). In China, the production of alcoholic drinks was historically focussed on products made from grain and rice. This resulted in many varieties of rice wines and liquors which were consumed with food and regularly featured at formal banquets and celebrations (they were considered to have cultural and spiritual values).

Ideas about food

People's ideas about food and meals and about what is and what is not acceptable vary according to:

- where and how they were raised
- the area in which they live and its social customs.

Different societies and cultures have conflicting ideas about what constitutes good cooking, a good chef and about the sort of food a good chef should provide. The French tradition of producing fine food and their chefs being highly regarded continues to this day, whereas other countries traditionally may have less interest in the art of cooking, and less respect for chefs.

Individuals' ideas of what constitutes a snack, a proper meal or a celebration will depend on their backgrounds, as will their interpretation of terms such as lunch or dinner. One person's idea of a snack may be another person's idea of a main meal; a celebration for some will be a visit to a hamburger bar, for others, a meal at a fashionable restaurant.

The idea of what is 'the right thing to do' when eating varies with age, social class and religion. To certain people it is right to eat with the fingers, while others use only a fork. Some will have cheese before the sweet course; others will have cheese after it. It is accepted that children and, sometimes, elderly people need to have their food cut up into small pieces and that people of some religions do not eat certain foods. These ideas usually originate from practical and hygiene reasons, although sometimes the origin is obscure.

Images of food

Fashions, fads and fancies affect foods and it is not always clear if catering creates or copies these trends. Nutritionists inform us about foods that are good and necessary in the diet, what the effect of particular foods will be on the figure and how much of each food we require. This helps to produce an 'image' of food. This image changes according to research, availability of food and what is considered to constitute healthy eating.

What people choose to eat says something about them as a person; it creates an image. We are what we eat, but why do we choose to eat what we do

when there is choice? One person will perhaps avoid trying snails because of ignorance of how to eat them or because the idea is repulsive, while another will select them deliberately to show off to other diners.

One person will select a dish because it is a new experience; another individual will choose it because they have previously enjoyed eating it. A glutton is someone who eats too much; a gourmand particularly likes eating; a gourmet is someone who is particularly interested in the quality aspects of eating and drinking.

Crop failure or distribution problems may make food scarce or not available at all. However, foods in season are now supplemented by imported foods so that foods out of season at home are now available much of the time. This means that there is a wide choice of food for the caterer and the customer.

Food is available through shops, supermarkets, cash and carry stores, wholesalers and direct suppliers. It is now possible for people at home and caterers to purchase, prepare, cook and present almost every food imaginable due to rapid air transport and food preservation. Food spoilage and wastage are minimised; variety and quality are maximised.

It is essential that food looks attractive, has a pleasing smell and tastes good, since individuals are less likely to eat food that does not meet these criteria, even if it is nutritious. Remember that people's views on what is attractive and appealing will vary according to their background and experience.

Money, time and facilities

Money, time and facilities affect what people eat – the economics of eating affect everyone. How much money an individual has available or decides to spend on food is crucial to their choice of what to eat. Some people will not be able to afford to eat out; others will be able to eat out only occasionally; while for others, eating out will be a frequent event. The money that individuals allocate for food will determine whether they:

- cook and eat at home
- use a takeaway (for example, fish and chips, Chinese)

- go to a pub; eat at a pizza restaurant or at an ethnic or other restaurant.

The amount of time people have to eat at work will affect whether they use any facilities provided, go out for a snack or meal during their lunch break, or take in their own food to the workplace. The ease of obtaining food, the use of convenience and frozen food, and the facility for storing foods have led to the availability of a wide range of foods in both the home and catering establishments. It is possible to freeze foods that are in season and use them throughout the year, so eliminating spoilage in the event of a glut of items.

The media

The media influences what we eat: television, radio, newspapers, magazines and literature of all kinds have an effect on our eating habits. Healthy eating, nutrition, hygiene and outbreaks of food poisoning are publicised; experts in all aspects of health, including those extolling exercise, diet and environmental health, tell us what they think should and should not be eaten.

Information given about the content of food in packets and the advertising of food influences our choice. The media contributes to our knowledge about eating and foods alongside our learning on this topic from the family, teachers, at school meals, at college, and through the experience of eating abroad.

The influences on our choice of food that are listed in Table 2.2 are separated for convenience but in reality they overlap. Only when sufficient food is available for survival can people begin to derive pleasure from eating food.

Contributions of individuals who have made a significant impact on professional cookery

Professional cookery has developed significantly over time and more rapidly in recent years than at any stage in history. With the major developments in transport, science, information communications technology and media, information is literally available at the touch of a screen or button.

In line with these developments, many individuals have made significant contributions to the industry, particularly over the last 150 years or so, and their contributions have helped to develop professional cookery in many new formats. For example, Alexis Soyer (1810–58) was a French chef who became the most celebrated cook in Victorian England, and was arguably the first celebrity chef.

Auguste Escoffier (1846–1935), another French chef, was a restaurateur and the writer of the legendry 'Repetoire de la cuisine'. Escoffier popularised and updated traditional French cooking methods and was responsible for the introduction of the 'partie' system as deployed in the modern kitchen. Fernand Point (1897–1955) another French restaurateur was later considered to be the father of modern French cuisine and today we still have Paul Bocuse in Lyon, France, continuing the era of such worldwide domination of French classical cuisine.

In current times, there are a number of major contributors to the industry including chefs such as Raymond Blanc (France), Gordon Ramsay (UK) and Thomas Keller, an American chef from California and the chef patron of the iconic French Laundry restaurant in California.

In line with the vast developments in and knowledge of products, processes and technologies, there are chefs challenging the rulebooks and defining new boundaries. Such chefs are using scientific principles, new technologies and equipment and are largely responsible for a new and exciting style of cooking which challenges the senses, whilst providing exasperating and highly impressive forms of presentation. Chefs such as Ferran Adria (Spain) and Heston Blumenthal (UK) are leading figures in these areas.

Futhermore, the media coverage of the catering industry has raised the profile of many chefs and restaurateurs to a point where they enjoy celebrity status alongside those from areas such as sports and entertainment. Some writers and food critics are now household names and there are many sources of information that the general public can use to research the quality and reputation of chefs and restaurants. For example, the Michelin

Table 2.2 Influences on choice of food

Media	Transport	Religion
TV	Transport of foods by sea, rail and air	Taboos
Books	Transport of people	Festivals
Newspapers		Pork, beef, shellfish, alcohol, halal, kosher
Journals		
Geographical	**Historical**	**Economic**
Climate	Explorations	Money to purchase
Indigenous fish, birds, animals, plant life	Invasions	Goods to exchange
Soil, lakes, rivers, seas, terrain	Establishment of trade routes	
Sociological	**Political**	**Cultural**
Family	Tax on food	Ethnic
School	Policies on 'food mountains'	Tribal
Workplace	Export and import restrictions	Celebrations
Leisure		
Fashion and trends		
Psychological	**Physiological**	**Scientific**
Appearance of food	Nutritional	Preservation
Smell	Healthy eating	Technology
Taste	Illness	Shorter ripening times
Aesthetics	Additives	Reduction in fat content in livestock
Reaction to new foods		Increased resistance to pests/disease
		Increased use of fertilisers
		Increased yields
		Increased shelf life
		GM foods
		Irradiated foods
		Intensive farming
		Ready meals – chilled, frozen, sous-vide

guide is produced every year to reflect the highest standards in the culinary world. Inspectors grade restaurants using a star-based system where even one star is a highly commendable achievement and the much rarer award of three stars shows a restaurant at the very top of the culinary ladder. There are a number of other schemes including the rosette scheme operated by the AA and nomination in the Good Food Guide, which is based on a percentage system.

Cultural influences on our choice of food

Cultural variety

The differing races and nations of the world represent a great variety of cultures, each with their own ways of cooking. Knowledge of this is essential for those working in the catering industry because:

- there has been a rapid spread of tourism, creating a demand for a broader culinary experience
- many people from overseas have opened restaurants using their own foods and styles of cooking

- the development of air cargo means perishable foods from distant places are readily available
- the media, particularly television, has stimulated an interest in worldwide cooking.

Religion and food and drink

- Christianity has little/no restrictions on diet, follows some rituals (fasting at Lent/no meat on Fridays).
- Judaism follows the law of Torah; diet has no pork or certain other animals including camels and horses; no molluscs or crustaceans; follows Kosher practices.
- Islam/Muslim (Middle East) follows the Prophet Muhammad and the Koran; pork is forbidden; animals are slaughtered according to ritual (Halal); during Ramadan eating is forbidden between dawn and dusk; no alcohol.
- Hinduism (India) believes in the principle of non-violence and therefore cannot harm other animals; vegetarian (not bulking the flesh by eating other beings) and no eggs (embryos); cook with ghee.
- Buddhism (East Asia – China, Japan, Thailand) love and compassion are central virtues; believes in the power of own conscience; avoid strong drinks.

Food and celebrations

Food is used every day to express a range of emotions: love, happiness, joy, satisfaction, and so on. Food is used in celebrations to convey these emotions in all parts of the world, regardless of culture or religion. Food can unite and strengthen community bonds and help to maintain a common identity amongst groups of people. Different countries use food in different ways to help celebrate special occasions such as Christmas, New Year, weddings and birthdays.

Christmas

In Britain it is traditional to serve roast turkey, Christmas pudding and mince pies as part of the festivities, but in other countries different foods are eaten. Some examples of different types of food eaten at Christmas include:

- France – black and white pudding, which is a sausage containing blood

- French Canada – desserts include doughnuts and sugar pie
- Germany – gingerbread biscuits and liqueur chocolates
- Nicaragua – chicken with a stuffing made from a range of fruits and vegetables including tomato, onion and papaya
- Russia – a feast of 12 different dishes, representing Christ's disciples.

New Year

Traditional New Year foods around the world include:

- Greece – a special sweet pastry baked with a coin inside it
- Japan – up to 20 dishes are cooked and prepared one week before the start of the celebrations; each food represents a New Year's wish (for example, seaweed asks for happiness in the year ahead)
- Scotland – haggis (sheep's stomach stuffed with a spiced mix of oatmeal and offal), gingerbread biscuits and scones
- Spain – 12 grapes, meant to be put into the mouth one at a time at each chime of the clock at midnight.

Lunar New Year

In many Asian countries, the New Year does not start on January 1, but with the first full moon in the first Chinese lunar month. Traditional Lunar New Year's food includes:

- China – fish, chestnuts and fried foods
- Korea – dumpling soup
- Vietnam – meat-filled rice cakes and shark fin soup.

Weddings

Around the world, weddings share common ground – no matter what the religion or culture, the typical wedding is a joint celebration for the families that involves a wedding cake and traditional foods. Foods that feature prominently in weddings include:

- China – roast suckling pig, fish, pigeon, chicken, lobster and a type of bun stuffed with lotus seeds (it is especially important to offer both lobster and chicken – the lobster represents the dragon and the chicken the phoenix so including both on the menu is thought to harmonise the Yin and Yang of the newly joined families)

- Indonesia – foods served depend on the region and religion but could include spicy rice dishes like nasi goreng, dim sum, sushi or even Western recipes like beef Wellington
- Italy – bowtie-shaped twists of fried dough, sprinkled with sugar (representing good luck), roast suckling pig or roast lamb is often the main dish, accompanied by pastas and fruits; traditional wedding cakes in different regions of Italy include one made from biscuits and another topped with fruit
- Korea – noodles are served, because they represent longevity
- Norway – the traditional wedding cake is made from bread topped with cream, cheese and syrup
- Britain – the honeymoon has been said to originate from a time when the father of the bride gave the groom a moon's (month's) worth of mead (alcoholic beverage made from honey) before the bride and groom left after the ceremony.

Birthdays

The custom of the birthday party originated in medieval Europe, when it was supposed that people were vulnerable to evil spirits on their birthdays. Friends, family members, festivities and presents were thought to ward off the spirits. Traditional birthday foods from around the world include:

- Australia – birthdays are often celebrated by sharing a decorated birthday cake with lit candles, which the person celebrating the birthday blows out while making a wish
- England – a cake may be baked containing symbolic objects which foretell the future (if your piece of cake has a coin, for example, you will one day be wealthy)
- Ghana – the child's birthday breakfast is a fried patty made from mashed sweet potato and eggs, and traditional birthday party fare includes a dish made from fried plantain (a kind of banana)
- Korea – for their first birthday, the child is dressed and sat before a range of objects including fruit, rice, calligraphy brushes and money; whichever item the child picks up predicts their future (for example, picking up the rice indicates material wealth); after this ceremony, the guests eat rice cakes

- Mexico – a papier-mâché container in the shape of an animal (piñata) is filled with lollies and other treats and the child who is celebrating his or her birthday is blindfolded and hits at the piñata until it breaks; the treats are shared among the guests
- Western Russia – the birthday boy or girl is given a fruit pie instead of a cake.

Consumer behaviour

Consumer behaviour describes the relationship between how individuals make decisions about how to spend their available resources on food, goods and services (their resources being their money, time and effort) and how producers such as the hospitality industry act to meet or create their needs and wants.

For example:
- who buys what
- when they buy
- why they buy
- how they buy
- where they buy
- how often they buy.

Consumer decision making

Here are some examples of the types of factors affecting the consumer purchasing process, that is, why consumers buy certain foods and services:
- cultural – cultural trends and norms, customs, religion, myths, symbolism, local, regional, national preferences, habits
- economic – income, prices, taxes
- marketing – advertising and promotion, distribution, restaurant/hotel location, size, product, portfolio and layout
- physiological – heredity, allergy, taste, food acceptability/intolerance
- political – EU legislation, food policy, Common Agricultural Policy
- psychographic – personality, self-concept, lifestyle, values, attitudes, beliefs, emotions, mood, preferences, significance of food
- social – social class, reference groups, household size, family, life-cycle stage, demography, educational level

- technical – food processing and preparation methods, cooking and storage options, packaging materials and type, nature of ingredients
- other – seasonality, perishability, portability.

(Source: Suzan Green, 'Consumer product management' in Proudlove, 2010, *The Science and Technology of Foods*, Forbes Publications)

Depending on individual products, dishes, types of restaurants and the circumstances of the consumer, all of these factors will have some bearing on what is selected, the quality, when and how often.

An example of four basic stages in choosing food is shown in Table 2.3.

Consumers may be conscious of all or none of these stages. Sometimes decision, purchases and consumption experiences are done on autopilot. Sometimes one stage may dominate – I'm hungry.

The consumer and society

Lifestyles are actual patterns of behaviour and are constructed by measuring consumers' activities, interests and opinions, which affect their food choice (see Table 2.4).

More people are consuming more varied food and drink products on more occasions than ever before. For an explanation of the relationship between cultural values and consumer food choice see Table 2.5.

Factors affecting the dining experience (sensory evaluation)

The quality of cooking and the food served is often the first thing that comes to mind when evaluating a dining experience. However, there are many other factors that contribute. The dining experience is a journey from the point of selection, the reservation process (if required) and the welcome on arrival through to the friendliness and professionalism of the staff, the quality of furnishings, the wine list, the technical ability of the staff (chefs, waiting staff, sommelier, etc.) right through to the end of the meal and the point of closure with payment and departure. It refers to the execution of the whole process.

Dining experiences come in many formats, both formally and informally. This could be across a

Table 2.3 Example of four basic stages in choice of food

1	Noticing	I'm hungry. That looks tasty.
2	Choosing	I feel like a snack. I like that brand/flavour.
3	Acting	I'll buy that to eat now. Just a small portion will do.
4	Assessing	I prefer the item I usually buy. That was good value for money.

Table 2.4 Factors affecting lifestyles

Activities	Work, hobbies, social events, vacation, entertainment, club membership, community, shopping, sports
Interests	Family, home, job, community, recreation, fashion, media, achievements
Opinions	Themselves, social issues, politics, business, economics, education, products, future, culture
Demographics	Age, education, income, occupation, family size, dwelling, geography, stage of life cycle

(Source: Reprinted with permission from *Journal of Marketing*, published by the American Marketing Association, Plummer, 38:1, 1974, 'The concepts and application of life style segmentation')

range of occasions including breakfast, lunch, afternoon tea, dinner or even when snacking. The formality of the occasion often leads to the development of perceptions and expectations; a business lunch would not have the same atmosphere or level of formality as a 21st birthday celebration, for example.

The venue also adds a level of variety to the type of food and service that is perceived. Chained restaurants are often identical in the way they standardise their food and service. Bistros and brasseries offer informal, social dining whereas fine dining establishments can be quite formal in their approach (although this sector of the industry is becoming less formal while being no less attentive to quality). There is also the staff-feeding industry to consider, which has changed significantly from traditional staff canteens to often thoughtful dining spaces with multiple food offerings in the form of sandwich and salad bars, live cooking stations, dessert bars and a range of international dishes.

Table 2.5 Cultural values and consumer food choice

Core cultural values	Food production
A more casual lifestyle with less formality	Destructuring of meal occasions and more individual autonomy over what is eaten
	Multiple product choices consumed at the same meal time by different people and less formal meals and meal times
Pleasure seeking and novelty – a desire for products and services which make life more fun	Constant innovation and product differentiation in all aspects – taste, texture, portion size, packaging, advertising, branding, product concepts, etc.
	Food as entertainment
Consumerism – increased concern over value for money with rising expectations about quality and performance	Rise in 'grocerant' products – restaurant-style food available to take home, for 'eating out, staying in'
	Increase in functional foods (e.g. energy drinks, vitamin-enriched products)
Instant gratification – living for today and intolerance of non-immediate availability	Rise in treats, indulgence, luxury items and super-premium lines
	More convenient access and availability through wider distribution of food
Simplification – a removal of time and energy spent on 'unnecessary' things or tasks	More pre-prepared, pre-packaged, processed and added-value lines for consumption at once or after microwaving to cut down effort in product selection, preparation, cooking and clearing away
Time conservation – time has to be used effectively	Pre-/part-prepared complete or partially ready meals
Concern with appearance and health, youth, keeping fit and looking good	Expansion/creation of calorie-light product meals
	Increase in low-fat, low-calorie, low-salt, high-fibre products and substitutes
	More product innovation in the areas of functional foods and nutraceuticals
	Meat reduction and meat substitutes (mycoprotein, soya, tofu)
	Eat yourself healthy campaign – 'five fruits and vegetables a day'
	Mediterranean 'superfoods' (e.g. garlic, olive oil, red wine, red peppers, sun-dried tomatoes, pasta, rice, fish and shellfish)

(Source: Proudlove, 2010, *The Science and Technology of Foods*, Forbes Publications)

The supply and use of commodities

The impact of the development of transport on food

The development of transport networks has completely changed the way in which food is consumed. Even going back just 40 years, the food available would be nothing like the offer available today. This is due to the development of transport and the conditions in which fresh food items, in particular, are distributed throughout the world. The development of the aircraft, in particular, and the extent to which air travel became accessible to trade networks changed the range of ingredients at a chef's disposal. This range became colossal in comparison, with foods from every corner of the world available to chefs through the various networks of supply.

Although this is fantastic in one sense, it does raise many issues including the use of seasonal and local produce. There are also many environmental and ethical considerations, including the carbon footprint created by food miles and fair-trade issues for farmers from poorer countries.

Suppliers

Suppliers of food include:

- Producers – the producers of the food, for example the farmers, also sell their produce direct through farm shops or to supermarkets or large catering companies.
- Wholesalers – wholesale suppliers buy from the producers, food manufacturers, and so on and sell to the caterer. Cash and carry wholesalers are an example of this type of operation.
- Retailers – large retail supermarkets stock a wide range of ingredients. This type of buying may be suitable for a small restaurant operation.

- Purchasing consortiums – organisations set up to negotiate prices for goods and commodities for hospitality companies to enable them to obtain the best possible price for the desired quality.

Selecting suppliers

The selection of suppliers is an important part of the purchasing process. First, consider how a supplier will be able to meet the needs of your operation. Consider:

- price
- delivery
- quality/standards.

You may obtain information on suppliers from other purchasers. Buyers should be encouraged to visit suppliers' establishments. When interviewing prospective suppliers, you need to question how reliable a supplier will be compared to the competition and how stable under varying market conditions.

It is important to ensure that suppliers are able to supply to the demand that the business requires. It is also essential to ensure that the supplier operates hygienically and to an HACCP (Hazard Analysis Critical Control Points) system to identify any critical control point within the food preparation chain. This will help to develop the supplier's reputation as a high quality and reliable business.

Knowing the market

Since markets vary considerably, in order to do a good job when purchasing commodities a buyer must know the characteristics of each market. A market is a place in which ownership of a commodity changes from one person to another. This exchange of ownership could occur while using the telephone, on a street corner, in a retail or wholesale establishment, or at an auction. It is important that a food and beverage purchaser has knowledge of the items to be purchased, such as:

- where they are grown
- seasons of production
- approximate costs
- conditions of supply and demand
- laws and regulations governing the market and the products

- marketing agents and their services
- processing
- storage requirements
- commodity and product, class and grade.

Buying tips

As a buyer you should:

- Make sure that you have an up-to-date and sound knowledge of all commodities, both fresh and convenience, to be purchased.
- Be aware of the availability of the different types and qualities of each commodity.
- When buying fresh commodities, be aware of part-prepared and ready-prepared items available on the market.
- Keep a sharp eye on price variations. Buy at the best price you can to ensure the required quality and also an economic yield. (The cheapest item may prove to be the most expensive if waste is excessive.) When possible, order by number and weight. For example, 10 kg pineapples could be 40 × 250 g or 20 × 500 g. It could also be 10 kg total weight of various sizes, which would make efficient portion control difficult.
- Organise an efficient system of ordering, ensuring that you keep copies of all orders for cross-checking, whether orders are given in writing, verbally or by telephone.
- Compare purchasing by retail, wholesale and contract procedures to ensure the best method is selected for your own particular organisation.
- Explore all possible suppliers: local or markets, town or country, small or large.
- Keep the number of suppliers to a minimum.
- At the same time, have at least two suppliers for every group of commodities, when possible. The principle of having competition for the caterer's business is sound.
- Issue all orders to suppliers fairly, allowing sufficient time for the order to be implemented efficiently.
- Request price lists as frequently as possible and compare prices continually to make sure that you buy at a good market price.
- Buy perishable goods when they are in full season, as this gives the best value at the cheapest price. To help with purchasing the correct quantities, it is useful to compile a purchasing

chart for 100 covers from which items can be divided or multiplied according to requirement. An indication of quality standards can also be incorporated in a chart of this kind.

- Ensure that all deliveries are checked against the orders given for quantity, quality and price. If any goods delivered are below an acceptable standard they must be returned, either for replacement or credit.
- Ensure that all containers are correctly stored, returned to the suppliers where possible and the proper credit given – containers can account for large sums of money.
- Check all invoices for quantities and prices.
- Check all statements against invoices and pass them swiftly to the office so that payment may be made in time to ensure maximum discount on purchases.
- Foster good relations with trade representatives because you can gain much useful up-to-date information from them.
- Keep up-to-date trade catalogues, visit trade exhibitions, survey new equipment and continually review the space, services and systems in use in order to explore possible avenues of increased efficiency.

Ethical and sustainable considerations

Here are some simple guidelines to follow to help take into account ethical and sustainable considerations when purchasing food:

- Use local, seasonal and available ingredients as standard to minimise goods transport, storage and energy use.
- Specify produce from farming systems that minimise harm to the environment, such as certified organic.
- Limit foods of animal origin such as meat products, dairy products and eggs, as livestock farming is one of the most significant contributors to climate change. Promote meals rich in fruit, vegetables, pulses and nuts.
- Ensure that meat, dairy and egg products are produced to high environmental and animal welfare standards.
- Exclude fish species identified as most 'at risk' by the Marine Conservation Society and specify fish only from sustainable sources.

- Buy fair-trade certified products and drinks imported from poorer countries to ensure a fair deal for disadvantaged producers.
- Avoid bottled water. Serve plain, filtered tap water. This will minimise transport and packaging waste.

(Source: Defra, *Putting it into Practice*, 2008, reproduced under the terms of the click-use licence)

Free trade

Free trade means that governments have to treat local and foreign producers in the same way, for example by not creating barriers to importing goods, services or people from other countries, or giving national businesses and farmers an advantage over foreign firms by offering them financial support. In practice, truly free trade has never existed and the reduction of trade barriers is always subject to intense political negotiation between countries of unequal power.

Ethical trade

This involves companies finding ways to buy their products from suppliers who provide good working conditions and respect the environment and human rights.

Fair trade

This encourages small-scale producers to play a stronger role in managing their relationship with buyers, guaranteeing them a fair financial return for their work. Some corporate buyers help to set up schools and health centres on the farms in countries where the food is produced.

World Trade Organization (WTO)

The WTO was created in 1944 to liberalise world trade through international agreements. Based in Geneva, the WTO has 140 member countries, some of which have much more power than others. Some poor countries cannot afford to keep any staff in Geneva.

Globalisation

This is the rapid integration of trade and culture between the world's nations. With a more open market, goods and, by association, cash can now travel across the globe much more freely. This greater global trade has been able to happen for the following reasons:

- governments have changed laws that, in the past, restricted economic trade
- new technologies have enabled faster communication
- travel and transport costs have been reduced
- western companies have looked abroad for investment.

Measuring success of sustainability

Key success indicators for sustainability are:

- percentage of food sourced locally, nationally and abroad
- decrease in food wastage
- reduction in food miles
- financial contribution to the local economy
- increase in recycling
- increase in food sales.

Technologies and 'food miles' bring with them implications for greenhouse gas emissions and their consequent effects on the climate.

Consider for example:

- Meat – according to figures from the United Nations, animal farming globally produces more greenhouse gas emissions than all of the cars', lorries' and planes' carbon emissions in the world put together.
- Food miles – this refers to the distance the food travels from producer to consumer. Eating according to the seasons has disappeared for most people in the developed world. You can buy asparagus and strawberries all year round thanks to refrigeration, heated greenhouses and, of course, global food transportation.

Assessing commodities and ingredients

When deciding to purchase a commodity or ingredient, it is important to assess whether it is suitable for your needs and whether it will satisfy the demands of your menu. Ensure that you check:

- Quality and flavour – is the commodity or ingredient affordable? Is it cost-effective? Will it give the number of portions required? Will it give the yield required?

- Terms of supply – what are the terms of supply, for example delivery times, payment requirements?
- Supply meets demand – is the supplier able to regularly supply to your requirements? Can they supply the quantity required on a regular basis? Is the supply sustainable?
- Hygiene, hazard analysis critical control points – visit the supplier before committing to purchase. Inspect the supplier premises for hygiene. Do they have a HACCP policy? How effective is the policy? Do they have appropriate records?
- Supplier's reputation – what reputation does the supplier have? Who are they already supplying to? What do these restaurants, hotels, catering establishments think of the supplier and the produce? Are they reliable? What type of packaging do they use? How efficient are they? Do they consider the environment and ethics? What are their policies regarding the use of packaging, local produce, fair trade, animal welfare, for example?

Provenance

Provenance refers to the origin of foods. As with the production of wine, geography has a huge part to play in the way food is grown and produced. The climate, soil type and general terrain dictate how successful crops will fare. Water and sun are also essential components that lead to the ripeness and sweetness of products such as fruit and vegetables. Similarly, livestock and fish will be affected by their surroundings. For example, cattle and sheep need land to graze, and fish need suitable food supplies in order to grow and survive.

The food chain relies on geography in so many ways. Water is an essential commodity as it is required by mammals, vegetation and is the home to fish in rivers, lakes and seas. The variety of species and varieties of foods are endless but they all depend on a certain habitat to survive and prosper. This summarises the importance that geography plays in the development of food sources and the respect that it deserves from the human race that rely on its produce.

Due to their provenance, certain food types are particularly well-suited to the natural conditions in which they grow. The cocoa pod, for example, is well-suited to the tropical conditions surrounding the equator. Most of the world's cocoa is grown in regions 10 degrees either side of the Equator. This is because cocoa trees grow well in humid, tropical climates with regular rainfall and a short dry season. The trees also need even temperatures between 21–23°C. Other foods are naturally associated with their regions. Cheeses such as Roquefort in France and Stilton in England, for example, are only produced in the regions with which they are associated.

With provenance and seasonality comes an opportunity for restaurants to market the fact that they are utilising top-quality, sustainably sourced and/or seasonal produce. Customers gain confidence with this knowledge and are often attracted to dishes where the use of such foods is brought to their attention. A restaurant's reputation can also be enhanced as a thoughtful and high quality business by utilising products in this way.

Test yourself

1 Reflect on your own taste preferences and how they have been developed.
2 Consider how you consume food and drink and have learnt how to use food and drink to build relationships.
3 Think about your own images of food and how they affect your food choice.
4 Explain how the media affects people's food choices.
5 Explain how food is used in religion to symbolise events, its meanings and its symbolic use in ceremonies.
6 Explain why a chef should understand the implications of different cultures when planning menus.
7 Using these cultural values and consumer food choice factors, devise a week's menus for a sports team of your choice for breakfast, lunch and dinner.
8 Research the local suppliers in your area for fresh fruit and vegetables. From your list, choose your preferred supplier and give reasons.
9 Design a poster or leaflet to explain a sustainable food policy.
10 Are your dairy products produced to high animal welfare standards?
11 Name five foods that are particularly well known and linked to their provenance.

3 Pastes, fillings, creams and sauces

This chapter covers:

→ **NVQ level 2 Prepare and cook and finish basic pastry products**
→ **VRQ level 2 Produce paste products**
→ **NVQ level 3 Prepare and cook and finish complex pastry products**
→ **NVQ level 3 Produce sauces, fillings and coatings for complex desserts**
→ **VRQ level 3 Produce paste products.**

In this chapter you will:

→ **Prepare and cook paste products using correct tools and equipment, and safe and hygienic practices (level 2)**
→ **Identify and prepare a range of fillings, creams and sauces that are used to finish paste products (levels 2 and 3)**
→ **Prepare and cook paste products to the recipe specifications, in line with current professional practice (level 3).**

Recipes in this chapter

It is very important to note that many of the basic items prepared in this chapter are also used as foundations in other areas, and are therefore referred to in other chapters. For example, a filling such as crème pâtissière is commonly used in paste products but is also utilised in the preparation of a variety of hot and cold desserts (Chapters 6 and 7), petits fours (Chapter 9) and dough products (Chapter 5). Another example would be ganache, which is also used in a variety of ways.

Introduction

This chapter is linked very closely to Chapter 4. It is focussed on the preparation of the various pastes, fillings, creams and preparations used to produce a wide range of pastry products, whereas Chapter 4 takes the use of these base products further to produce a variety of pastry goods.

The range of pastes includes simple pastes such as shortcrust and sweet pastes, as well as pastes produced in a variety of different ways such as choux paste and puff paste.

The specialist area of paste production requires close attention to the following key points:
- Check all weighing scales for accuracy.
- Follow recipes carefully.
- Check all storage temperatures are correct.
- Always work in a clean, tidy and organised way; clean all equipment after use.
- Always store ingredients correctly: eggs should be stored at 12°C, flour in a bin with a tight-fitting lid, sugar and other dry ingredients in closed storage containers.
- Keep equipment clean and dry.

Hygienic working practices are essential in all areas of the pastry kitchen and it is paramount that food safety practices are followed throughout all stages of preparation and cooking. Items must be stored appropriately using the following guidelines:
- chilled items in a refrigerator between 1 and 4°C
- frozen items in a freezer between –18 and –22°C.

When storing items, they should be:
- clearly labelled
- dated
- covered or wrapped
- positioned on an appropriate shelf
- rotated with other stock items.

Health and safety !

Patisserie items are often viewed as an indulgence or a treat and people allow themselves to enjoy items that are often high in sugar and fat and therefore calories. However, there are opportunities to produce pastry products with low fat and low sugar options and also the opportunity to use wholemeal flour in place of white flour (e.g. in short pastry).

In terms of the health and welfare of customers, it is important to identify the ingredients that make up patisserie products. Flour-based products, of which there are countless types, would not be suitable for a customer with a gluten intolerance or coeliac disease. Another regularly used ingredient in the production of pastry products is sugar. Products of this type would be unsuitable for a diabetic customer.

Techniques in pastry work

Adding fat to flour

Fats act as a shortening agent in short and sweet pastes. The fat has the effect of shortening the gluten strands in flour, which are easily broken when eaten, making the texture of the product more crumbly. Short and sweet pastes are commonly used to produce the lining for the base of tarts and flans, filled with creams, purées and custards, such as crème pâtissière.

In contrast to the production of such pastes, the development of gluten in puff pastry is very important as it is needed to support the expanding steam during the baking process – this is what makes the paste rise. Therefore, depending on the type of paste being produced, the type of flour to be used will differ according to the way

Note

Service of frozen items can involve removal of the product from the freezer before serving. This will allow the product to acclimatise and the temperature will rise to its ideal service temperature of between –8 and –10°C.

in which the gluten will act in the preparation and cooking processes. Pastes such as short and sweet use soft flour, low in gluten, whereas choux and puff pastes require strong, high gluten, flour. As a general rule, any pastry that rises during baking is made from strong flour.

Fat can be added to pastes in a number of ways depending on the type of paste being produced. Fat can be rubbed into flour, melted in water or layered between an existing paste, to produce a range of pastes with different textures and uses. Generally, the more fat in ratio to flour, the richer the paste will become and, in the case of short and sweet pastes, the more shortening properties (light and crumbly) it will possess. However, as fat softens as temperature rises, pastes with a high ratio of fat to flour will become increasingly more difficult to handle. Some chefs will therefore choose to use a paste with a lower ratio of fat to flour in warmer conditions.

The various methods of adding fat to flour and examples of their use are described below.

Rubbing in

Rubbing in (for example for short pastry) can be carried out by machine or by hand.

With this method it is better to work with the fat if it is cold – it will be much easier to produce a fine crumb.

Creaming

In the production of sweet pastry, the fat can be creamed with sugar, followed by egg before the flour is added. The method of creaming can be achieved by machine or by hand.

When creaming, it is better to work with fat that is 'plastic' (at room temperature). This will make it easier to cream.

Remember to always cream the fat and sugar well, before adding the liquid.

Note

When the liquid in a paste is egg, rather than water, it is much less likely to activate the gluten within the flour, and will therefore produce a short, crumbly and light pastry product.

Lamination

Puff pastry is produced using the lamination method, by making a series of alternating layers of a flour-based paste and a fat of the same texture. This is done using a series of either single or double turns (see diagram) during the preparation of the paste.

Single turn

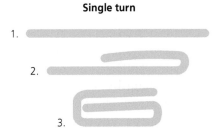

1.

2.

3.

At step 3, roll out and repeat five more times to give six turns in total.

Double turn

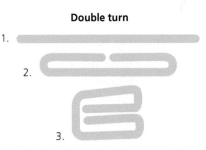

1.

2.

3.

At step 3, roll out and repeat three more times to give four turns in total.

Puff pastry is very versatile in both the patisserie and the savoury kitchen. Its texture is light and crisp and it is buttery and crunchy to the palate. It can be combined with all types of food in sweet and savoury dishes. One of the differences of making it is in the fat used. The taste and texture of a puff pastry made with pastry butter is considered the finest in comparison to those made with pastry margarine.

With this method the flour paste and the fat are laid in successive folds and rolled between each turn, rather than kneaded, so the two elements do not bind completely. The fat forms a separating layer which, when cooked, retains the steam generated by the water in the dough and produces the layer-separation effect. The flour paste, which includes part of the fat, becomes crunchy and takes on a pleasant golden tone rather than becoming hard and dry.

Boiling

Choux pastry is produced by placing butter in water and melting it by bringing it to the boil. Once boiled and melted, flour is mixed to the liquid and cooked until a smooth paste (panada) is produced. This is then cooled before beaten eggs are mixed in to produce a paste of piping consistency.

During the baking process, the moisture within the paste produces steam and the eggs and starch in the flour form a coating or case in which the steam is captured. As the paste cooks, it naturally aerates to form a hollow centre in which fillings such as crème Chantilly or crème pâtissière can be piped once the paste is cooked. The most well-known choux pastry products are profiteroles and chocolate éclairs.

Handling pastry

Techniques used to work pastry include:

- Folding: folding the initial paste when making puff pastry to create its layers, as in a vol-au-vents or gâteau pithiviers.
- Kneading: using your hands to work dough or puff pastry in the first stage of making.
- Relaxing: keeping pastry covered with a damp cloth, cling film or plastic to prevent a skin forming on the surface and to help prevent the pastry from shrinking during the baking process.
- Shaping: when producing flans, tartlets, barquettes and other goods with short, sweet or lining paste; this also refers to the crimping with the back of a small knife when using the finger and thumb technique.
- Docking: this is the piercing of raw pastry with small holes to prevent it from rising during baking, as when cooking tartlets blind (without a filling).

Rolling

- Roll the pastry on a lightly floured surface; turn the pastry regularly but delicately to prevent it sticking. Keep the rolling pin lightly floured and free from the pastry.
- Always roll with care, handling the pastry lightly – never apply too much pressure.
- Always apply even pressure when using a rolling pin.
- Handle as lightly and quickly as possible.

In a commercial environment, a pastry break is often used for rolling, due to the quantity of product.

Cutting

- Always cut with a sharp, damp knife.
- When using cutters, always flour them before use by dipping in flour. This will give a sharp, neat cut.
- Only use a lattice cutter on firm pastry; if the pastry is too soft, you will have difficulty lifting the lattice.

Glazing

A glaze is something that gives a product a smooth, shiny surface. Examples of glazes used for pastry dishes are as follows:

- Hot clear gel, which is produced from a pectin source obtainable commercially for finishing flans and tartlets; always use this while it is still hot. Cold gel is exactly the same except that it is used cold. Both gels give a sheen to the products and keep out oxygen, which might otherwise cause discoloration.
- Apricot glaze, produced from apricot jam, acts in the same way as a hot gel.
- Egg wash, applied prior to baking, produces a rich glaze during the cooking process.
- Icing sugar dusted on the surface of the product caramelises in the oven or under the grill.
- Fondant gives a rich sugar glaze, which may be flavoured and/or coloured.
- Water icing gives a transparent glaze, which may also be flavoured and/or coloured.

Finishing and presentation

It is essential that all products are finished according to the recipe requirements. Finishing and presentation are key stages in the process, as failure at this point can affect sales. The way products are presented is an important part of the sales technique. Each product of the same type must be of the same shape, size, colour and finish. The decoration should be attractive, delicate and in keeping with the product range. All piping should be neat, clean and tidy.

Some methods of finishing and presentation are as follows:

- dusting: a light sprinkling of icing sugar on a product using a fine sugar dredger or sieve, or muslin cloth
- piping: using fresh cream, chocolate or fondant
- filling: with fruit, cream, pastry cream, etc. (be careful never to overfill as this will often give the product a clumsy appearance and may be problematic for the customer to eat).

Piping fresh cream

- The piping of fresh cream is a skill; and like all other skills it takes practice to become proficient. Finished items should look attractive, simple, clean and tidy, with neat piping.
- Modern practice is to use a disposable plastic piping bag. If using a washable piping bag, it should be sterilised and dried after each use.
- Make sure that all the equipment you need for piping is hygienically cleaned before and after use to avoid cross-contamination.

Other considerations when preparing pastry items

- Ensure all cooked products are cooled before finishing.
- Always plan your time carefully.
- Understand why pastry products react in different ways according to the production process. Understand why pastry items must be rested or relaxed and docked. This will prevent excessive shrinkage in the oven, and docking will allow the air to escape through the product, preventing any unevenness.
- Use silicone paper or specialist silicone mats for baking in preference to greaseproof.

Sauces, creams and fillings used in patisserie

Sauces, creams and fillings used in pastry work have changed dramatically over the past 15 years and have gone from the classic anglaise, Chantilly and coulis preparations to what we see in modern restaurants today, such as foams, oils, cold creams stabilised with gelatine, syrups and convenience fruit purées.

Cold sauces, creams and fillings include:

- **coulis** – made from various soft fruit such as strawberries, raspberries, blackberries, etc.
- **crème anglaise** – also used as a base for desserts such as bavarois, ice cream and oeufs à la neige
- **crème pâtissière** – which can be transformed into crème chiboust, crème diplomat or crème mousseline and used for filling pastry products
- **cold set creams** – made from cream which is sweetened and flavoured, set with gelatine and when cold and set is whisked and spooned/dragged on the dessert plate
- **crème Chantilly** – this is sweetened cream lightly whipped to piping consistency and flavoured with vanilla
- **crème fouettée** – this is lightly whipped cream with no flavourings or sweetener, and is used to enrich and aerate certain pastry products such as mousses, parfaits and cold soufflés
- **crémeux** – literally means creamy/smooth and is made from a crème anglaise base emulsified onto chocolate, which is whipped when cold to a smooth shiny chocolate cream used for filling pastry products
- **pâte à bombe** – this is made by whipping egg yolks until aerated, slowly pouring on sugar cooked to 121°C and whisking until cold: it is used in the base of chocolate mousses, parfait glace, bombe glace and fruit gratins
- **lemon cream/curd** – this is made by whisking egg yolks, lemon juice/zest and sugar over a simmering bain-marie of water until the mixture thickens and reaches a temperature of 80°C; once thickened, melted butter is added and the mixture is then chilled down and piped into pastry cases or used as a filling for petits fours
- **ganache** – this is made by heating two parts double cream to 80°C and emulsifying onto one part couverture using a stem blender (ratios may vary depending on the use of the ganache)
- **sabayon** – a mixture of whole egg or egg yolk with caster sugar, whisked over a pan of simmering water to form a cooked, aerated mass; alcohol may be added, e.g. Marsala for a zabaglione.

E'spumas and foams

The word e'spuma directly translates from Spanish into 'foam' or 'bubbles'. An e'spuma is created using a thermo-whip (classic cream-whipper), which is a stainless steel vessel fitted with a screw top and a non-return valve which you charge with nitrogen dioxide (which constitutes 78 per cent of the air we breathe). This has minimum water solubility, therefore it will not affect the product that is being charged.

The principle role of the gas is to force the liquid out of the canister under pressure through two nozzles, making the cream more voluminous (increasing its volume) due to the mechanical disturbance of the fats. (Although this statement may seem quite complicated it is necessary to explain the mechanics and function of this equipment.)

In simple terms, the canister, once charged, will whip cream the same way as a whisk. The key factors that are essential to a successful preparation are detailed below.

Cold fat-based

In a litre canister, 750 g is the maximum amount of product to be placed inside. Depending on the viscosity required, one or two charges can be used – for low viscosity (thin mixtures) use two charges; for high viscosity (thicker mixtures) use one charge.

Once the product has been charged, it will need to be treated like any fat-based product that has been aerated, and not stored at room temperature as the aeration will be reduced dramatically.

Warm fat-based

In a litre canister, 600 g is the maximum amount of product to be placed inside.

Warm products tend to need two charges to ensure good aeration.

50–55°C is the optimum temperature to have the canister charged and ready for use. Any hotter and the expansion in the canister will be too great and uncontrollable when the trigger is pressed. If the canister is too cold, the fat molecules will tend to coat the tongue and not give optimum flavour.

Gelatine-based

In a litre canister, 750 g is the maximum amount of product to be placed inside.

The product will be liquid when it is poured into the canister. It will need to be charged immediately, placed in the fridge and shaken every 10–15 minutes to prevent total setting.

This preparation will give you a purer flavour as there is little or no fat involved. Fat coats the tongue, therefore the absence of fat in this preparation will increase flavour.

Why use e'spumas?

The boundaries of gastronomy have changed dramatically over the last 20 to 30 years and will no doubt continue to do so for the next 20 to 30, but the current approach is 'less volume, more flavour'.

By offering more flavours, the dining experience will be heightened; by reducing the volume that is taken, more flavour combinations can be offered – e'spumas are excellent vehicles to achieve such a result.

However, this is a technique that should be used in moderation as too much on one menu will become repetitive to the palate, and what was initially a motivation for using them will become the norm.

1 Short paste (pâte à foncer)

Short paste, rough puff paste (Recipe 7) and sweet paste (Recipe 2)

Makes (approximately) >	400g	850g
Flour (soft)	250 g	500 g
Salt	pinch	large pinch
Butter or block/cake margarine	125 g	250 g
Water	40–50 ml	80–100ml

1 Sieve the flour and salt.

2 Rub in the fat to achieve a sandy texture.

3 Make a well in the centre.

4 Add sufficient water to make a fairly firm paste.

5 Handle as little and as lightly as possible. Refrigerate until firm before rolling.

Try something different

- For wholemeal short pastry use half to three-quarters wholemeal flour in place of white flour.
- Short pastry is used in fruit pies, Cornish pasties, etc.
- Short pastry for sweet dishes such as baked jam roll may be made with self-raising flour.
- Lard can be used in place of some or all of the fat (butter or pastry margarine). Lard has excellent shortening properties and would lend itself, in terms of flavour, to savoury products, particularly meat-based ones. However, many people view lard as an unhealthy product, being very high in saturated fat. It is also unsuitable for anyone following a vegan or vegetarian diet as it is an animal product.

Key point

The amount of water used varies according to:

- the type of flour (a very fine soft flour is more absorbent)
- the degree of heat (for example, prolonged contact with hot hands, and warm weather conditions).

Different fats have different shortening properties. For example, paste made with a high ratio of butter to other fat will be harder to handle.

Faults

Possible reasons for faults in short pastry are detailed below.

Hard:
- too much water
- too little fat
- fat rubbed in insufficiently
- too much handling and rolling
- over-baking.

Soft-crumbly:
- too little water
- too much fat.

Blistered:
- too little water
- water added unevenly
- fat not rubbed in evenly.

Soggy:
- too much water
- too cool an oven
- baked for insufficient time.

Shrunken:
- too much handling and rolling
- pastry stretched whilst handling.

Correct, blistered and shrunken short pastry

2 Sweet (sugar) paste (pâte à sucre)

Makes (approximately) >	400g	1kg
Sugar	50 g	125 g
Butter or block/cake margarine	125 g	300 g
Egg	1	2–3
Flour (soft)	200 g	500 g
Salt	pinch	large pinch

Method 1 – sweet lining paste (rubbing in)

1 Sieve the flour and salt. Lightly rub in the margarine or butter to achieve a sandy texture.
2 Mix the sugar and egg until dissolved.
3 Make a well in the centre of the flour. Add the sugar and beaten egg.
4 Gradually incorporate the flour and margarine (or butter), and lightly mix to a smooth paste. Allow to rest before using.

Professional tip

Sugar pastry is used for products such as flans, fruit tarts and tartlets. The higher the percentage of butter, the shorter and richer the paste will become. However, as the butter will soften and melt during handling, the paste will become softer and more difficult to work with. Therefore chilling and light, quick handling is required when using a sweet paste with a high butter content.

This also applies to the working environment. For example, in a particularly warm kitchen, it will be more difficult to work with a paste of this structure than in a cooler kitchen. The butter in this recipe could be reduced from 125g to 100g to make handling easier.

Measure out the sugar and cut the butter into small chunks

Cream the butter and sugar together

Add the beaten egg in stages, thoroughly mixing each time

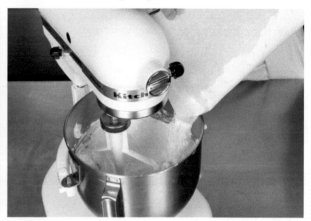

Incorporate the flour and salt

Press into a tray and leave to chill

The paste will need to be rolled out before use in any recipe

Method 2 – traditional French sugar paste (creaming)

1 Taking care not to over-soften, cream the butter and sugar.

2 Add the beaten egg gradually, and mix for a few seconds.

3 Gradually incorporate the sieved flour and salt. Mix lightly until smooth.

4 Allow to rest in a cool place before using.

3 Lining paste

Makes (approximately) >	450 g/3 × 15 cm tarts
Soft flour	250 g
Caster sugar	10 g
Salt	5 g
Butter	125 g
Water	40 ml
Egg	1

1 Combine the dry ingredients together; rub in the butter.
2 Add the water and egg to gently form a dough.
3 Wrap in cling film and leave to rest in the refrigerator for several hours before using.

Note

Lining paste may be used in place of sweet paste. Lining paste is not as rich and sweet as sweet (sugar) paste.

4 Sablé paste

Makes (approximately) >	500g
Egg	1
Caster sugar	75 g
Butter or block/cake margarine	150 g
Soft flour	200 g
Salt	Pinch
Ground almonds	75 g

1 Lightly cream the egg and sugar without over-softening.
2 Lightly mix in the butter – do not over-soften.
3 Incorporate sieved flour, salt and the ground almonds.
4 Mix lightly to a smooth paste.
5 Chill in the refrigerator before use.

Note

Sablé paste may be used for petits fours, pastries and as a base or platform for other desserts. 'Sablé' means a sandy texture.

Variation

Sablé paste may also be made with a creaming method. An example of this is the paste used in gateau MacMahon (see Chapter 7, Recipe 55).

5 Choux paste

Makes (approximately) >	750 g	1.5 kg
Water	250 ml	500 ml
Sugar	pinch	large pinch
Salt	pinch	large pinch
Butter or block/cake margarine	100 g	200 g
Flour (strong)	150 g	300 g
Eggs	4–5	8–10

1 Bring the water, sugar, salt and fat to the boil in a saucepan. Remove from heat.

2 Add the sieved flour and mix in with a wooden spoon (50 per cent, 70 per cent or 100 per cent wholemeal flour may be used).

3 Return to a moderate heat and stir continuously until the mixture leaves the sides of the pan. (This is known as a panada.)

4 Remove from the heat and allow to cool.

5 Gradually add the beaten eggs, beating well. Do not add all the eggs at once – check the consistency as you go. The mixture should just flow back when moved in one direction (it may not take all the egg).

Note

Choux paste is used to make products such as éclairs, profiteroles and gâteaux Paris-Brest.

Faults

Greasy and heavy paste:
- basic mixture over-cooked.

Soft paste, not aerated:
- flour insufficiently cooked.
- eggs insufficiently beaten in the mixture
- oven too cool
- under-baked.

Split or separated mixture:
- egg added too quickly.

The choux buns on the left are light and well risen; those on the right are poorly aerated.

Cut the butter into cubes and then melt them in the water

Add egg until the mixture is the right consistency – it should drop from a spoon under its own weight

Add the flour

Pipe the paste into the shape required – these rings can be used for Paris-Brest (Chapter 4, Recipe 36)

When the panada is ready, it will start to come away from the sides

A selection of shapes in raw choux paste

6 Puff paste (French method)

Makes (approximately) >	1.5 kg
Flour (strong)	560 g
Salt	12 g
Pastry butter or pastry margarine	60 g
Water, ice-cold	325 ml
Pastry butter or pastry margarine	500 g
Lemon juice or ascorbic or tartaric acid or white vinegar	a few drops

1 Sieve the flour and salt.

2 Rub in the 60g of butter/pastry margarine.

3 Make a well in the centre.

4 Add the water and lemon juice or acid (to make the gluten more elastic), and knead well into a smooth dough in the shape of a ball.

5 Relax the dough in a cool place for 30 minutes.

6 Cut a cross halfway through the dough and pull out the corners to form a star shape.

7 Roll out the points of the star square, leaving the centre thick.

8 Knead the remaining butter/pastry margarine to the same texture as the dough. This is most important – if the fat is too soft it will melt and ooze out, if too hard it will break through the paste when being rolled.

9 Place the butter or margarine on the centre square, which is four times thicker than the flaps.

10 Fold over the flaps.

11 Roll out to 30 cm × 15 cm, cover with a cloth or plastic and rest for 5–10 minutes in a cool place.

12 Roll out to 60cm × 20 cm, fold both the ends to the centre, fold in half again to form a square. This is one double turn.

13 Allow to rest in a cool place for 20 minutes.

14 Half-turn the paste to the right or the left.

15 Give one more double turn; allow to rest for 20 minutes.

16 Give two more double turns, allowing to rest between each.

17 Allow to rest before using.

Rub the first batch of butter into the flour

Mix in the water and lemon juice

Knead into a smooth dough

Roll the dough out into a cross shape

Knead the remaining butter in a plastic bag, then place it on the centre of the dough

Fold over each flap

Roll into a neat rectangle, then take each end and fold to meet in the centre

Fold again from the top end of the paste to the bottom

These photos have shown one double turn. When resting the turned and folded paste, leave an indented finger mark on the surface to show the number of turns completed

Key points

- Care must be taken when rolling out the paste to keep the ends and sides square.
- When rolling between each turn, always roll with the folded edge to the left.
- The addition of lemon juice (acid) helps to strengthen the gluten in the flour, thus helping to make a stronger dough so that there is less likelihood of the fat oozing out; 3 g (7.5 g for 10 portions) ascorbic or tartaric acid may be used in place of lemon juice.
- The rise is caused by the fat separating layers of paste during rolling. When heat is applied by the oven, steam is produced, causing the layers to rise and give the characteristic flaky formation. This is aeration by lamination.

Faults

The pastry on the left is unevenly laminated. Possible reasons for this:
- Paste was not folded equally
- Paste was rolled too thinly
- Re-used scraps of paste were used, instead of making up a virgin paste.

Note

There are different methods of making puff pastry, incorporating the fat in different ways. This recipe is the French method; Recipe 7 shows another method.

7 Rough puff paste (Scottish method)

Makes (approximately) >	475 g	1.2 kg
Flour (strong)	200 g	500 g
Salt	2 g (large pinch)	4 g (2 large pinches)
Butter or block/cake margarine (lightly chilled)	150 g	375 g
Water, ice-cold	125 ml	300 ml
Lemon juice, ascorbic or tartaric acid	10 ml	25 ml

1 Sieve the flour and salt.

2 Cut the fat into small pieces and lightly mix them into the flour without rubbing in.

3 Make a well in the centre.

4 Add the liquid and mix to a dough. The dough should be fairly tight at this stage.

5 Turn on to a floured table and roll into an oblong strip, about 30 × 10 cm, keeping the sides square.

6 Give one double turn (as for puff pastry).

7 Allow to rest in a cool place, covered with cloth or plastic for 30 minutes.

8 Give three more double turns, resting between each. (Alternatively, give six single turns.) Allow to rest before using.

Make a well in the centre of the flour and butter, and add the liquid

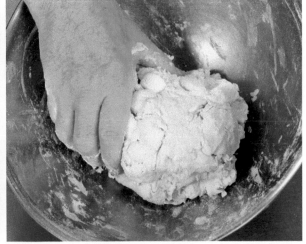

Mix to a fairly stiff dough

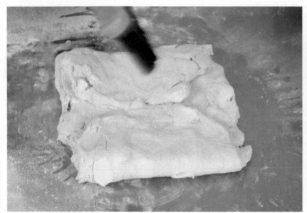

Roll out and fold the ends to the middle

Keep rolling, folding and turning

The finished paste, ready to rest and then use

Video: rough puff pastry,
http://bit.ly/16PT5fX

Note

Each time you leave the paste to rest, gently make finger indentations, one for each turn you have given the paste. This will help you to keep track.

8 Strudel paste

Makes (approximately) >	1.25 kg
Strong flour	680 g
Eggs, whole	3
Egg yolks	3
Oil	3 tbsp
Salt	7 g
Water, cold	To make up to 575 ml

1 Sift the flour and place into a mixer.

2 Place the eggs, egg yolks, oil and salt into a measuring jug. Add cold water to the 575 ml mark.

3 Add the liquid to the flour and mix with a hook attachment, to make a smooth dough. If the dough is very sticky, add more flour.

4 Divide the dough into 4 equal pieces. Leave to rest in a cool area between oiled plates.

5 Cover a free-standing table, away from the wall, with a large, clean cloth. Dust the cloth with flour.

6 Roll out the paste as far as possible across the table with a rolling pin.

7 Stretch and pull the paste out by hand.

Professional tip

Make sure you have all the ingredients and equipment needed for the strudel, including the filling, before you start rolling out the paste. As strudel paste is so fine and delicate, it is important that the process of making the strudel is completed quickly to avoid the paste from either drying out (if left uncovered for too long) or becoming soggy (if in contact with a filling and not baked immediately).

Half this recipe will be enough for ten portions, but when making strudel it is advisable to make more than you need in case of mistakes being made at the stretching stage.

Pulling the dough

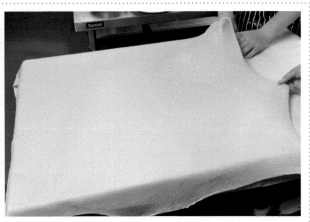

Fully stretched paste

9 Hot water paste

Makes (approximately) >	500 g	1.2 kg
Flour (strong)	250 g	625 g
Salt	5 g	12 g
Lard, butter or block/cake margarine	125 g	300 g
Water	125 ml	312 ml

1 Sift the flour and salt into a basin.

2 Make a well in the centre.

3 Boil the fat with the water and pour immediately into the centre of the flour.

4 Mix with a wooden spoon until cool.

5 Mix to a smooth paste and use while still warm.

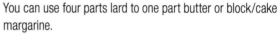

Professional tip

You can use four parts lard to one part butter or block/cake margarine.

Note

Hot water paste is used to produce savoury pies and the lining for terrines. Not all pies are cooked in moulds. Instead they are hand-raised using a hot water paste. A well-known example is a pork pie.

10 Suet paste

Makes (approximately) >	400 g	1 kg
Flour (soft) or self-raising flour	200 g	500 g
Baking powder	10 g	25 g
Salt	pinch	large pinch
Prepared beef or vegetarian suet	100 g	250 g
Water	125 ml	300 ml

1 Sieve the flour, baking powder and salt.

2 Mix in the suet. Make a well. Add the water.

3 Mix lightly to a fairly stiff paste.

Faults

Possible reasons for faults in suet paste:
- paste is heavy and soggy – it may be that the cooking temperature was too low
- paste is tough – it may have been handled too much or over-cooked.

Note

Suet paste is used for steamed fruit puddings, steamed jam rolls, steamed meat puddings and dumplings. Vegetarian suet is also available to enable products to be meat free.

Self-raising flour already contains baking powder so this element could be reduced by half if using self-raising flour.

11 Chantilly cream

Makes >	500 ml
Whipping cream	500 ml
Caster sugar	100 g
Vanilla arome/fresh vanilla pod	a few drops to taste/seeds from 1 vanilla pod

1 Place all ingredients in a bowl. Whisk over ice until the mixture forms soft peaks. If using a mechanical mixer, stand and watch until the mixture is ready – do not leave it unattended as the mix will over whip quickly, curdling the cream.

2 Cover and place in the fridge immediately.

12   Pastry cream (crème pâtissière)

Pastry cream, crème diplomat (Recipe 14) and crème chiboust (Recipe 15)

	Makes >	Approx. 750 ml
Milk		500 ml
Vanilla pod		1
Egg yolks		4
Caster sugar		125 g
Soft flour		75 g
Custard powder		10 g

1   Heat the milk with the cut vanilla pod and leave to infuse.

2   Beat the sugar and egg yolks together until creamy white. Add the flour and custard powder.

3   Strain the hot milk, gradually blending it into the egg mixture.

4   Strain into a clean pan and bring back to the boil, stirring constantly.

5   When the mixture has boiled and thickened, pour into a plastic bowl, sprinkle with caster sugar and cover with cling film.

6   Chill over ice and refrigerate as soon as possible. Ideally, blast chill.

7   When required, knock back on a mixing machine with a little kirsch.

Professional tip

At step 4, the microwave may be used effectively. Pour the mixture into a plastic bowl and cook in the microwave for 30-second periods, stirring in between, until the mixture boils and thickens.

Recipes 13, 14 and 15 are based on the recipe for crème pâtissière, with additional ingredients.

13 Crème mousseline

Beat in 100g of soft butter (a pomade). The butter content is usually about 20 per cent of the volume but this can be raised to 50 per cent depending on its intended use.

14 Crème diplomat

When the pastry cream is chilled, fold in an equal quantity of whipped double cream.

15 Crème chiboust

When the pastry cream mixture has cooled slightly, fold in an equal quantity of Italian meringue (Recipe 20).

Variations

Additional flavourings can also be added to crème pâtissière, crème diplomat or crème chiboust.

16 Butter icing

Makes >	350 g
Icing sugar	150 g
Butter	200 g

1 Sieve the icing sugar.
2 Cream the butter and icing sugar until light and creamy.
3 Flavour and colour as required.

Try something different

Variations include:
- rum – add rum to flavour and blend in
- chocolate – add melted chocolate, sweetened or unsweetened according to taste.

17 Boiled buttercream

Makes >	750 ml
Eggs	2
Icing sugar	50 g
Granulated sugar or cube sugar	300 g
Water	100 g
Glucose	50 g
Unsalted butter, cut into cubes	400 g

1 Beat the eggs and icing sugar until at ribbon stage (sponge).

2 Boil the granulated or cube sugar with water and glucose to 118°C.

3 Gradually add the sugar at 118°C to the eggs and icing sugar at ribbon stage, whisk for 2–3 minutes.

4 Gradually add the unsalted butter while continuing to whisk until a smooth cream is obtained.

Try something different

Possible flavours for buttercream include:

- chocolate and rum
- whisky and orange (using a flavour compound)
- strawberry and vanilla
- brandy and praline
- coffee and hazelnut.

Buttercream may also be made by adding unsalted butter to either a softened fondant or a cold, enriched crème anglaise.

Whisk the eggs

Add the boiling sugar and water

Add the butter

18 Frangipane (almond cream)

	Makes >	300 g
Butter		100 g
Caster sugar		100 g
Eggs		2
Ground almonds		100 g
Flour		10 g

1 Cream the butter and sugar until aerated.

2 Gradually beat in the eggs.

3 Mix in the almonds and flour (mix lightly).

4 Use as required.

Cut the butter into small pieces and add to the sugar

Cream the butter and sugar together

Beat in the eggs (before adding to the flour)

Try something different

Try adding lemon zest or vanilla seeds to the recipe.

19 Ganache

Makes >	750 g
Version 1 (for decoration)	
Double cream	300 ml
Couverture, cut into small pieces	350 g
Unsalted butter	85 g
Spirit or liqueur	20 ml
Makes >	**1 kg**
Version 2 (for a filling)	
Double cream	300 ml
Vanilla pod	½
Couverture, cut into small pieces	600 g
Unsalted butter	120 g

1 Boil the cream (and the vanilla for Version 2) in a heavy saucepan.

2 Gradually pour the cream over the couverture. Whisk with a fine whisk to form a shiny, emulsified ganache.

3 Whisk in the butter (and the liqueur for Version 1).

4 Stir over ice until the mixture has the required consistency.

20 Italian meringue

Makes >	250 g	625 g
Granulated or cube sugar	200 g	500 g
Water	60 ml	140 g
Cream of tartar	pinch	large pinch
Egg whites	4	10

1 Boil the sugar, water and cream of tartar to hard-ball stage of 121°C. (To ensure the sugar is not heated beyond this point, it is advisable to remove from the heat at 115°C as the sugar will continue to rise in temperature, and this will provide a little time to ensure the egg whites are whipped to the correct point.)

2 While the sugar is cooking, beat the egg whites to full peak and, while stiff, beating slowly, pour on the boiling sugar.

3 Use as required.

Boil the sugar

Combine with the beaten egg whites

The mixture will stand up in stiff peaks when it is ready

21   Apple purée (marmalade de pomme)

Makes >	400 g	1 kg
Cooking apples	400 g	1 kg
Butter	10 g	25 g
Sugar	50 g	125 g

1 Peel, core and slice the apples.

2 Place the butter in a thick-bottomed pan; heat until melted.

3 Add the apples and sugar, cover with a lid and cook gently until soft.

4 Drain off any excess liquid and pass through a sieve or liquidise.

22   Lemon curd

Makes >	450 ml	1.5 litres
Granulated sugar	450 g	1.125 kg
Zest of lemon, grated	2	5
Freshly squeezed lemon juice	240 ml	600 ml
Eggs, large	8	20
Egg yolks, large	2	5
Unsalted butter, cut into pieces	350 g	875 g

1 Place the sugar into a bowl. Grate the lemon zest into it and rub together.

2 Strain the lemon juice into a non-reactive pan. Add the eggs, egg yolks, butter and zested sugar. Whisk to combine.

3 Place over a medium heat and whisk continuously for 3–5 minutes, until the mixture begins to thicken.

4 At the first sign of boiling, remove from the heat and strain into a bowl and cool.

23 Pâte à bombe

Makes >	700 ml
Caster sugar	300 g
Water	200 ml
Glucose	20 g
Egg yolks, large	10

Note

This is used as a base for parfaits, iced soufflés and fruit gratins.

1 Boil the sugar, water and glucose together in a heavy-based pan. Continue to gently boil. Wash down the sides of the pan to prevent crystallisation.

2 Meanwhile put the eggs yolks in a food mixer and start mixing, whilst watching the sugar.

3 When the sugar reaches 121°C pour on to the eggs in a steady stream.

4 Carry on whisking until the mixture has increased in volume and is cold.

5 Use as required.

24 Stock syrup

Makes >	750 ml	1.5 litres
Water	500 ml	1.25 litres
Granulated sugar	250 g	625 g
Glucose	50 g	125 g

1 Boil the water, sugar and glucose together.

2 Strain and cool.

Professional tip

Glucose helps to prevent crystallising.

25 Apricot glaze

Makes >	150 ml
Apricot jam	100 g
Stock syrup (Recipe 24) or water	50 ml

1 Prepare by boiling apricot jam with a little syrup or water.
2 Pass through a strainer. The glaze should be used hot.

Professional tip

A flan jelly (commercial pectin glaze) may be used as an alternative to apricot glaze. This is usually a clear glaze to which food colour may be added.

26 Sabayon sauce (sauce sabayon)

Makes (approximately) >	450–500ml
Egg yolks, pasteurised	6
Caster or unrefined sugar	100 g
Dry white wine	250 ml

1 Whisk the egg yolks and sugar in a 1-litre pan or basin until white.
2 Dilute with the wine.
3 Place the pan or basin in a bain-marie of boiling water.
4 Whisk the mixture continuously until it increases to four times its bulk and is firm and frothy.

Note

Sauce sabayon may be offered as an accompaniment to any suitable hot sweet (e.g. pudding soufflé or soufflés).

Try something different

A sauce sabayon may also be made using milk in place of wine, which can be flavoured according to taste (for example, vanilla, nutmeg, cinnamon).

A classic Italian dessert, **zabaglione**. is made from this recipe with the use of Marsala wine. It is served with biscuits à la cuillère (Chapter 8, Recipe 6).

27 Fresh egg custard sauce (sauce à l'anglaise)

1 Mix the yolks, sugar and vanilla in a bowl.
2 Whisk in the boiled milk and return to a thick-bottomed pan.
3 Place on a low heat and stir with a wooden spoon until it coats the back of the spoon. Do not allow the mix to boil or the egg will scramble. A probe can be used to ensure the temperature does not go any higher than 85°C.
4 Put through a fine sieve into a clean bowl. Set on ice to seize the cooking process and to chill rapidly.

Try something different

Other flavours may be used in place of vanilla, for example:
- coffee
- curaçao
- chocolate
- Cointreau
- rum
- Tia Maria
- brandy
- whisky
- star anise or cardamom seeds
- kirsch
- orange flower water.

Makes >	300 ml	700 ml
Egg yolks, pasteurised	40 ml	100 ml
Caster or unrefined sugar	25 g	60 g
Vanilla extract or vanilla pod (seeds)	2–3 drops/½ pod	5–7 drops/1 pod
Milk, whole or skimmed, boiled	250 ml	625 ml

28 Custard sauce

Makes (approximately) >	250–275 ml	600–650 ml
Custard powder	10 g	25 g
Milk, whole or semi-skimmed	250 ml	600 ml
Caster or unrefined sugar	25 g	65 g

1 Dilute the custard powder with a little of the milk.
2 Boil the remainder of the milk.
3 Pour a little of the boiled milk on to the diluted custard powder.
4 Return to the saucepan.
5 Stir to the boil and mix in the sugar.

29 Fruit coulis (cooked)

Makes >	1.4 litres
Fruit purée	1 litre
Caster sugar	500 g
Lemon juice	10 g

1 Warm the purée.

2 Boil the sugar with a little water to soft-ball stage (121°C).

3 Pour the soft-ball sugar into the warm fruit purée while whisking vigorously. Add the lemon juice. Bring back to the boil.

4 This will then be ready to store.

Professional tip

The reason the soft-ball stage needs to be achieved when the sugar is mixed with the purée is that this stabilises the fruit and prevents separation once the coulis is presented on the plate.

Adding lemon juice brings out the flavour of the fruit.

30 Strawberry sauce (raw)

Makes (approximately) >	225 ml	600 ml
Strawberries (fresh or puréed)	200 g	500 g
Icing sugar	50 g	125 g
Lemon juice	10ml	20ml

Blend all the ingredients together (using a liquidiser if using fresh fruit) and strain through a fine sieve.

Variations

Alternative fruits or purées that can be used include peach, apricot, mango, pawpaw, blackberry and raspberry. For peach sauce, for example, proceed as above, substituting peach purée for the strawberries.

A commercial product, Ultratex, may be used to thicken and stabilise a sweetened fruit purée, like this sauce.

31 Orange, lemon or lime sauce

Makes (approximately) >	300–325 ml	700–800 ml
Sugar, caster or unrefined	50 g	125 g
Water	250 ml	625 ml
Cornflour or arrowroot	10 g	25 g
Oranges, lemons or limes	1–2	4–5

1 Boil the sugar and water.
2 Add the cornflour (or arrowroot) diluted (slaked) with a little cold water, stirring continuously.
3 Re-boil until clear; strain.
4 Add blanched julienne of zest and the strained juice from the fruit being used.

Variations

A little curaçao or Cointreau may be added for additional flavour.

32 Chocolate sauce (sauce au chocolat)

Makes (approximately) >	300 ml	750 ml
Method 1		
Double cream	150 ml	375 ml
Butter	25 g	60 g
Milk or plain chocolate couverture callets	180 g	420 g
Method 2		
Caster sugar	40 g	100 g
Water	120 ml	300 ml
Dark chocolate couverture (75 per cent cocoa solids)	160 g	400 g
Unsalted butter	25 g	65 g
Single cream	80 ml	200 ml

Method 1

1 Place cream and butter in a saucepan and gently bring to a simmer.
2 Add the chocolate and stir well until the chocolate has melted and the sauce is smooth.

Method 2

1 Dissolve the sugar in the water over a low heat.
2 Remove from the heat. Stir in the chocolate and butter.
3 When everything has melted, stir in the cream and gently bring to the boil.

33 | Butterscotch sauce

Makes (approximately) >	300 ml	750 ml
Double cream	250 ml	625 ml
Butter	62 g	155 g
Demerara sugar	100 g	250 g

1 Boil the cream, then whisk in the butter and sugar.

2 Simmer for 3 minutes.

34 | Caramel sauce

Makes >	1 litre
Granulated sugar	400 g
Oranges	4
Lemons	4
Apricot purée or jam	170 g
Crème de cacao	100 ml

1 Place the sugar in a large, heavy saucepan on a very low gas to melt and caramelise. Do not stir until the sugar has melted.

2 While the sugar melts, grate the orange and lemon zests and squeeze the juice. Place the zest, juice and apricot purée in a saucepan and heat slowly.

3 When the sugar has caramelised, add the fruit mixture and stir carefully until dissolved.

4 Pass through a chinois into a suitable container. Once cold, stir in the crème de cacao liqueur.

5 Chill thoroughly before using.

35 Apricot sauce (sauce abricot)

Makes >	4 portions
Apricot jam	200 g
Water	100 ml
Lemon juice	2–3 drops
Cornflour	10 g

1 Boil the jam, water and lemon juice together.
2 Adjust the consistency with a little cornflour (or arrowroot) diluted with water.
3 Reboil until clear and pass through a conical strainer.

36 Melba sauce (sauce Melba)

Makes >	4 portions
Raspberry jam	400 g
Water	125 ml

Variations
Raspberry coulis (see Recipe 29) can also be used as Melba sauce.

Boil ingredients together and pass through a conical strainer.

37 Syrup sauce

Makes >	8 portions
Golden syrup	200 g
Water	125 ml
Grated zest and juice of lime or orange	1
Cornflour or arrowroot	10 g

Variations
This is a traditional sauce served with the classic British golden syrup pudding. It is also used with steamed puddings or chilled and served over ice cream.

The sauce may be flavoured with white rum or whisky.

1 Bring the syrup, water and lemon juice to the boil and thicken with diluted cornflour (or arrowroot).
2 Boil for a few minutes and strain.

38 Water icing (glacé icing)

1 Take 400 g of icing sugar and start to add 60 ml (4 tbsp) of hot water; add the water until the icing is thick enough to coat the back of a spoon.

2 If necessary, add more water or icing sugar to adjust the consistency.

Try something different

Water may be replaced with other liquids to add flavour to the icing – for example, orange juice, mango juice, lemon juice, apple juice, lime juice, grape juice, passionfruit juice – or use a combination of juices with Cointreau, kirsch, Grand Marnier, rum, calvados, etc.

39 Royal icing

Makes >	400 g
Icing sugar	400 g
Whites of egg, pasteurised	3
Lemon, juice of	1
Glycerine	2 tsp

1 Mix well together in a basin the sieved icing sugar and the whites of egg, with a wooden spoon.

2 Add a few drops of lemon juice and glycerine and beat until stiff.

Professional tip

Modern practice is to use egg white substitute or dried egg whites. Always follow the manufacturer's instructions for quantities.

40 Chocolate glaze

Makes >	600 ml
Double cream	188 g
Water	175 g
Caster sugar	225 g
Cocoa powder	75 g
Gelatine, soaked in cold water	4½ leaves (10 g)

1 Bring the cream, water, sugar and cocoa to the boil slowly in a heavy-bottomed saucepan.

2 Simmer for 2 to 3 minutes, then remove from the heat.

3 Drain the gelatine and add it to the mixture. Stir until dissolved.

4 Pass, then cool by stirring over ice.

5 Store in a plastic container with cling film pressed directly onto the surface.

6 If using to glaze a frozen product, warm the glaze in the microwave before use.

An example of a product finished with this glaze is Opera (see Chapter 8, Recipe 17).

41 Crémeux

Makes >	15 portions
Base anglaise	
35 per cent UHT whipping cream	208 g
Milk	208 g
Egg yolk	83 g
Sugar	42 g
Cremèux	
Base anglaise	500 g
Plain couverture	220 g

1 Boil the milk and cream and add the mix of sugar and yolks. Cook to 84°C. Strain and weigh 500 g.

2 Part melt the couverture and slowly pour on the warm anglaise.

3 Emulsify with a stem blender.

4 Refrigerate.

5 When firm beat to a piping consistency.

42 Caramel e'spuma

Makes (approximately) >	600 ml
Yoghurt	250 g
Cream	50 g
Milk	100 g
Caramel sauce	166 g
Icing sugar	104 g

1 Mix all ingredients together.

2 Pour into an e'spuma gun and charge with two gas cartridges.

Professional tip

An e'spuma is usually dispensed from the gun into individual glasses. The e'spuma gun is also known as a siphon.

43 Hot chocolate e'spuma

Makes (approximately) >	650 g
Milk chocolate couverture	300 g
Dark chocolate couverture	50 g
White chocolate couverture	100 g
Hot water	200 g

1 Melt the three types of chocolate over a bain-marie until at 45°C.

2 Add the hot water and whisk until smooth.

3 Pour mix into an e'spuma gun and charge with two cartridges.

4 Place in a bain-marie to keep warm.

44 Pernod foam

Makes (approximately) >	800 ml
Skimmed milk	450 ml
Sugar	170 g
Leaf gelatine, pre-soaked	3 leaves
Pernod	160 g

1 Boil the milk and sugar. Add the pre-soaked gelatine leaves and allow them to dissolve.

2 Allow the mixture to cool.

3 Add the Pernod.

4 Pour into an e'spuma siphon and charge with two cartridges. Refrigerate.

5 Once chilled, shake well before using.

45 Raspberry foam

Makes (approximately) >	300 ml
Still mineral water	100 ml
Raspberry purée	200 ml
Sugar	25g
Gelatine leaves	2

1 Mix together the mineral water and the raspberry purée.

2 Soak the gelatine in cold water.

3 Warm a quarter of the raspberry purée mixture. Squeeze the excess water from the gelatine and dissolve in the warmed raspberry mixture.

4 Strain this into the remaining raspberry mixture.

5 Pour into an e'spuma siphon, charged with one cartridge, and refrigerate for 30 minutes.

6 Once chilled, shake well before using.

46 Coconut foam

Makes >	400 ml
Coconut purée	250 ml
Sugar	50 g
Double cream	100 g

1 Blitz the ingredients together.

2 Place in an e'spuma gun and chill.

3 Shake well before using.

Test yourself

Level 2

1 What type of nut is used to make frangipane?

2 What will happen to crème anglaise if it is allowed to boil?

3 At what temperature is the sugar cooked when making Italian meringue?

4 Describe two conditions that could lead to short pastry shrinking during the cooking process.

Level 3

5 Describe the difference between crème chiboust and crème mousseline, providing an example of their use for each.

6 What is Ultratex used for in the pastry kitchen?

7 Which gas is used to aerate foams when using a thermo-whip?

8 When laminating pastry, describe the difference between a single turn and a double turn.

4 Pastry products

This chapter covers:
- → **NVQ level 2 Prepare and cook and finish basic pastry products**
- → **VRQ level 2 Produce paste products**
- → **NVQ level 3 Prepare and cook and finish complex pastry products**
- → **VRQ level 3 Produce paste products.**

In this chapter you will:
- → **Produce a variety of pastry goods using the correct techniques and equipment, and prepare a range of fillings, creams and sauces (level 2)**
- → **Produce, finish and present a variety of pastry products to the recipe specifications, in line with current professional practice (level 3).**

Recipes in this chapter ······················

Introduction

This chapter continues on from Chapter 3 to illustrate the various types of pastry products that are commonly produced in the pastry kitchen. There are many further examples of products made from the foundations introduced in Chapter 3, and the intention here is to provide a variety of products using the base pastes, fillings, creams and sauces that are introduced in that chapter. This chapter is broken into sections, the first looking at examples of products made from short paste and the various forms of sweet paste. The sections following this provide further examples of products made from puff paste and choux paste. The chapter concludes with examples of products less commonly made in the pastry kitchen with examples of hot water paste and suet paste. An example of using a manufactured paste such as filo or spring roll paste is provided in Chapter 6, Recipe 20 (Almond and apricot samosas).

1 Lining a flan case

1 Prepare the pastry as per recipe (short, sweet, sablé, etc., as detailed in Chapter 3).

2 Grease the flan ring and baking sheet, or use a non-stick liner such as a non-stick silicon baking mat (for example, Silpat).

3 Roll out the pastry so that the circumference is 2 cm larger than the flan ring. The pastry may be rolled between cling-film, greaseproof or silicone paper.

4 Place the flan ring on the baking sheet.

5 Carefully place the pastry on the flan ring, by rolling it loosely over the rolling pin, picking up and unrolling it over the flan ring.

6 Press the pastry into shape without stretching it, being careful to exclude any air.

7 Allow a 0.5 cm ridge of pastry on top of the flan ring.

8 Cut off the surplus paste neatly.

Baking blind

Some recipes call for the flan case to be baked blind, which means that it is baked before the filling is added. Line the pastry case with cling film and fill with baking beans or dried beans. Bake at 190°C.

Professional tips

The rim can be left straight or moulded using the edge with the thumb and forefinger, pressing the pastry neatly to form a corrugated pattern. The back of a small, paring knife can also be used to press a rim between the forefinger and thumb.

If baking blind, an alternative approach is to trim the paste so that the rim is slightly higher than the flan ring. Partly bake blind, then remove from the oven. Remove the beans. Carefully cut back the rim of the pastry with a sharp, thin-bladed knife, so that it is level with the top of the flan ring. Return to the oven to fully bake.

Place the pastry into the flan ring

Firm the pastry into the bottom of the ring

Bake blind, filled with beans, if the recipe requires

2 Quiche Lorraine (cheese and ham savoury flan)

Makes >	4 portions	10 portions
Short paste	100 g	250 g
Ham, chopped	75 g	150 g
Cheese, grated	50 g	125 g
Egg	1	2
Milk	125 ml	300 ml
Cayenne		
Sea-salt (e.g. Maldon)		

Try something different

- The filling can be varied by using lightly fried lardons of bacon (in place of the ham), chopped cooked onions and chopped parsley.
- A variety of savoury flans can be made by using imagination and experimenting with different combinations (for example, stilton and onion; salmon and dill; sliced sausage and tomato).

1 Lightly grease an appropriately sized flan ring or barquette, or tartlet moulds if making individual portions. Line thinly with pastry (as per Recipe 1).

2 Dock the bottom of the paste. Prepare for baking blind.

3 Cook in a hot oven at 200°C for 3–4 minutes or until the pastry is lightly set. Reduce the oven temperature to 160°C.

4 Remove from the oven and remove the flan beans; press the pastry down if it has tended to rise.

5 Add the chopped ham and grated cheese.

6 Mix the egg, milk, salt and cayenne thoroughly. Strain over the ham and cheese.

7 Return to the oven at 160°C and bake gently for approximately 20 minutes or until nicely browned and the egg custard mix has set.

3 | Fruit pie

Makes >	4–6 portions	10–15 portions
Fruit (see note)	400 g	1.5 kg
Sugar	100 g	250 g
Short paste	200 g	500 g

1 Prepare the fruit, wash and place in a bowl.

2 Add the sugar and mix gently. Partially cook on the stove until semi-soft, depending on the type of fruit.

3 Line an appropriately sized flan ring with two-thirds of the pastry, leaving a third for the top.

4 Fill the lined ring with the fruit mixture. (Place a clove in an apple pie.)

5 Dampen the edge of the pastry with a little water.

6 Roll out the reserved piece of pastry to cover the fruit.

7 Carefully lay the pastry on top of the fruit and firmly seal to the edge of the lined pastry. Cut off any surplus pastry. Spare pastry can be used to create a lattice effect by running strips of pastry in diagonal lines across the pie, cutting off any excess pastry at the edges (see picture).

8 Sprinkle the top of the pie with caster sugar.

9 Place the pie on a baking sheet and bake at 200°C for 10 minutes, reduce the oven to 180°C and continue to bake for another 25–30 minutes. If the pastry colours too quickly cover with a sheet of paper.

10 Once cooked, allow to cool slightly before removing the flan-ring. Serve with custard, fresh cream or ice cream.

Preparation of fruit for pies

- Apples – peeled, quartered, cored, washed, cut in slices
- Cherries – stalks removed, washed, stones removed
- Blackberries – stalks removed, washed
- Gooseberries – stalks and tails removed, washed
- Damsons – picked and washed, stones removed
- Rhubarb – leaves and root removed, tough strings removed, cut into 2 cm pieces, washed.

The pie may be filled with a single fruit or a combination such as blackberry and apple or damson and apple.

Treacle tart

Makes >	4 portions	10 portions
Short paste	125 g	300 g
Treacle	100 g	250 g
Water	1 tbsp	2.5 tbsp
Lemon juice	3–4 drops	8–10 drops
Fresh white bread or cake crumbs	15 g	50 g

1 Lightly grease an appropriately sized flan ring or barquette, or tartlet moulds if making individual portions.

2 Line with pastry.

3 Warm the treacle, water and lemon juice; add the crumbs.

4 Place into the pastry ring and bake at 170°C for about 20 minutes.

Try something different

This tart can also be made in a shallow flan ring. Any pastry debris can be rolled and cut into 0.5 cm strips and used to decorate the top of the tart before baking.

Try sprinkling with vanilla salt as a garnish.

5 Egg custard tart

	Makes >	8 portions (1 × 20 cm flan-ring)
Sweet paste		250 g
Egg yolks		9
Caster sugar		75 g
Whipping cream, gently warmed and infused with 2 sticks of cinnamon		500 ml
Nutmeg, freshly grated		

1 Roll out the pastry on a lightly floured surface, to 2 mm thickness. Use it to line a flan ring, placed on a baking sheet.

2 Line the pastry with food-safe cling film or greaseproof paper and fill with baking beans. Bake blind in a preheated oven at 190°C, for about 10 minutes or until the pastry is turning golden brown. Remove the cling film and beans, and allow to cool. Turn the oven down to 130°C.

3 To make the custard filling, whisk together the egg yolks and sugar. Add the cream and mix well.

4 Pass the mixture through a fine sieve into a saucepan. Heat to 37°C.

5 Fill the pastry case with the custard to 5 mm below the top. Place it carefully into the middle of the oven and bake for 30–40 minutes or until the custard appears to be set but not too firm.

6 Remove from the oven and cover liberally with grated nutmeg. Allow to cool to room temperature.

	Makes >	1 × 20 cm flan ring
Sweet paste		250 g
Fruit (e.g. strawberries, raspberries, grapes, blueberries, etc.)		500 g
Pasrry cream (Chapter 3, Recipe 12)		
Glaze		5 tbsp

1 Line an appropriately sized flan ring, or tartlet or barquette moulds with paste and cook blind at 190°C (see Recipe 5). Allow to cool.

2 Pick and wash the fruit, then drain well. Wash and slice/segment, etc. any larger fruit being used.

3 Pipe pastry cream into the flan case, filling it to the rim. Dress the fruit neatly over the top. Coat with the glaze.

4 Use a glaze suitable for the fruit chosen; for example, with a strawberry tart, use a red glaze.

Professional tip

Brush the inside of the pastry case with melted couverture before filling. This forms a barrier between the pastry and the moisture in the filling.

Faults

Although this strawberry tart may appear to be fine in this picture, the husks of the strawberries are visible. To improve this, it would be better to present the strawberries with their tops pointing upwards or in a way that presented the strawberries sliced and overlapping.

There is also quite a wide gap between the rows of strawberries, showing the crème pâtissière underneath. This should also be avoided.

The second photo shows the importance of ensuring that fillings are prepared and/or cooked properly. In this case, the crème pâtissière has not been cooked sufficiently or prepared accurately as the filling is not structured sufficiently to support the fruit once the tart has been cut.

For tartlets

1 Roll out pastry 3 mm thick.

2 Cut out rounds with a fluted cutter and place them neatly in greased tartlet moulds. If soft fruit (such as strawberries or raspberries) is being used, the pastry should be cooked blind first.

3 After baking and filling (or filling and baking) with pastry cream, and dressing neatly with fruit, glaze the top.

Professional tip

Certain fruits (such as strawberries and raspberries) are sometimes served in boat-shaped moulds (**barquettes**). The preparation is the same as for tartlets.

Tartlets and barquettes should be glazed and served, allowing one large or two small per portion.

French apple flan (flan aux pommes)

1 Line an appropriately sized flan ring with the sweet paste. Pierce the bottom several times with a fork.

2 Pipe a layer of pastry cream into the bottom of the flan.

3 Peel, core, halve and wash an apple.

4 Cut into thin slices and lay carefully on the pastry cream, in overlapping slices starting at the outside and working towards the centre. Ensure that each slice points to the centre of the flan, then no difficulty should be encountered in joining the pattern up neatly.

5 Sprinkle a little sugar on the apple slices and bake the flan at 200–220°C for 30–40 minutes.

6 When the flan is almost cooked, remove the flan ring carefully, return to the oven to complete the cooking. Mask with hot apricot glaze or flan gel.

Makes >	4 portions	10 portions
Sweet paste	100 g	250 g
Pastry cream	250 ml	625 ml
Dessert apples, large	3	8
Sugar	50 g	125 g
Apricot glaze	2 tbsp	6 tbsp

Pipe the filling neatly into the flan case

Slice the apple very thinly for decoration

Arrange the apple slices on top of the flan

Complete the arrangement of apple slices

8 Lemon tart (tarte au citron)

Makes >	8 portions
Sweet paste	200 g
Lemons	juice of 3, zest from 4
Eggs	8
Caster sugar	300 g
Double cream	250 ml

Professional tip

If possible, make the filling one day in advance. The flavour will develop as the mixture matures.

Try something different

Limes may be used in place of lemons. If so, use the zest and juice of 5 limes or use a mixture of lemons and limes.

1 Prepare 200 g of sweet paste, adding the zest of one lemon to the sugar.

2 Line a 20 cm flan ring with the paste.

3 Bake blind at 190°C for approximately 15 minutes.

4 Prepare the filling: mix the eggs and sugar together until smooth, add the cream, lemon juice and zest. Whisk well.

5 Seal the pastry, so that the filling will not leak out. Pour the filling into the flan case, bake for 30–40 minutes at 150°C until just set. (Take care when almost cooked as overcooking will cause the filling to rise and possibly crack.)

6 Remove from oven and allow to cool.

7 Dust with icing sugar and glaze under the grill or use a blowtorch. Portion and serve.

8 The mixture will fill one 16 × 4 cm or two 16 × 2 cm flan rings. If using two flan rings, double the amount of pastry and reduce the baking time when the filling is added.

Lemon meringue pie

	Makes >	2 × 20 cm flan rings
Sweet paste flan cases, pre-baked		2
Granulated sugar		450 g
Lemon zest, grated		2
Fresh lemon juice		240 ml
Eggs, large		8
Large egg yolks		2
Unsalted butter, cut into small pieces		350 g
For the meringue		
Egg whites		6
Caster sugar		600 g

1 Place the sugar into a bowl and grate the zest of lemon into it, rubbing together.

2 Strain the lemon juice into a non-reactive pan. Add the eggs, egg yolks, butter and zested sugar. Whisk to combine.

3 Place over a medium heat and whisk continuously for 3–5 minutes, until the mixture begins to thicken.

4 At the first sign of boiling, remove from the heat. Strain into the pastry cases. Leave to set.

5 Make the meringue (see Chapter 3, Recipe 20). Pipe it on top of the filled pie.

6 Colour in a hot oven at 220°C.

10 Pear and almond tart

Makes >	1 × 20cm ring (8 portions)
Sweet paste	200 g
Apricot jam	25 g
Almond cream	350 g
Poached pears	4
Apricot glaze	
Flaked almonds	
Icing sugar	

1 Line a buttered 20 cm flan ring with sweet paste. Trim and dock.

2 Using the back of a spoon, spread a little apricot jam over the base.

3 Pipe in almond cream until the flan case is two-thirds full.

4 Dry the poached pears. Cut them in half and remove the cores and string.

5 Score across the pears and arrange on top of the flan.

6 Bake in the oven at 200°C for 25–30 minutes.

7 Allow to cool, then brush with apricot glaze.

8 Sprinkle flaked almonds around the edge and dust with icing sugar.

11 Pecan pie

Makes >	1 × 20 cm flan (8 portions)
Sweet or lining paste	200 g
Eggs	4
Light brown sugar	185 g
Golden syrup	85 g
Salt	2 g
Vanilla pod/vanilla extract (e.g. vanilla arome)	1/2.5 ml
Whisky	30 ml
Unsalted butter, melted	42 g
Pecan nuts	285 g
Apricot glaze	
Chocolate for piping (optional)	

1 Make up the paste and place a 20 cm flan ring on a baking tray. Line it with the paste and bake the flan case blind.

2 Whisk the eggs a little, to break them up.

3 Mix in the sugar, syrup, salt, vanilla and whisky.

4 Stir in the butter and pecan nuts.

5 Pour the pecan mixture into the case.

6 Bake at 180°C for approx. 30 minutes, until firm.

Professional tip

Allow to cool completely. Brush with apricot glaze. Remove from the flan ring and pipe a chocolate trellis over the flan.

12 Bakewell tart

Makes >	1 × 20 cm ring (8 portions)
Sugar paste	200 g
Raspberry jam	50 g
Eggwash	
Apricot glaze	50 g
Icing sugar	35 g
Frangipane (almond cream)	250 g

1 Roll out the paste to 2 mm thick. Line a flan ring.

2 Pierce the bottom with a fork.

3 Spread with jam and the frangipane.

4 Bake in a moderately hot oven at 200–210°C for 30–40 minutes. Brush with hot apricot glaze.

5 When cooled brush over with very thin water icing.

13 Mince pies

Makes >	12 small pies
Sweet or short paste	200 g
Mincemeat (see below)	200 g
Eggwash	
Icing sugar	

1 Roll out the pastry 3 mm thick.

2 Cut half the pastry into fluted rounds 8 cm in diameter.

3 Line buttered tartlet cases with the rounds of pastry.

4 Moisten the edges. Fill each with mincemeat.

5 Cut the remainder of the pastry into fluted rounds, 6 cm in diameter.

6 Cover the mincemeat with pastry and seal the edges. Brush with eggwash.

7 Bake at 210°C for approximately 20 minutes.

8 Sprinkle with icing sugar and serve warm. Accompany with a suitable sauce (such as custard, brandy sauce, brandy cream).

Try something different

Puff pastry versions of these pies are also made.

Piped Viennese biscuit may be used for the top of the tarts.

Various toppings can also be added, such as crumble mixture or flaked almonds and an apricot glaze.

To make the mincemeat

Makes >	800 g
Suet, chopped	100 g
Mixed peel, chopped	100 g
Currants	100 g
Sultanas	100 g
Raisins	100 g
Apples, chopped	100 g
Barbados sugar	100 g
Mixed spice	5 g
Lemon, grated zest and juice of	1
Orange, grated zest and juice of	1
Rum	60 ml
Brandy	60 ml

1 Mix the ingredients together.

2 Seal in jars and use as required.

Swiss apple flan

	Makes >	1 × 20 cm flan (6 portions)
Sweet paste		250 g
Apricot jam		20 g
Frangipane		200 g
Dessert apples, small		5
Apricot glaze		100 ml
Flaked almonds, roasted		50 g
Neige-décor or icing sugar, for dusting		

1 Make up the sweet paste and place a 20cm flan ring on a baking tray. Line it with the sweet paste. Dock the pastry base.

2 Beat the jam until smooth. Spread a very thin layer over the flan base, using the back of a spoon.

3 Spread the frangipane inside the pastry case; the case should be no more than half full.

4 Peel and core the apples. Cut them in half and score them. Place them onto the flan, evenly spaced. The ideal arrangement is six halves around the edge and one in the centre.

5 Bake at 185°C for 25–35 minutes.

6 Allow to cool. Brush with boiling apricot glaze. Place flaked almonds around the edge of the flan and dust the almonds with neige-décor.

Note

Neige-décor is made from dextrose, wheat starch and vegetable fat. It is used to finish and decorate pastries and desserts. Unlike icing sugar, it does not melt easily and will stay where it is placed more robustly.

15   Plum and soured cream tart

Makes >	1 × 20 cm flan	3 × 20 cm flans
Sweet paste	200 g	600 g
Filling		
Eggs, beaten	1	4
Plums	6	18
Unsalted butter	45 g	130 g
Caster sugar	60 g	180 g
Grated nutmeg	Small pinch	Pinch
Soured cream	140 ml	400 ml
Semolina	15 g	50 g
Zest and juice of lemon	½	1
Topping		
Soft flour	75 g	200 g
Butter	50 g	150 g
Demerara sugar	25 g	70 g

1 Line 20cm flan rings with the pastry and chill in the fridge.

2 Bake blind at 190°C for 15 minutes until golden brown. Egg wash the tart cases and cook for 2 minutes then lower the oven to 150°C.

3 Cut the plums in half, remove the stones, and arrange them cut side up over the base of the tart. Cream the butter and the sugar together until light and fluffy.

4 Gradually beat in the remaining egg, and then stir in the nutmeg, soured cream, semolina, lemon zest and juice. Pour the mixture over the plums and bake for 25 minutes until lightly set.

5 Meanwhile, for the topping, rub the flour and butter together until it resembles fine breadcrumbs, stir in the demerara sugar.

6 After 25 minutes sprinkle the crumble mixture over the top of the tart and bake for another 25 minutes.

7 Decorate with sliced plums.

16 Baked chocolate tart

Makes >	1 × 20 cm flan
Sweet paste	200 g
Filling	
Eggs	3
Egg yolks	3
Caster sugar	60 g
Butter	200 g
Chocolate callets (55% cocoa, unsweetened)	300 g

1 Roll out the sweet paste and line a 20 cm flan ring. Bake the flan case blind at 190°C.

2 For the filling, whisk the eggs, yolks and sugar together to make a sabayon.

3 Bring the butter to the boil, remove and mix in the chocolate callets until they are all melted.

4 Once the sabayon is light and fluffy, fold in the chocolate and butter mixture, mixing very carefully so as not to beat out the air.

5 Pour into the cooked flan case and place in a deck oven at 150°C until the edge crusts (approximately 5 minutes). Chill to set.

6 Once set, remove from fridge and then serve at room temperature.

17 White chocolate and citrus meringue tart

Note

Recipes 17 and 18 are examples of tarts with inserts of different flavours and textures. In addition to the eating qualities, the inserts also add an extra element to the presentation of the tarts, providing a layered visual effect.

1 Wash and zest the lemons.

2 Mix the lemon juice with the sugar, zest and the eggs.

3 Cook slowly over a low heat until the mixture thickens.

4 Remove from the heat at 85°c. Add the softened gelatine.

5 Pour the lemon cream over the melted chocolate and cocoa butter

6 Emulsify using a stem blender.

Makes >	1 × 20 cm flan
1 × 20 cm flan case, baked blind using sable paste	
White chocolate lemon cream	
Lemon juice	250 g
Lemon zest	1
Caster sugar	80 g
Whole eggs	5
Gelatine leaves	2
White couverture	200 g
Cocoa butter	10 g
Lime meringue (insert/filler)	
Gelatine (approximately 4 leaves)	10 g
Caster sugar	115 g
Water	135 g
Lime juice	125 g

For the insert

1 Soak the gelatine in cold water.

2 Boil the water, sugar and lime juice and add the gelatine.

3 Allow to set in the fridge until solid.

4 Partially melt in the microwave and whip up at full speed to aerate like a meringue.

5 Pour into a tray and freeze.

6 Once frozen cut out a disc slightly smaller than the baked pastry case.

Assembly

1 Half fill the pastry case with warm lemon cream.

2 When almost set push in the frozen lime meringue disc.

3 Top up level to the rim of the pastry case with the remaining cream and refrigerate.

4 Cut out a wedge and decorate the top with raspberry and passion fruit coulis thickened with Ultratex (refer to the Appendix on specialist ingredients), green pistachio nuts and squares of lime meringue.

18 Chocolate and ginger tart with insert of mango jelly and praline

Makes >	1 × 20 cm flan
1 × 20 cm flan case, baked blind using flan case	
1 × disc of mango jelly (Chapter 9, Recipe 26)	
1 × disc of praline	
Chocolate and ginger ganache	
Whipping cream	200 g
Fresh ginger	22 g
Preserved ginger	22 g
Plain couverture	250 g
Unsalted butter	50 g

1 Boil the cream with the finely grated ginger and allow to infuse for 10 minutes.

2 Add the chopped preserved ginger. Strain, and pour onto the melted chocolate and butter. Stem blend to emulsify.

Assembly

1 Half fill the tart case with the ganache, allow to partially set, and place on top the discs of mango jelly and praline feuilletine, cut slightly smaller than the flan case.

2 Top with the remaining ganache to fill the pastry case, refrigerate.

3 Once set, cut a wedge using a warm knife and just before service run the flame of a blow torch over the top of the ganache to give a shine. Decorate with gold leaf.

4 Serve with pear water ice.

The praline disc

Makes >	1 disc
Melted plain couverture, tempered	100 g
Praline paste (commercial product – see Appendix)	100 g
Pâte à feuilletine	100 g

1 Warm the praline paste in a microwave.

2 Mix with the melted couverture.

3 Add the pâte à feuilletine.

4 Roll out thinly between silicone paper.

5 Refrigerate until solid.

6 Cut out a disc using a flan ring narrower than the baked pastry case.

19 Puff pastry cases (bouchées, vol-au-vents)

Makes >	12 bouchées/6 vol-au-vents
Puff pastry	200 g

1. Roll out the pastry approximately 5 mm thick.
2. Cut out with a round, fluted cutter of the size required.
3. Place on a greased, dampened baking sheet; eggwash.
4. Dip a plain cutter of a slightly smaller diameter into hot fat or oil and make an incision 3 mm deep in the centre of each.
5. Allow to rest in a cool place.
6. Bake at 220°C for about 20 minutes.
7. When cool, remove the caps or lids carefully and remove all the raw pastry from inside the cases.

Note

Bouchées are filled with a variety of savoury fillings and are served hot or cold. They may also be filled with creams or curds as a pastry. Larger versions are known as vol-au-vents. They may be produced in one-, two-, four- or six-portion sizes; a single-sized vol-au-vent would be approximately twice the size of a bouchée. When preparing one and two-portion sized vol-au-vents, the method for bouchées may be followed. When preparing larger-sized vol-au-vents, it is advisable to have two layers of puff pastry each 0.5 cm thick, sealed together with eggwash. One layer should be a plain round, and the other of the same diameter with a circle cut out of the centre.

20 Cheese straws (paillettes au fromage)

Makes >	8–10 portions	16–20 portions
Puff paste or rough puff paste	100 g	250 g
Cheese, grated	50 g	125 g
Cayenne pepper		

1. Roll the pastry to 60 × 15 cm, 3 mm thick.
2. Sprinkle with the cheese and cayenne pepper.
3. Roll out lightly to embed the cheese.
4. Cut the paste into thin strips by length.
5. Twist each strip to form rolls in the strip.
6. Place on a silicone mat.
7. Bake in a hot oven at 230–250°C for 10 minutes or until a golden brown. Cut into lengths as required.

21 Sausage rolls

Makes >	12 × 8 cm rolls
Puff pastry	200 g
Sausage meat	400 g
Eggwash	

1. Roll out the pastry 3 mm thick into a strip 10 cm wide.
2. Shape the sausage meat into a roll 2 cm in diameter.
3. Place on the pastry. Moisten the edges of the pastry.
4. Fold over and seal. Cut into 8 cm lengths.
5. Mark the edge with the back of a knife. Brush with eggwash.
6. Place on to a greased, dampened baking sheet.
7. Bake at 220°C for approximately 20 minutes.

22 Gâteau pithiviers

Video: gâteau pithiviers,
http://bit.ly/17lm9ec

Makes >	2 × 22 cm gâteaux
Puff pastry	1 kg
Pastry cream	60 g
Frangipane	600 g
Eggwash (yolks only)	
Granulated sugar	

1 Divide the paste into four equal pieces. Roll out each piece in a circle with a 22 cm diameter, 4 mm thick.

2 Rest in the fridge between sheets of cling film, preferably overnight.

3 Lightly butter two baking trays and splash with water. Lay one circle of paste onto each tray and dock them.

4 Mark a 16 cm diameter circle in the centre of each.

5 Beat the pastry cream, if desired, and mix it with the frangipane.

6 Using a plain nozzle, pipe the cream over the inner circles, making them slightly domed. (The paste may be brushed with apricot glaze first, if desired.)

7 Eggwash the outer edges of the paste. Lay one of the remaining pieces over the top of each one, smooth over and press down hard.

8 Mark the edges with a round cutter. Cut out a scallop pattern with a knife, or use a cut piping nozzle as shown in the photo sequence.

9 Eggwash twice. Mark the top of both with a spiral pattern.

10 Bake at 220°C for 10 minutes. Remove from the oven and sprinkle with granulated sugar or icing sugar. Turn the oven down to 190°C and bake for a further 20 to 25 minutes.

11 Glaze under a salamander.

Adding the filling to the rolled base

Trimming the edge

Marking the top

23 Palmiers

Palmiers are usually made from leftover or off-cuts of puff-pastry. These can be used because a biscuit-like finish is desired, and less rise is needed than for products such as bouchées and vol-au-vents.

1. Roll out puff pastry on sugar into a square 3 mm thick.
2. Sprinkle liberally with caster sugar on both sides and roll into the pastry.
3. Fold into three from each end so as to meet in the middle; brush with eggwash and fold in two.
4. Cut into strips approximately 2 cm thick; dip one side in caster sugar.
5. Place on a greased baking sheet, sugared side down, leaving a space of at least 2 cm between each.
6. Bake in a very hot oven for about 10 minutes.
7. Turn with a palette knife, cook on the other side until brown and the sugar is caramelised.

Professional tip

Puff pastry trimmings are suitable for this recipe. Palmiers may be made in a wide variety of sizes. Two joined together with a little whipped cream may be served as a pastry, or small ones for petits fours secs. They may be sandwiched together with soft fruit, whipped cream and/or ice cream and served as a sweet.

24 Pear jalousie

Makes >	8–10 portions
Puff pastry	200 g
Frangipane	200 g
Pears, poached or tinned (cored and cut in half lengthways)	5

1. Roll out two-thirds of the pastry 3 mm thick into a strip 25 × 10 cm and place on a greased, dampened baking sheet.
2. Pierce with a docker. Moisten the edges.
3. Pipe on the frangipane, leaving 2 cm free all the way round. Place the pears on top.
4. Roll out the remaining one-third of the pastry to the same size. Chill before cutting.
5. Cut the dough with a trellis cutter to make a lattice.
6. Carefully open out this strip and neatly place on to the first strip.
7. Trim off any excess. Neaten and decorate the edge. Brush with eggwash.
8. Bake at 220°C for 25–30 minutes.
9. Glaze with apricot glaze. Dust the edges with icing sugar and return to a very hot oven to glaze.

25 Puff pastry slice (mille-feuilles)

	Makes >	10 portions
Puff pastry trimmings		600 g
Pastry cream		400 ml
Apricot glaze		
Fondant		350 g
Chocolate		100 g

1 Roll out the pastry 2 mm thick into an even-sided square.

2 Roll up carefully on a rolling pin and unroll onto a greased, dampened baking sheet.

3 Dock well.

4 Bake in a hot oven at 220°C for 15–20 minutes; turn the strips over after 10 minutes. Allow to cool.

5 Using a large knife cut each to form three even-sized rectangles.

6 Keeping the best strip for the top, pipe the pastry cream on one strip.

7 Place the second strip on top and pipe with pastry cream.

8 Place the last strip on top, flat side up. Press gently. Brush with boiling apricot glaze to form a key.

To decorate by feather-icing

See photos on next page.

1 Warm the fondant to 37°C (warm to the touch) and correct the consistency with sugar syrup if necessary.

2 Pour the fondant over the mille-feuilles in an even coat.

3 Immediately pipe the melted chocolate over the fondant in strips, ½ cm apart.

4 With the back of a small knife, wiping after each stroke, mark down the slice at 2 cm intervals.

5 Quickly turn the slice around and repeat in the same direction with strokes in between the previous ones.

6 Allow to set and trim the edges neatly.

7 Cut into even portions with a sharp thin-bladed knife, dip into hot water and wipe clean after each cut.

For a traditional finish, crush the pastry trimmings and use them to coat the sides.

Pipe cream between layers of pastry

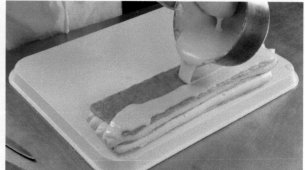

Ice the top with fondant

Decorate with chocolate

Variations

- Whipped fresh cream may be used as an alternative to pastry cream. Spread lightly with jam before adding the cream.
- The pastry cream or whipped cream may also be flavoured with a liqueur if so desired, such as curaçao, Grand Marnier or Cointreau.

26 | Mille-feuilles Napoleon

Makes >	10 portions
Puff paste	600 g
Crème diplomat (see below)	400 ml
Strawberries, small or halves	225 g
Icing sugar	20 g
Crème diplomat	
Double cream, whipped	300 ml
Pastry cream	300 g

1 Roll out the puff paste in a very thin sheet. Dock well.

2 Rest the pastry with a clean baking sheet on top to prevent it from rising. Bake at 220°C.

3 Prepare the crème diplomat by folding the whipped double cream into the chilled pastry cream.

4 Cut the paste neatly into rectangles, approximately 9 × 5 cm. Allow 3 pieces per portion.

5 Take one rectangle of paste and pipe on bulbs of crème diplomat placing strawberry halves between the bulbs.

6 Place a second rectangle on top and layer with crème diplomat and strawberries as before.

7 Dust the third rectangle of paste with icing sugar and sear in diagonal lines with a hot poker. (This technique is referred to as 'quadrillage'.) Place gently on top of the dessert.

8 Serve with strawberry coulis or crème anglaise.

27 Fruit slice (bande aux fruits)

Makes >	8–10 portions
Puff pastry	250 g
Fruit (see note)	400 g
Pastry cream	250 ml (approximately)
Apricot glaze	2 tbsp

1 Roll out the pastry 2 mm thick in a strip 12 cm wide.

2 Place on a greased, dampened baking sheet.

3 Moisten two edges with eggwash; lay two 1.5 cm-wide strips along each edge.

4 Seal firmly and mark with the back of a knife. Prick the bottom of the slice.

5 Depending on the fruit used, either put the fruit (such as apple) on the slice and cook together, or cook the slice blind and afterwards place the pastry cream and fruit (such as tinned peaches) on the pastry. Glaze and serve as for flans.

Note

Fruit slices may be prepared from any fruit suitable for flans/tarts. Berries and soft summer fruits are often used.

Variations

Alternative methods are:
- to use sweet pastry in a slice mould
- to use short of sweet pastry in a tranche mould (rectangular flan case).

28 Cream horns

Makes >	16 horns
Puff pastry	250 g
Eggwash	
Icing sugar, to sprinkle	
Jam	50 g
Caster sugar	50 g
Vanilla pod/vanilla extract (e.g. vanille arome)	Seeds from 1/ few drops
Cream	500 ml

1 Roll out the pastry 2 mm thick, 30 cm long.
2 Cut into 1.5 cm-wide strips. Moisten on one side.
3 Wind carefully round lightly greased cream horn moulds, starting at the point and carefully overlapping each round slightly.
4 Brush with eggwash on one side and place on a greased baking sheet.
5 Bake at 220°C for about 20 minutes.
6 Sprinkle with icing sugar and return to a hot oven for a few seconds to glaze.
7 Remove carefully from the moulds and allow to cool.
8 Place a little jam in the bottom of each.
9 Add the sugar and vanilla to the cream and whip until firm peaks form.
10 Place in a piping bag with a star tube and pipe a neat rosette into each horn.

Variations

These may also be partially filled with pastry cream, to which various flavourings or fruit may be added. For example:

- praline
- chocolate
- coffee
- lemon
- raspberries
- strawberries
- mango
- orange segments.

29   Eccles cakes

Makes >	12 cakes
Puff pastry or rough puff pastry	300 g
Egg white, to brush	
Caster sugar, to coat	
Filling	
Butter	50 g
Raisins	50 g
Demerara sugar	50
Currants	200
Mixed spice (optional)	pinch

1 Roll out the pastry 2 mm thick.

2 Cut into rounds 10–12 cm diameter. Damp the edges.

3 Mix together all the ingredients for the filling and place a tbsp of the mixture in the centre of each round.

4 Fold the edges over to the centre and completely seal in the mixture.

5 Brush the top with egg white and dip into caster sugar. Press flat and score the top.

6 Place on a greased baking sheet.

7 Cut two or three incisions with a knife so as to show the filling.

8 Bake at 220°C for 15–20 minutes.

30   Apple turnovers (chausson aux pommes)

Makes >	12 turnovers
Puff pastry	300 g
Dry, sweetened apple purée	250 g
Egg white, to brush	
Caster sugar, to coat	

1 Roll out the pastry 2 mm thick. Rest.

2 Cut into 8 cm diameter rounds.

3 Roll out slightly oval, 12 × 10 cm.

4 Moisten the edges, place a little apple purée in the centre of each.

5 Fold over and seal firmly.

6 Brush with egg white and dip in caster sugar.

7 Place sugar side up on a dampened baking sheet.

8 Bake in a hot oven, 220°C for 15–20 minutes.

Variations

Other types of fruit may be included in the turnovers, such as apple and mango, apple and blackberry, apple and passion fruit, and apple, pear and cinnamon.

31 Strawberry soleil

Makes >	1 (8 portions)
Puff pastry	250 g
Pastry cream	400 g
Slice of sponge	1
Grand Marnier or rum stock syrup (Chapter 10, Recipe 21)	20 ml
Strawberries, hulled	2 punnets
Strawberry glaze	200 ml

Professional tip: the soleil shape

- Cut out a 30 cm circle of pastry.
- Mark (but do not cut) an inner circle with a 22 cm diameter.
- Starting from the inner circle, make 16 cuts, regularly spaced.
- Fold over each piece so that edge a aligns with edge b (see diagram), forming points.

1 Roll the pastry. Rest it, then cut and fold it into a soleil (sun) shape. Dock the central area, eggwash the points and press down.

2 Bake blind at 205°C for about 15 minutes. Dock the base again at least once during cooking.

3 Beat the pastry cream until smooth. Pipe it in the centre of the pastry to within 2 cm of the inside edge. Pipe in a cone shape so that it is domed towards the centre.

4 Cut a slice of sponge to fit and lay it on top of the cream. Brush with the Grand Marnier stock syrup.

5 Spread a little more pastry cream on top of the sponge.

6 Arrange whole strawberries all over the sponge, pointing outwards. Brush with boiling strawberry glaze, making sure any gaps are filled in.

32 Cream buns (choux à la crème)

Makes >	10 buns
Choux paste	150 ml
Chopped almonds and/or nib sugar	35 g
Whipped cream/Chantilly cream	300 ml
Icing sugar, to serve	

1 Place the choux paste into a piping bag with a 1 cm plain tube.

2 Pipe out on to a lightly greased, baking sheet into pieces the size of a walnut.

3 Using a wet finger, gently press down any spikes or peaks of paste to make round bulbs of paste, and then eggwash.

4 Bake at 200°C for approximately 20 minutes.

5 Allow to cool. Make a hole in the base of each bun using a small, round piping nozzle.

6 Fill with sweetened, vanilla-flavoured whipped cream (Chantilly cream) using a piping bag and small tube.

7 Sprinkle with icing sugar and serve with cream and fresh fruits such as strawberries or raspberries.

33 Profiteroles and chocolate sauce (profiteroles au chocolat)

Makes >	10 portions
Choux paste	200 ml
Chocolate sauce (Chapter 3, Recipe 32)	250 ml
Chantilly cream	250 ml
Icing sugar, to serve	

Variations

Alternatively, coffee sauce may be served and the profiteroles filled with non-dairy cream. Profiteroles may also be filled with chocolate-, coffee- or rum-flavoured pastry cream.

1 Spoon the choux paste into a piping bag with a plain nozzle (approx. 1.5 cm diameter).

2 Pipe walnut-sized balls of paste onto the greased baking sheet, spaced well apart. Level the peaked tops with the tip of a wet finger.

3 Bake for 18–20 minutes at 200°C, until well risen and golden brown. Remove from the oven, transfer to a wire rack and allow to cool completely.

4 Make a hole in each and fill with Chantilly cream.

5 Dredge with icing sugar and serve with a sauceboat of cold chocolate sauce, or coat the profiteroles with the sauce.

34 Chocolate éclairs (éclairs au chocolat)

Makes >	12 éclairs
Choux paste	200 ml
Whipped cream/Chantilly cream	250 ml
Fondant	100 g
Chocolate couverture	25 g

Note

Traditionally, chocolate éclairs were filled with chocolate pastry cream.

1. Place the choux paste into a piping bag with a 1 cm plain tube.
2. Pipe into 8 cm lengths onto a lightly greased baking sheet.
3. Bake at 200–220°C for about 30 minutes.
4. Allow to cool. Make two small holes in the base of each.
5. Fill with Chantilly cream (or whipped cream), using a piping bag and small tube. The continental fashion is to fill with pastry cream.
6. Warm the fondant to 37°C, add the finely cut chocolate, allow to melt slowly, adjust the consistency with a little sugar and water syrup if necessary. Do not overheat or the fondant will lose its shine.
7. Glaze the éclairs by dipping them in the fondant; remove the surplus with the finger. Allow to set.

Variations

For **coffee éclairs** (éclairs au café) add a few drops of coffee extract to the fondant instead of chocolate; coffee éclairs may also be filled with pastry cream (Chapter 3, Recipe 12) flavoured with coffee.

Pierce the éclair with a cream horn mould or similar

Pipe in the filling

Dip the éclair in fondant; wipe the edges to give a neat finish

Gâteaux Paris-Brest

Makes >	1 large or 8 individual
Choux paste	200 ml
Crème diplomat	400 ml
Praline	
Flaked almonds, hazelnuts and pecans (any combination)	200 g
Granulated sugar	250 g

To make the praline

1 Place the nuts on a baking sheet and toast until evenly coloured.

2 Place some of the sugar in a large, heavy, stainless steel saucepan. Set the pan over a low heat and allow the sugar to caramelise. Do not over-stir, but do not allow the sugar to burn. Gradually feed in the rest of the sugar.

3 When the sugar is evenly coloured, remove from the heat and stir in the warm nuts.

4 Immediately deposit the mixture on a Silpat mat. Place another mat over the top and roll as thinly as possible.

5 Allow to go completely cold. Break up and store in an airtight container. A food processor such as a robot-coupe can be used to break up the praline. This will produce quite a fine mix, particularly if ground multiple times.

For the Paris-Brest

1 Pipe choux paste into rings (approximately 8cm), sprinkle with flaked almonds (optional) and bake.

2 Once cooled, slice each ring in half and fill by piping with a mixture of crème diplomat and praline.

3 Dust with icing sugar.

36 Choux paste fritters (beignets soufflés)

Shape the fritter with two spoons

Lower the fritters into hot oil on a strip of greaseproof paper

Lift them out with a spider

Makes >	10 fritters
Choux paste	150 ml
Caster sugar	
Fruit-based sauce to serve	125 ml

1 Using two spoons of the same size, shape the paste into quenelles.

2 Place onto strips of lightly greased, greaseproof paper.

3 Lower the pieces into moderately hot deep fat at 170°C. Allow to cook gently for 10–15 minutes.

4 Drain well and roll in the caster sugar. The sugar can be flavoured at this point. (Ground cinnamon is often used at this stage.)

5 Serve with a sauceboat of hot fruit-based sauce.

38 | Gâteau St Honoré

Makes >	1 (10 portions)
Choux paste	500 g
Puff paste disc	16 cm
Pastry cream	300 ml
Kirsch	30 ml
Double cream, whipped	300 ml
Caramel	1 kg

1 Pipe the choux paste in a ring around the edge of the docked puff paste disc, and also pipe a set of profiteroles.

2 Bake at 220°C for 10 minutes, then at 165°C for 15 to 20 minutes.

3 Beat the pastry cream, flavour it with kirsch and fold in the whipped double cream.

4 Dip the profiteroles in caramel. Fill the profiteroles with the cream.

5 Stick them to the ring of choux paste on the base.

6 Fill the centre with quenelles of the cream using a St Honoré piping nozzle.

Note

Traditionally, this gateau is flled with crème chiboust (crème St Honoré), which is pastry cream with leaf gelatine added (6g per 250ml) while hot, with Italian meringue folded through.

38 Croquembouche

Water	400 ml
Granulated sugar	1 kg
Glucose	200 g
Profiteroles, piped and baked	
Crème diplomat	
Nougatine	

Note

Quantities required depend on the size of croquembouche required.

1 Boil the water, sugar and glucose to make a caramel.
2 Dip each profiterole in caramel and allow to cool.
3 When cool, fill with crème diplomat.
4 Using caramel as glue, assemble the profiteroles around a croquembouche mould (see Appendix).
5 Assemble the nougatine in shapes to form a base.
6 Once the caramel has set, turn the profiteroles out of the mould onto the nougatine base.
7 Decorate with cut-out nougatine shapes and pulled sugar.

Note

This dish is traditionally served at weddings in France. The embellishment techniques used to create Croquembouche (such as pulled sugar, etc.) can be found in Chapter 10.

39 Veal and ham pie

Hot water paste is commonly used in the production of savoury items, rather than sweet. However, the pastry chef may be required to produce the paste for such items or even to produce such items in their entirety. One of the most traditional products made from hot-water paste is the classic hand-raised pork pie. This veal and ham pie is another example.

	Makes >	1 pie/terrine (10 portions)
Ham or bacon		375 g
Lean veal (shoulder, leg or loin)		625 g
Salt, pepper		to season
Parsley and thyme		1 tsp
Lemon, grated zest of		2
Stock or water		5 tbsp
Bread, soaked in milk		125 g
Hot water paste (see Chapter 3, Recipe 9)		375 ml
Hard-boiled egg		2
Gelatine		12.5 g

1. Cut the veal and ham/bacon into small even pieces and combine with the rest of the main ingredients.

2. Keep one-quarter of the paste warm and covered. Roll out the remaining three-quarters and carefully line a well-greased raised pie or terrine mould. Ensure that there is a thick rim of pastry.

3. Add half of the filling and press down firmly. Place the shelled egg in the centre of the mixture and top with the remaining mixture.

4. Roll out the remaining pastry for the lid, and eggwash the edges of the pie.

5. Add the lid, seal firmly, neaten the edges, cut off any surplus paste; decorate if desired.

6. Make a hole 1 cm in diameter in the centre of the pie; brush all over with eggwash.

7. Bake in a hot oven (230–250°C) for approximately 20 minutes.

8. Reduce the heat to moderate (150–200°C) and cook for 1½–2 hours in total.

9. If the pie colours too quickly, cover with greaseproof paper or foil. Remove from the oven and carefully remove tin. Eggwash the pie all over and return to the oven for a few minutes.

10. Remove from the oven and fill with approximately 125 ml of good hot stock in which 5 g of gelatine has been dissolved.

11. Serve when cold, garnished with picked watercress and offer a suitable salad.

Partly fill the pie, then place eggs into the centre

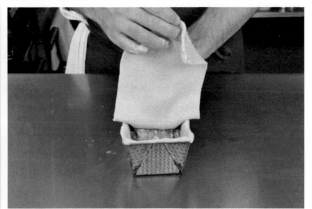
Add the pastry lid and seal it firmly

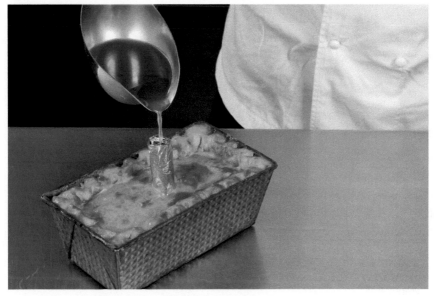
Pour gelatine dissolved in stock into the pie after baking

40 Steamed fruit pudding

Suet paste is used for steamed fruit puddings, steamed jam rolls, steamed meat puddings and dumplings. Vegetarian suet is also available to enable products to be meat free. Here is an example of a steamed pudding using apple.

Makes >	10 individual or 1 large
Suet paste (Chapter 3, Recipe 10)	300 g
Fruit	1 kg
Sugar	200 g
Water	60 ml

1 Grease a basin or individual moulds.
2 Line the moulds, using three-quarters of the paste.
3 Add the prepared and washed fruit and the sugar. (Add between one and two cloves in an apple pudding.)
4 Add water. Moisten the edge of the paste.
5 Cover with the remaining quarter of the pastry.
6 Seal firmly.
7 Cover with greased greaseproof paper, a pudding cloth or foil.
8 Steam for about 1½ hours (large basin) or 40 minutes (individual moulds). Serve with custard.

Fillings

Steamed fruit puddings can be made with apple, apple and blackberry, rhubarb, rhubarb and apple, and so on.

Test yourself

Level 2

1 What is the difference between a bouchée and a vol-au-vent?
2 Which paste is used to make the following products?
   a Cheese straws
   b Lemon tart
   c Cream buns
   d Fruit pies
3 What is the difference between a Swiss apple flan and a French apple flan?

Level 3

4 Name two products that can be made using off-cuts of (or secondary rolled) puff pastry?
5 Which paste(s) is used to make the following products?
   a Steamed fruit puddings
   b Gâteau St Honoré
   c Pork pies
   d Gâteau pithiviers
6 What is the name of the pastry product that is often served at weddings in France?

5 Fermented goods

This chapter covers:
→ **NVQ level 2 Prepare, cook and finish basic bread and dough products**
→ **VRQ level 2 Produce fermented dough products**
→ **NVQ level 3 Prepare, cook and finish complex bread and dough products**
→ **VRQ level 3 Produce fermented dough and batter products.**

This chapter will help you to:
→ **Identify, prepare and cook different types of fermented dough products using the correct techniques and equipment (level 2)**
→ **Identify, prepare and finish a variety of fermented dough products to the recipe specifications, in line with current professional practice (level 3).**

Recipes in this chapter

Introduction

Bread, as we would recognise it, first appeared in Ancient Egypt around 4000 BC, and was the result of a happy accident between a brewery and a bakery. Froth from the fermenting beer blew over from the brewery and landed on some dough; the result was a lighter product. This raised (or leavened) bread was later helped by adding some of the previous day's dough, a method still used in the production of sourdough. It is generally accepted that the longest development time, with the minimum amount of yeast, will give bread the best flavour and character.

Although some methods have changed little over thousands of years, techniques are continually evolving and rules once written on tablets of stone are being broken as knowledge and understanding increase. For instance, it was taught that the liquid must be tepid, but in some modern recipes iced water is used (not just very cold water, but water with ice in it). A cold dough is easier to shape. Also, improvers, used to speed up the production of factory bread, are now being used by artisan bakers, whose main interest is to improve the quality of the finished product.

Bread making involves few ingredients but is a complex science. Here we can only scratch the surface, but a thorough understanding of the reactions and effects of ingredients and processes is the difference between producing a good product and an excellent one.

Food allergies and intolerance

It is estimated that up to 45 per cent of the population suffer from some sort of food intolerance which, while not life threatening, can be very uncomfortable for sufferers. The symptoms vary from person to person and are numerous.

An allergy is an allergic reaction, it is much more serious and can be life threatening.

An increasing number of people are intolerant to gluten (the protein found in wheat, barley and rye) and to a lesser extent, yeast. This is known as coeliac disease, a digestive condition which damages the lining of the small intestine. Sufferers of coeliac disease must avoid wheat- and flour-based products.

It is important that staff are well briefed on the ingredients used in fermented goods and that gluten/yeast-free products are kept separate to avoid contamination.

The main ingredients of fermented goods

Strong flour

This is the essential ingredient when making fermented products. Strong flour is a white wheat flour which has been processed to remove the outer skin (bran), the husk and germ. It has a high gluten or protein content. The gluten provides increased water absorption and elasticity, which is essential for allowing the dough to expand by trapping the CO_2 gas produced by the yeast. Legislation requires that four specific nutrients in specific quantities are added to wheat flour; they are: iron, vitamin B1, nicotinic acid (niacin) and calcium carbonate.

Wholemeal and wheatmeal flour

As the names would suggest, these flours contain the whole wheat grain. Nothing is added or taken away, and they are considered to be a healthier alternative to white flour. Because the nutrients mentioned above are naturally present, these flours do not require any additions.

Rye flour

Rye is a type of grass which grows in harsh climates and is associated with Northern and Eastern Europe, where there is a tradition of rye breads. Rye was looked upon as being inferior to wheat. It has a low gluten content and will add flavour and texture when used with strong flour.

Spelt

A grain related to wheat but with less gluten and a nutty flavour.

Rice cones and rice flour

Coarsley ground rice can be sprinkled on the baking sheet before placing on the baking tray and on top of the bread, which adds texture and crunch to the products. Rice flour is ground much finer and is used with gram flour to make the gluten-free loaf in Recipe 15.

Gram flour

This is flour made from chickpeas.

Yeast

Yeast is a living organism which, when fed (on sugar), watered and kept warm, will multiply and produce carbon dioxide gas and ethyl alcohol. Yeast is essential to lighten or leaven fermented products. It comes compressed in a block (fresh) or dried, sometimes with the addition of ascorbic acid (vitamin C) which is an improver. Most recipes will say to dissolve yeast in the (usually warm) liquid, but more bakers are moving away from this and feel it is better to rub the yeast into the flour rather than 'drown' it in the liquid. Dried yeast is concentrated so remember to use half the quantity if using it in place of fresh yeast.

Using too much yeast can affect flavour and the products will stale more quickly. The best results are achieved by using the minimum quantity of yeast and allowing it to prove and develop over a longer time.

Improvers

Available in powder form, these usually have a vitamin C or ascorbic acid base, which speeds up the action of the yeast. This eliminates the need for bulk proving or BFT (bulk fermentation time) – see the ADD method below. Fast action dried yeast is a combination of dried yeast with an improver added.

Salt

In the last few years salt has had a bad press mainly because of its overuse in processed foods, but in truth we cannot live without salt in our diet, as a lack of it can lead to dehydration.

It is best to use sea or rock salt because unlike table salt, where most of the mineral content is destroyed by the high temperatures used in its production, sea and rock salts are relatively natural products.

As salt plays such a huge role in fermented goods, it must be measured carefully. It helps to stabilise the fermentation and strengthen the gluten, improves the crust texture, colour and flavour and lengthens the shelf life of products.

Note

Salt must not come into direct contact with the yeast as it will slow, or at worst kill the yeast and stop the fermentation.

Sugar

Sugar has also gained a bad reputation and is believed to make a significant contribution to obesity. Much of this is due to the food manufacturing industry using it in excessive quantities in their processed food. Although yeast feeds on sugar there is plenty naturally present in flour. The sugar added to a recipe is for flavour (it does, however, help speed up the fermentation).

Note

Like salt, sugar should not come into direct contact with the yeast. You may come across recipes which recommend creaming the yeast and sugar together. Be aware that this is bad practice.

Liquid

Some bakers recommend using bottled water but this is by no means essential unless your tap water is heavily chlorinated. Milk, beer or buttermilk can also form part of the liquid content when making different products. The liquid is usually added at around body temperature, unless the air temperature is very hot or you want to delay the fermentation, in which case it is added cold. The quantity of liquid added may vary depending on the strength of the flour and the time of year (the flour tends to need more liquid during the winter).

Milk produces a softer dough. However, because milk contains an enzyme which inhibits yeast activity, they are not natural partners. Bringing the milk to the boil before adding it will neutralise this enzyme. Often, milk is added in the form of dried powder, as in several recipes featured in this chapter.

Fat

Butter is the most common fat in fermented goods recipes, but oils are also used. The advantage of using oil is that it does not have to melted or rubbed in. Fat shortens the gluten strands, making the dough less elastic, so in most bread recipes only a little fat (or sometimes none at all) is used. In brioche, where the butter content is high (anything from 25–100 per cent of the flour weight) the texture is much softer and more like a cake.

Eggs

Eggs are used in enriched doughs. Good quality ingredients will result in good quality products (free range hens are thought to produce eggs which contain a stronger albumen and have a deeper coloured yolk), and the fresher the egg the better. Most recipes will be based on a size 3 egg, but often the egg content is expressed as a liquid measurement which helps achieve greater accuracy. Pasteurised whole egg, yolks and whites are also available in litre cartons.

Classification

Fermented doughs can be divided into four categories: simple, enriched, laminated and batters.

Simple doughs

These include most types of bread. They consist of flour, salt, yeast, water and sometimes a small amount of fat.

Enriched doughs

Enriched doughs use the same ingredients as simple doughs, but are enriched with eggs and butter. Sometimes milk is used instead of water, and they may also include dried fruits, spices and a larger quantity of sugar.

Effects of adding ingredients which enrich a dough:
- increased nutritional content
- change in colour
- softer texture and a finer crumb
- may retard (slow down) the yeast
- increased shelf life.

Laminated doughs

These are either a simple dough which is layered with butter (such as croissant paste) or an enriched dough layered with butter (such as Danish pastry). This raising agent is a combination of yeast fermentation and lamination of the butter.

Batters

Batters, as the name would suggest, are much wetter than other doughs. They cannot be moulded or shaped and have to be cooked in moulds. Examples include savarin and blinis.

Methods

Straight dough

This is the simplest method and is often used when making simple bread doughs. All the ingredients are added at once and mixed together.

Sponge and dough

This method is often used for making enriched doughs (those which contain larger quantities of fat, sugar and eggs). As these additional ingredients slow down yeast activity, this method involves making a batter or 'sponge' with the yeast, liquid, and some of the flour to enable the yeast to start working before the other ingredients are added, giving the fermentation a 'kick start'.

Ferment and dough

This is very similar to the sponge and dough method and is used for the same reason. Here the yeast is mixed with the liquid and added to a well in the flour, a little of the flour is mixed in and more sprinkled over to cover. As this batter ferments it 'erupts' through the flour crust, indicating it is ready to go to the next stage. This method is most often used when making very slack doughs or batters such as savarin and blinis.

Activated/accelerated dough development (ADD)

This process is primarily associated with the industrial manufacture of bread made by commercial methods. Such methods are concerned with saving time and reducing labour costs, which in turn saves money and maximises profit. Industrial bread produced by these methods is now often looked upon as contributing to an unhealthy diet – it is believed that high-speed milling reduces the nutritional content. Also, two or three times the usual amount of yeast is used, and the hydrogenated fats are often replaced with a fractionated variety. Breads produced using these methods have been linked to the increase in coeliac disease, gluten intolerance, yeast intolerance and irritable bowel syndrome.

The ADD method

The 'accelerator' referred to in this method is known as an 'improver' or conditioner, the main ingredient of which is more often than not ascorbic acid (Vitamin C). The second feature of this process is the speed and length of mixing (very fast for a longer time). This generates heat in the dough, which speeds up the fermentation. This is also assisted by the 'improver' which develops the gluten much faster. The advantage of beating the dough in this way is that proving in bulk (see BFT, below), as used in traditional methods, is not required, thus considerably reducing the production time. It may reduce the flavour. This method is now being adapted and developed by smaller bakeries, where the emphasis is on quality, rather than time and cost savings.

Retardation

By holding the dough at temperatures between 2°C and 4°C, yeast activity is stopped. This enables the dough to be made and held, then baked as needed. For example, a dough can be made, shaped and held overnight then baked in the morning.

This method is best undertaken using a piece of equipment known as a retarder-prover. A timer can be set, allowing the dough to be proved slowly and be ready to bake at a specific time.

Such pieces of equipment go hand in hand with modern developments and are contributing to quality improvements. For instance, using an 'improver' with less yeast and a much longer final proving can result in a better flavour, character and shelf life – more like a sourdough where the proving is long and the fermentation natural.

Quality points

Although individual products will vary in terms of texture and crumb size, depending on the ingredients and methods used, they should all be:

- consistent in size and shape
- evenly coloured
- correctly finished according to type.
- fresh.

The main equipment used

Planetary mixer

Every professional kitchen will have a planetary mixer. They are used when making anything other than the smallest quantities. The speed and length of mixing time are often stated in the recipes and affect the elasticity and development of the gluten.

> **Health and safety**
>
> Large mixers can be dangerous. All operatives must be over 18 and must be trained before using them. Some useful points to remember are:
> - Do not use large equipment if working alone.
> - If the mixer is table mounted, always make sure it is secure.
> - Use the correct attachment.
> - All guards must be in place.
> - Always start on the slow speed.
> - Disconnect when not in use.

Prover

A prover provides the ideal conditions required to encourage yeast activity. It provides warmth and moisture in a controlled environment. The optimum temperature is 24–26°C for most products. If no prover is available then leave in a warm place covered with a large plastic bag. (See also Proving and Retardation.)

Bakers' oven

Professional bakers' ovens are decked (separately controlled ovens stacked on top of each other) and often come with top/bottom heat controls and steam injection.

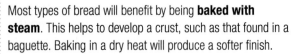

> **Professional tip**
>
> Most types of bread will benefit by being **baked with steam**. This helps to develop a crust, such as that found in a baguette. Baking in a dry heat will produce a softer finish.

> **Health and safety**
>
> Care must be taken when removing baking sheets from the oven.
> - Always place the trays in the oven with the open end facing out.
> - Have a cooling oven rack next to the oven.
> - Always use two dry folded oven cloths to remove hot baking sheets.
> - Make others aware when removing hot baking sheets.

Digital scales

Accurate measurement is essential, particularly when the quantities are small. Salt, sugar and yeast all have a big effect on the speed and development of the fermentation.

Small equipment

It is expected that the kitchen will have a good supply of general items, such as:

- mixing bowls in various sizes
- sturdy baking sheets
- silicone mats
- cooling racks
- loaf/baking tins
- flour sieves
- flour brushes
- plastic scrapers
- measuring jugs
- rubber spatulas, etc.

Definitions

Mixing and kneading

More often than not, unless very small quantities are required, doughs will be mixed mechanically. They should be mixed slowly on speed 1 to start with to make sure the flour is properly hydrated, and then mixed at speed 2 to develop the gluten. The dough should come away from the bowl, leaving the sides clean.

Kneading is a process done by hand when making smaller quantities. It consists of pulling and stretching the dough with the heel of the hand, changing the structure and creating a smooth, uniform elastic mixture.

Biga/poolish/starter/levain

These are all terms for a soft ferment made with yeast, water, flour and sometimes other ingredients such as honey, yoghurt or raisins which all boost yeast activity and encourage the dough to ripen.

Proving

Most doughs are proved twice, first in bulk as explained below and then a second time just before baking. During this process the yeast feeds on the sugars and ferments, producing carbon-dioxide gas and ethyl alcohol. The gas is trapped by the gluten strands, which stretch and allow the dough to increase in size. This process develops character, texture and flavour. The alcohol gives the bread its distinctive flavour and aroma.

Bulk fermentation time (BFT)

This is the period of time after the dough is made and before it is scaled and shaped. It is the time in which the dough is allowed to prove, developing the flavour and texture. The optimum time is usually one hour at 24–26°C.

Knocking back

After the bulk proving (BFT), the dough is 'knocked back' to expel the old gas and bring the yeast back into contact with the dough so it can prove a second time.

Scaling

Scaling is the process of dividing the dough by weight before shaping.

Baking

All fermented goods are baked in a hot to very hot oven which kills the yeast and stops the fermentation. Enriched doughs are baked at a lower temperature. Bread in particular will benefit from the addition of steam, which helps to develop the crust and colour. As stated earlier, modern bakers' ovens will often have a steam injection facility.

Eggwash

Many fermented products will be eggwashed before baking. Be consistent – eggwash is best made to a recipe in a reasonable quantity, not just as you need it. For the best results use the yolks only, add 10 per cent water and a good pinch of salt. This will give a deep rich glaze to the finished products.

Storage

If storing uncooked dough it should be flattened, placed inside a plastic bag and frozen. Some dough such as brioche can be left in the fridge overnight, but generally if doughs are stored in the fridge for too long the yeast will expire.

Cooked doughs can be kept at an ambient temperature for a short time but are best frozen if keeping for longer. Never store cooked dough products in the fridge as this will speed up staling.

When storing, all products should be placed in plastic bags or wrapped in cling film, labelled and dated. Cooked items should be kept separate from raw doughs and the stock rotated (using FIFO – first in, first out).

The Maillard reaction

The Maillard reaction was discovered by the French chemist Louis Maillard around the turn of the twentieth century. It is the name given to describe the effect of baking – the reaction between carbohydrates (sugar) and amino acids (protein) which occurs when subjected to high temperatures. When heat is applied (baking), chemical changes take place on the surface, and the result is browning. This gives the products colour, aroma, flavour and texture. This process will not happen below 150°C.

Faults in fermented products

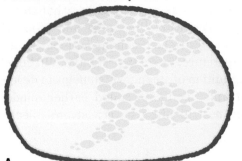

A

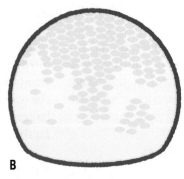

B

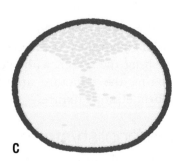

C

Underripe dough, sometimes referred to as a 'green' dough, means the dough is under-proved and will be small in volume and tough (C). When baked it will have a high crust colour, will possibly split at the side and have a very close texture.

Causes:
- not enough yeast and/or not enough time
- too much sugar, salt or spice.

Overripe dough has been over-proved, resulting in flat shapes where the products have risen and dropped back (A). When cooked they will have a loose, open texture and an anaemic colour.

Causes:
- too much yeast, too much proving
- lack of salt.

Video: fermentation in bread,
http://bit.ly/14CDgJW

1 Seeded bread rolls

Makes (approximately) >	30 rolls
Strong flour	1 kg
Yeast	30 g
Water at 37°C	600 ml
Salt	20 g
Caster sugar	10 g
Milk powder	20 g
Sunflower oil	50 g
Eggwash	
Poppy seeds	
Sesame seeds	

Method: straight dough

1 Sieve the flour onto paper.

2 Dissolve the yeast in half the water.

3 Dissolve the salt, sugar and milk powder in the other half.

4 Add both liquids and the oil to the flour at once and mix on speed 1 for 5 minutes or knead by hand for 10 minutes.

5 Cover with cling film and leave to prove for 1 hour at 26°C.

6 'Knock back' the dough and scale into 50 g pieces.

7 Shape and place in staggered rows on a silicone-paper-covered baking sheet.

8 Prove until the rolls almost double.

9 Eggwash carefully and sprinkle with seeds.

10 Bake immediately at 230°C with steam for 10–12 minutes.

11 Break one open to test if cooked.

12 Allow to cool on a wire rack.

Professional tip

Instead of weighing out each 50 g piece of dough, weigh out 100 g pieces and then halve them.

Placing bread rolls in staggered rows means they are less likely to 'prove' into each other. The spacing allows them to cook more evenly and more will fit on the baking sheet.

Variations

Try using different seeds such as sunflower, linseed or pumpkin.

For a beer glaze, mix together 150 ml beer with 100 g rye flour and brush on before baking.

2　Wholemeal loaf

Makes >	2 standard loaves
Strong flour	125 g
Wholemeal flour	625 g
Yeast	25 g
Water at 37°C	500 ml
Sunflower oil	60 g
Honey	40 g
Salt	10 g

Method: straight dough

1　Sieve the flours onto paper.

2　Dissolve the yeast in half the water.

3　Mix the honey, salt and oil with the rest of the water.

4　Add both liquids to the flours and mix well for 5 minutes or knead by hand for 10 minutes to achieve a soft and slightly sticky dough.

5　Place in a clean oiled bowl, cover with cling film and prove at 26°C for 1 hour.

6　Knock back and divide in half, roll and shape, place into a prepared loaf tin and leave to prove until double.

7　Bake at 220°C for 35–40 minutes. When cooked, the bread should sound hollow when tapped on the base.

8　Leave to cool on a wire rack.

Professional tip

When adding walnuts, first boil and then dry them before adding, or they will discolour the dough.

Try something different

For walnut and sultana bread, add 50 g walnuts and 50 g sultanas to the recipe.

3 Parmesan rolls

Use the bread recipe for seeded rolls (Recipe 1).

Makes (approximately) >	30 rolls
Grated Parmesan cheese	approx. 200 g

Method: straight dough

1 Follow method for seeded rolls (Recipe 1) up to step 5.
2 Lightly flour work surface and roll the dough into a rectangle until 3 cm thick.
3 Make sure the dough is not stuck to the surface.
4 Brush with water and cover with Parmesan.
5 Using a large knife cut into squares 6 × 6 cm.
6 Place on a silicone-paper-covered baking sheet and leave to prove until almost double in size.
7 Bake at 230°C for 10–12 minutes with steam.
8 Cool on a wire rack.

Professional tips

● When making bread that requires rolling out as opposed to being individually shaped (Recipes 3 and 4), it is helpful to decrease the liquid content by 10 per cent so it will be easier to process.
● To ensure the squares are all the same size, mark a grid using the back of the knife before cutting.

4 Red onion and sage rolls

Use the bread recipe for seeded rolls (Recipe 1).

Makes (approximately) >	30 slices
Red onion	1 small
Fresh sage	few leaves

Method: straight dough

1 Follow method for seeded rolls up to step 5.
2 Finely dice the onion and gently cook with a little oil, then add the chopped sage. Leave to cool.
3 Lightly flour work surface and roll the dough to a rectangle 8 cm wide, 2 cm thick and as long as it will go.
4 Shake, cover and rest.
5 Spread the onion over ⅞ of the dough and eggwash the far edge.
6 Roll up as you would a Swiss roll and pinch along the edge to seal.

7 Cut into 50 g slices and lay flat on a silicone-paper-covered baking sheet.
8 Prove until almost double in size, carefully eggwash and bake at 230°C for approximately 10 minutes.
9 Cool on a wire rack.

Try something different

Try adding some grated smoked cheese before rolling up.

5 Olive bread

Makes >	4 loaves
Starter dough	
Yeast	40 g
Water at 37°C	180 ml
Strong flour	225 g
Sugar	5 g
Dough	
Strong flour	855 g
Sugar	40 g
Salt	20 g
Water at 37°C	450 ml
Olive oil	160ml
Green olives (cut into quarters)	100 g

Method: sponge and dough

1 For the starter, dissolve the yeast in the water, add the flour and sugar, mix well, cover and leave to ferment for 30 minutes.

2 For the dough, sieve the flour, sugar and salt into a mixing bowl, add the water followed by the starter dough and start mixing slowly.

3 Gradually add the oil and continue mixing to achieve a smooth dough.

4 Cover with cling film and prove for 1 hour or until double in size.

5 Knock back, add the olives and divide the dough into four.

6 Roll into long shapes and place on a baking sheet sprinkled with rice cones, return to the prover and leave until double in size.

7 Brush with olive oil and bake at 220°C for 20–25 minutes.

8 When cooked the bread should sound hollow when tapped on the base.

9 Leave to cool on a wire rack.

Make up the starter

Starter ready for use after proving

Start mixing in the ingredients for the main dough, tearing up the starter

Continue mixing in the ingredients and working the dough

Shape the dough

Divide and roll into loaves

6 | Bagels

Makes (approximately) >	10–12 bagels
Strong flour	450 g
Yeast	15 g
Warm water	150 ml
Salt	10 g
Caster sugar	25 g
Oil	45 ml
Egg yolk	20 g
Milk	150 ml
Poppy seeds	

Method: ferment and dough

1. Sieve the flour, place in a mixing bowl.
2. Make a well and add the yeast which has been dissolved in the water.
3. Mix a little of the flour into the yeast to form a batter, sprinkle over some of the flour from the sides and leave to ferment.
4. Mix together the salt, sugar, oil, egg yolk and milk.
5. When the batter has fermented, add the rest of the ingredients and mix to achieve a smooth dough.
6. Cover and prove for 1 hour (BFT).
7. Knock back and scale at 50 g pieces, shape into rolls and make a hole in the centre using a small rolling pin.
8. Place on a floured board and prove for 10 minutes.
9. Carefully drop into boiling water and simmer until they rise to the surface.
10. Lift out and place on a silicone-covered-baking sheet, eggwash, sprinkle or dip in poppy seeds and bake at 210°C for 30 minutes.

Use a rolling pin to make a hole in the centre of each bagel

Poach the bagels in water

Eggwash the bagels and sprinkle with seeds before baking

7 Baguette

Method: sponge and dough

1 For the starter dough, dissolve the yeast into the water.
2 Combine the two flours in a bowl, make a well, add the liquid and mix to a paste.
3 Cover the bowl and leave to ferment for 6 hours.
4 For the dough, add the cold water to the starter dough and mix well.
5 Place the flour in a mixing bowl, add the salt to one side and the yeast to the other, add the starter dough and mix slowly for 5 minutes.
6 Scrape down and continue to mix for 7 minutes on a medium speed until the dough is smooth and elastic. (A faster speed will generate heat and encourage fermentation.)
7 Prove until double in size.
8 Knock back, scale into 320 g pieces and roll into long sticks.
9 Score by making 7 diagonal cuts with a sharp knife.
10 Bake at 250°C with steam for 20 minutes.
11 Cool on a wire rack.

Note

In France a baguette must weigh 320g and have 7 cuts along the top!

Makes >	6 baguettes
Starter dough	
Yeast	5 g
Water	135 ml
Strong flour	100 g
Rye flour	100 g
Dough	
Cold water	680 ml
Strong flour	1070 g
Fine sea salt	15 g
Yeast	22 g

8 Ciabatta

Makes	4 loaves
Starter dough	
Yeast	10 g
Water	180 g
Strong flour	350 g
Dough	
Strong flour	450 g
Yeast	10 g
Water	340 g
Salt	20 g
Olive oil	50 g
Coarse semolina or rice cones	

Method: sponge and dough

1 For the starter dough, dissolve the yeast in the water, add to the flour and mix to a dough. Place in a bowl, cover with cling film and leave for 24 hours.

2 For the dough, sieve the flour, rub in the yeast, break the starter dough into small pieces and add.

3 Add the water, salt and oil and mix on a slow speed for 5 minutes.

4 Place in an oiled bowl, cover and prove for 1 hour at 22°C.

5 Knock back the dough and divide into four pieces.

6 Roll into long cylinders and place on a baking sheet dusted with rice cones/semolina, brush with water and sprinkle over more rice cones/semolina.

7 Prove until double in size and bake at 230°C for 18–20 minutes.

8 Cool on a wire rack.

9 Foccacia

Makes >	8 × 15cm
Starter dough	
Yeast	40 g
Water at 37°C	180 ml
Strong flour	225 g
Sugar	15 g
Dough	
Strong flour	855 g
Sugar	30 g
Salt	20 g
Water at 37°C	480 ml
Olive oil	180 ml
Salamoia	
Water at 50°C	200 g
Olive oil	200 g
Salt	20 g

1 For the starter dough, dissolve the yeast in the water, add the flour and sugar, mix well, cover and leave to ferment for 30 minutes.

2 For the dough, sieve the flour, sugar and salt into a mixing bowl, add the water followed by the starter dough and start mixing slowly.

3 Gradually add the oil and continue mixing to a smooth dough.

4 Cover with cling film and prove for 1 hour until double in size.

5   Brush individual round baking plates with olive oil.

6   Divide the dough into 250 g pieces, roll into balls and then roll out on a lightly floured surface, keeping them round.

7   Place on baking plates, brush with olive oil and sprinkle over garnish.

8   Push your fingers into the dough to create dimples.

9   Prove for 30 minutes or until dough has risen slightly.

10 Bake at 230°C for 25–30 minutes

11 Whisk together the salamoia ingredients to fully emulsify and brush over the cooked bread immediately it comes out of the oven.

Garnishes

- rosemary
- chopped olives
- thyme
- sun dried tomatoes
- pesto.

10   Pain de campagne

Makes >	2 large or 4 smaller loaves
Starter dough	
Strong flour	200 g
Dark rye flour	50 g
Yeast	5 g
Water	175 g
Salt	5 g
Dough	
Strong flour	500 g
Dark rye flour	100 g
Yeast	5 g
Water at 40°C	400 ml
Salt	7 g

Method: sponge and dough

1   For the starter dough, mix the two flours and rub in the yeast, add the water and salt and mix.

2   Place in an oiled bowl, cover with cling film and leave in the fridge overnight.

3   For the dough, place the flours in a mixing bowl and rub in the yeast, add the starter dough, water and salt and mix slowly for 5 minutes.

4   Cover and prove for 1 hour.

5   Knock back and divide, shape into round loaves.

6   Score the tops with a sharp knife, dust with flour and prove until double in size.

7   Bake at 220°C for 30–40 minutes.

11 Rye bread

Method: sponge and dough

1. Dissolve the yeast in the water, add 50g of the strong flour, cover and leave to ferment.
2. In a bowl combine the oil, salt, treacle and lager.
3. Place the two flours in a mixing bowl, add the yeast batter and the rest of the ingredients, mix slowly for 5 minutes.
4. Place in an oiled bowl, cover with cling film and prove until double in size. (This may take up to 2 hours.)
5. Knock back and shape into round or oval loaves.
6. Sprinkle rice cones on a baking sheet and place the loaves on it, allowing space to prove.
7. Prove until double in size.
8. Dust with flour and bake at 200°C for approximately 1 hour.
9. To test that loaves are cooked, tap the base and they should sound hollow.
10. Allow to cool on a wire rack.

Makes >	2 medium sized loaves
Yeast	30 g
Water at 40°C	120 ml
Strong flour	350 g
Dark rye flour	500 g
Sunflower oil	30 ml
Salt	20 g
Black treacle	30 g
Lager	500 ml
Rice cones	

Try something different

- Caraway seeds can be added to the dough and/or sprinkled on top before baking.
- Make into rolls and brush with beer wash.

12 Soda bread

Makes >	2 medium loaves
Plain flour	575 g
Salt	10 g
Baking powder	10 g
Bicarbonate of soda	10 g
Sugar	12 g
Buttermilk	45 g
Sunflower oil	56 g
Water	360 ml

1 Sieve the flour, salt, baking powder and bicarbonate of soda twice to thoroughly disperse.

2 Make a well, add all other ingredients and lightly mix to produce a slightly tacky dough.

3 On a lightly floured surface, shape into a round without overworking.

4 Place on a silicone-paper-covered baking sheet and dust with flour.

5 Score the top to make a cross.

6 Place in oven immediately and bake at 190°C for about 25 minutes.

7 Cool on a wire rack.

Note

Soda bread is suitable for those who suffer from yeast intolerance.

13 Wholemeal and black treacle soda bread

Makes >	3 medium loaves
Plain flour	330 g
Wholemeal flour	330 g
Salt	15 g
Baking powder	15 g
Bicarbonate of soda	15 g
Sunflower oil	75 ml
Buttermilk	300 ml
Black treacle	60 g
Water	225 ml

1 Sieve the flour, salt, baking powder and bicarbonate of soda twice to thoroughly disperse.

2 Mix all other ingredients, add to the flours and lightly mix to produce a slightly tacky dough.

3 On a lightly floured surface shape into a round without overworking.

4 Place on a silicone-paper-covered baking sheet, dust with flour and cut a cross in the top.

5 Place in oven immediately and bake at 190°C for about 25 minutes.

6 Cool on a wire rack.

14 Curry rolls

Makes >	30 rolls
Strong flour	500 g
Butter	50 g
Yeast	20 g
Water at 40°C	300 g
Curry powder	16 g
Salt	10 g
Currants	70 g
Flaked almonds	30 g

Method: straight dough

1 Sieve the flour and rub in the butter.

2 Dissolve the yeast in half the water.

3 Dissolve the curry powder and salt in the rest of the water.

4 Add both liquids to the flour and mix to a smooth dough.

5 Place in an oiled bowl, cover with cling film and prove at 26°C for 1 hour.

6 Knock back and add the currants and almonds.

7 Scale into 50 g pieces and shape into rolls.

8 Prove until double in size, dredge with a mix of flour and curry powder.

9 Bake at 230°C for 12–15 minutes.

10 Cool on a wire rack.

Professional tip

Make sure staff are thoroughly briefed that this recipe contains nuts, which can cause an **allergic reaction**. Keep separate from other products to avoid contamination.

15 Red onion and chive gluten-free loaf

Makes >	2 loaves
Red onions	1
Chives	1 small bunch
Eggs	4
Milk	400 ml
Butter	80 g
Corn meal or gram flour	350 g
Bicarbonate of soda	10 g
Cream of tartar	10 g
Rice flour	150 g
Sugar	20 g
Rock salt	
Black pepper	

1 Butter and line the loaf tins with silicone paper.
2 Finely shred the onions and chop the chives.
3 Whisk the eggs and milk together and melt the butter.
4 Sieve together all the dry ingredients and make a well.
5 Add the liquid and melted butter and mix.
6 Mix in the onions and chives.

7 Divide between the two loaf tins.
8 Sprinkle with rock salt and black pepper.
9 Rest for 15 minutes then bake at 180°C for 25 minutes.
10 Allow to cool slightly before unmoulding onto a cooling rack.

Try something different
The top can be covered with grated cheddar cheese.

Note

This recipe is suitable for those with a wheat allergy or gluten intolerance.

16 Sourdough

When it comes to bread, sourdough is the gold standard. It has a distinctive sour taste and a strong chewy crust.

The recipe looks long and complicated, but the different stages are necessary to get the process underway by making a 'sour' or starter dough. It is possible to have a 'sour dough' that is very old and has been passed down from one generation to the next. This will have a well-developed and distinctive character that only comes with age.

A routine must be established to keep the dough alive. Even if you do not intend to bake, you must **refresh** the dough regularly by taking out what you would use and adding an equal quantity back. At least every two days take out 500 g dough. If you are not going to use it to bake then discard it and add back in 500g flour with 250 g water, mix, cover and place in the fridge.

It is important not to sanitise bowls and surfaces with something that destroys all known germs as the wild airborne yeasts that you rely on for fermentation will not survive.

Day 1

	Makes >	2 loaves around 1 kg each
Sour		
Strong white flour		1 kg
Warm water		750 g

Mix together in a bowl, cover with a plastic bag and secure with an elastic band and leave to ferment at room temperature.

Day 2

Strong white flour	1 kg
Warm water	750 g

Add the flour and water to the day 1 mixture, mix well, cover and leave as before. It is important to use the same bowl and the same plastic bag.

Day 3

Strong white flour	500 g
Warm water	250 g

Repeat as above.

Days 4 and 5

Repeat as for day 3 and leave for a further two days before using.

The sourdough loaf

White sour dough	
Strong white flour	1 kg
Yeast	20 g
Salt	20 g
Mother	500 g
Olive oil	20 g
Water	500 g

1 Rub the yeast into the sieved flour and place in a mixing bowl.

2 Add the salt to one side and add the sour dough, oil and water.

3 Mix on the slow speed for 5 minutes, then turn up the speed and continue mixing for another 5 minutes.

4 Cover and rest for 4 hours.

5 Divide in half and shape into rounds.

6 Leave to prove until double in size.

7 Dust with flour and slash the top.

8 Bake at 230°C with steam for 30–35 minutes.

Note

Sourdoughs do not usually contain any additional yeast but rely solely on natural airborne yeasts for fermentation. This recipe does include a small amount of additional yeast to boost fermentation and shorten the production time.

A good-quality sourdough will have:
- a slightly acidic taste with characteristic sour notes
- a very strong crunchy crust
- an open-textured dense crumb
- a dark colour with distinctive cuts
- excellent keeping qualities.

Professional tip

Allow the 'sour dough' to come up to room temperature before adding to the other ingredients.

Variations

- Rye sourdough – use the above recipe but replace half the white flour with dark rye.
- Wholemeal sourdough – use the white starter dough but use 25% white flour to 75% wholemeal, and reduce the water by 10%.

17 Bun dough and varieties

Makes approx. >	12 buns	24 buns
Strong flour	500 g	1 kg
Yeast	25 g	50 g
Milk (scalded and cooled to 40°C)	250 ml	500 ml
Butter	60 g	120 g
Eggs	2	4
Salt	5 g	10 g
Sugar	60 g	120 g

Method: sponge and dough

1 Sieve the flour.

2 Dissolve the yeast in half the milk and add enough of the flour to make a thick batter, cover with cling film and place in the prover to ferment.

3 Rub the butter into the rest of the flour.

4 Beat the eggs and add the salt and sugar.

5 When the batter has fermented add to the flour together with the liquid.

6 Mix slowly for 5 minutes to form a soft dough.

7 Place in a lightly oiled bowl, cover with cling film and prove for 1 hour at 26°C.

8 Knock back the dough and knead on the table, rest for 10 minutes before processing.

Bun wash

Milk	250 ml
Caster sugar	250 ml

Bring both ingredients to the boil and brush over liberally as soon as the buns are removed from the oven. The heat from the buns will set the glaze and prevent it from soaking in, giving a characteristic sticky coat.

Sift the flour

Rub in the fat

Make a well in the flour, and pour in the beaten egg

Pour in the liquid

Fold the ingredients together

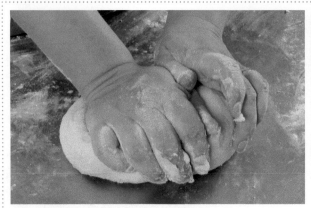

Knead the dough

Before and after proving: the same amount of dough is twice the size after it has been left to prove

Chelsea buns, hot cross buns and Bath buns

Chelsea buns

Makes >	12 buns	24 buns
Basic bun dough	1 kg	2 kg
Melted butter	60 g	120 g
Demerara sugar	50 g	100 g
Currants	100 g	200 g
Sultanas	100 g	200 g
Mixed peel	30 g	60 g

1 Roll out the dough on a lightly floured surface into a rectangle 25 cm deep.

2 Brush with melted butter and sprinkle over caster sugar followed by the dried fruit and mixed peel.

3 Eggwash the far edge and roll up lengthways like a swiss roll, pinch to seal.

4 Brush outside with melted butter and cut into 3 cm wide slices.

5 Line a deep-sided baking tray with silicone paper and lay in the slices so they are touching.

6 Allow to prove.

7 Bake at 220°C for 15–20 minutes.

8 Brush with bun wash as soon as they come out of the oven and break to separate.

Hot cross buns

Makes >	12–14 buns	24 buns
Basic bun dough	1 kg	2 kg
Currants	75 g	150 g
Sultanas	75 g	150 g
Mixed spice	5 g	10 g
Crossing paste		
Strong flour	125 g	250 g
Water	250 ml	500 ml
Oil	25 ml	50 ml

1 Add the dried fruit and spice to the basic dough, mix well.

2 Scale into 60 g pieces and roll.

3 Place on a baking sheet lined with silicone paper in neat rows opposite each other and eggwash.

4 Mix together the ingredients for the crossing paste. Pipe it in continuous lines across the buns.

5 Allow to prove.

6 Bake at 220°C for 15–20 minutes.

7 Brush with bun wash as soon as they come out of the oven.

Bath buns

Makes >	12–14 buns
Basic bun dough	1 kg
Bun spice	20 ml
Sultanas	200 g
Sugar nibs	360 g
Egg yolks	8

1 Mix the bun spice into the basic dough and knead.

2 Add the sultanas, two-thirds of the sugar nibs and all the egg yolks.

3 Using a plastic scraper, cut in the ingredients (it is usual for the ingredients not to be fully mixed in).

4 Scale into 60 g pieces.

5 Place on a paper-lined baking sheet in rough shapes.

6 Sprinkle liberally with the rest of the nibbed sugar.

7 Allow to prove until double in size.

8 Bake at 200°C for 15–20 minutes.

9 Brush with bun wash as soon as they come out of the oven.

Variation

To make **fruit** buns, proceed as for hot cross buns without the crosses.

Swiss buns

Makes >	12–14 buns
Basic bun dough	1 kg
Fondant	500 g
Lemon oil	5 ml

1 Scale the dough into 60 g pieces.

2 Roll into balls then elongate to form oval shapes.

3 Place on a baking sheet lined with silicone paper, eggwash.

4 Allow to prove.

5 Bake at 220°C for 15–20 minutes.

6 Allow to cool then dip each bun in lemon-flavoured fondant.

Doughnuts

	Makes >	12 doughnuts
Basic bun dough		1 kg
Caster sugar		500 g
Raspberry jam		250 g

1 Scale the dough into 60 g pieces.
2 Roll into balls and make a hole in the dough using a rolling pin.
3 Prove on an oiled paper-lined tray.
4 When proved carefully place in a deep fat fryer at 180°C.
5 Turn over when coloured on one side and fully cook.
6 Drain well on absorbent paper.
7 Toss in caster sugar.
8 Make a small hole in one side and pipe in the jam.

Try something different
The caster sugar can be mixed with ground cinnamon.

Health and safety
As a fryer is not a regular piece of equipment found in a patisserie, a portable fryer is often used. Always make sure it is on a very secure surface in a suitable position. Never attempt to move it until it has completely cooled down. In addition, extreme care must be taken to avoid serious burns.
- Only use a deep fat fryer after proper training.
- Make sure the oil is clean and the fryer is filled to the correct level.
- Pre-heat before using but never leave unattended.
- Always carefully place the products into the fryer – never drop them in. Use a basket if appropriate.
- Never place wet products into the fryer.

18 Savarin dough and derivatives

	Makes >	35 individual items
Basic dough		
Strong flour		450 g
Yeast		15 g
Water at 40°C		125 ml
Eggs		5
Caster sugar		60 g
Salt		pinch
Melted butter		150 g

Professional tip
Savarin paste is notorious for sticking in the mould. Always butter moulds carefully and then flour. After use do not wash the moulds but wipe clean with kitchen paper.

Method: ferment and dough
1 Sieve the flour and place in a bowl. Make a well.
2 Dissolve the yeast in the water and pour into the well.
3 Gradually mix the flour into the liquid, forming a thin batter. Sprinkle over a little of the flour to cover, then leave to ferment.
4 Whisk the eggs, sugar and salt.
5 When the ferment has erupted through the flour, add the egg mixture and mix to a smooth batter, cover and leave to prove until double in size.
6 Add the melted butter and beat in.
7 Pipe the batter into prepared (buttered and floured) moulds one-third full.
8 Prove until the mixture reaches the top of the mould and bake at 220°C for 12–20 minutes, depending on the size of the mould.
9 Unmould and leave to cool.

Cream the yeast in milk to make a ferment

Add the dissolved yeast to the flour, and sprinkle a little flour over it

The mixture after fermentation

Add beaten eggs, sugar and salt

The dough after proving

Add the butter

After proving, pipe into moulds

After proving for the final time

Note

Savarin and savarin-based products are never served without first soaking in a flavoured syrup. They are literally dry sponges and should be:

- golden brown in colour with an even surface
- smooth, with no cracks, breaks or tears
- evenly soaked without any hard or dry areas
- sealed by brushing with apricot glaze after soaking.

Cooked products can be stored in the fridge overnight, but if left for too long they will dry out and cracks will appear. They are best wrapped in cling film, labelled and stored in the freezer.

Savarin syrup

It is important that savarin syrup is at the correct density. It should measure 22 Baumé on the saccharometer. If the syrup is too thin the products are likely to disintegrate. If it is too dense it will not fully penetrate the product and leave a dry centre.

Makes >	3 litres
Oranges	2
Lemons	2
Water	2 litres
Sugar	1 kg
Bay leaf	2
Cloves	2
Cinnamon sticks	2

1 Peel the oranges and lemons and squeeze the juice.

2 Add all the ingredients into a large pan and bring to the boil, simmer for 2–3 minutes and pass through a conical strainer.

3 Allow to cool and measure the density (it should read 22 Baume), adjust if necessary. (More liquid will lower the density, more sugar will increase it.)

4 Re-boil then dip the savarins into the hot syrup until they swell slightly, check they are properly soaked before carefully removing and placing onto a wire rack with a tray underneath to drain.

5 When cooled, brush with boiling apricot glaze.

Savarin with fruit

A savarin is baked in a ring mould, either large or individual.

Before glazing, sprinkle with kirsch and fill the centre with prepared fruit, serve with a crème anglaise or raspberry coulis.

Marignans Chantilly

A marignan is baked in an individual boat-shaped mould or barquette.

Split the glazed marignans and fill with crème Chantilly after glazing.

Once split and before filling, they can be sprinkled with rum or Grand Marnier.

Blueberry baba

For a baba, currants are usually added to the basic savarin dough before baking.

As for marignans above, split the baba and fill with crème diplomat and blueberries or crème Chantilly.

19 | Brioche

The quantity of butter added to a brioche dough can vary from 25–100 per cent of the flour weight, and is identified in the list below. The higher the butter content the more difficult it is to handle. (The standard is to use 50 per cent butter.)

- Brioche commune – uses 25 per cent butter
- Brioche à tête – uses 50 per cent butter
- Brioche mousseline – uses 75 per cent butter
- Brioche surfine – uses 100 per cent butter

Makes >	2 × 500 g loaves or 16 individual
Basic dough	
Strong flour	500 g
Yeast	15 g
Milk (scalded and cooled)	70 ml
Eggs	4
Salt	15 g
Sugar	30 g
Diced butter	250 g

Brioche à tête (front) and brioche Nantaise

Method: straight dough

1 Sieve the flour onto paper and place in a mixing bowl.
2 Dissolve the yeast in the milk.
3 Beat the eggs, sugar and salt.
4 Add both liquids to the flour and mix for 5 minutes on speed 1 with the dough hook, followed by 5 minutes on speed 2.
5 Replace the dough hook with a beater and gradually add the butter and mix until smooth.
6 Rest in the fridge for 1 hour before processing.

Brioche à tête

This makes 16 brioches.

1 Scale the dough into 50 g pieces.
2 First roll into balls and then using the side of the hand, almost remove the top third, drop into buttered brioche moulds and using a finger, push down so the head sits neatly in the centre.
3 Prove at 22°C until double in size.
4 Eggwash carefully and bake at 210°C for 15–20 minutes.
5 Unmould immediately and cool on a wire rack.

Brioche Nantaise

This makes three standard-sized loaves.

1 Scale the dough into 40 g pieces
2 Roll each piece into a ball.
3 Place the balls side by side in a buttered loaf tin until full.
4 Prove at 22°C until mixture reaches the top of the tin.
5 Eggwash carefully and bake at 200°C for 25–30 minutes.
6 Unmould immediately and leave to cool on a wire rack.

Brioche mousseline

This makes four brioches.

The butter content can be increased to 75 per cent of the flour weight.

1. Divide the dough into four pieces and shape into balls.
2. Prepare the tins by buttering then lining with silicone paper. (Washed A2½ tins can be used in place of a specialist mould – just make sure the paper comes about 4 cm above the top of the tin.)
3. Elongate the brioche and carefully drop into the tin, snip the top with a pair of scissors to form a cross.
4. Prove at 22°C until mixture reaches the top of the tin.
5. Bake at 210°C for 20–25 minutes.
6. Unmould immediately and cool on a wire rack.

Try something different

Croûte Bostock: Slice the brioche mousseline into 2 cm rounds, spread each with almond cream or frangipane, slightly domed in the centre, and sprinkle with flaked almonds. Place on a paper-lined baking sheet and bake for 12–15 minutes at 210°C. When baked, dust with icing sugar and glaze under the salamander, serve with crème anglaise.

Brioche en tresse

This makes two loaves.

1. Divide the dough into half.
2. Divide each piece into three or five.
3. Roll each piece into long strands and plait.
4. Prove at 22°C until double in size.
5. Eggwash and bake at 210°C for 20–25 minutes.
6. Cool on a wire rack.

Brioche couronne (crown)

This makes two loaves.

1. Divide the dough into two pieces and shape into round balls
2. Using a rolling pin make a hole in the centre and moving the rolling pin in circles, make the hole bigger until the brioche is around 20 cm in diameter.
3. Place on a silicone-paper-covered baking sheet, carefully eggwash and snip around the top edge with scissors.
4. Prove slowly at 22°C until almost double in size, carefully re-eggwash around the base and bake at 205°C for 20–25 minutes.
5. Cool on a wire rack.

Note

A good-quality brioche will have:
- a soft crust with a deep golden colour
- a fine crumb and a short cake-like texture.

20 Croissants

Makes (approximately) >	35 pieces
The détrempe or basic paste	
Strong flour	960 g
Yeast	25 g
Cold water	600 ml
Salt	12 g
Sugar	30 g
Milk powder	30 g
Pastry butter (beurre sec)	500 g

1 Sieve the flour onto paper and place in a mixing bowl.

2 Dissolve the yeast in half the water.

3 Dissolve the salt and sugar in the rest of the water and whisk in the milk powder. Add both liquids to the flour and mix slowly for 5 minutes.

4 Take a large tray (approximately 30 cm × 60 cm) and cover with cling film.

5 Roll the dough into a large rectangle the same size as the tray, lay on the tray, cover with cling film and rest for 25 minutes.

6 Process the butter by rolling out to two-thirds the size of the dough inside a plastic bag.

7 Place the butter on the dough, leaving one third uncovered, seal the edges.

8 Fold over the uncovered dough and then fold over the other third, giving three layers of dough and two layers of butter.

9 Roll out into a rectangle the same size as before and shake before folding into three.

10 Rest in a cool place for 30 minutes.

11 Repeat points 9 and 10 twice more, giving three single turns in total.

12 Roll out into a rectangle 20 cm deep and 3 mm thick, shake well.

13 Cut into isosceles triangles, make a small cut in the short edge and roll away from yourself as tightly as possible and stretching the dough at the same time, seal the tip with eggwash.

14 Lay on a silicone-papered baking sheet and bend the ends around to meet in the middle to form the crescent shape, arrange in staggered rows, allowing the croissants room to expand.

15 Prove at 22°C for 35–40 minutes until double in size, carefully eggwash and bake at 230°C for 15 minutes, cool on a wire rack.

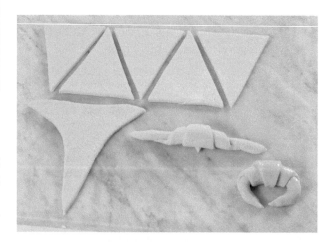

Try something different

Before rolling and shaping into crescents, they can be filled with Parma ham or gruyère cheese, or for a sweet variation try almond cream and sprinkle the top with a few flaked almonds after eggwashing.

Pain aux raisins

Makes >	35
Crème pâtissière	500 g
Raisins pre-soaked in rum	50 g

1 Follow the instructions for croissants up to point 12.

2 Spread the dough with the softened pastry cream and sprinkle with the raisins.

3 Roll up as for a Swiss roll, seal the edge with eggwash.

4 Cut into 1–2cm slices and lay flat on a silicone-papered baking sheet, spacing as for croissants.

5 Prove at 22°C for 35–40 minutes until almost double in size, carefully eggwash and bake at 220°C for 20 minutes, cool on a wire rack.

6 Brush with boiled apricot glaze.

Pain au chocolat

Makes >	30
Chocolate batons (purpose-made)	60 (approx.)

1 Follow the instructions for croissants up to point 12.

2 Cut into rectangles 5 × 8cm.

3 Lay on two batons (or one if preferred) of chocolate, eggwash and fold over the paste.

4 Space as before on a papered baking sheet and prove at 22°C until double in size.

5 Carefully eggwash and bake at 230°C for 15 minutes, cool on a wire rack.

21 Danish pastries

Makes >	20 pieces
The détrempe or basic dough	
Strong flour	450 g
Yeast	20 g
Cold milk (previously scalded)	200 ml
Eggs	2
Sunflower oil	50 ml
Salt	5 g
Sugar	70 g
Pastry butter (beurre sec)	250 g

Method: straight dough and lamination

1 Sieve the flour onto paper and place in a mixing bowl.

2 Dissolve the yeast in the milk.

3 Whisk together the eggs, oil, salt and sugar.

4 Add both liquids to the flour and mix to a soft dough.

5 Take a large tray (approximately 30 cm × 60 cm and cover with cling film.

6 Roll the dough into a large rectangle the same size as the tray, lay on the tray, cover with cling film and rest. (A word of caution – Danish pastry is a soft dough and needs careful handling, the work surface should be lightly dusted with flour at regular intervals.)

7 Process the butter by rolling out to two-thirds the size of the dough inside a plastic bag.

8 Place the butter on the dough, leaving one-third uncovered, seal the edges.

9 Fold over the uncovered dough and then fold over the other third, giving three layers of dough and two layers of butter.

10 Roll out into a rectangle the same size as before and shake before folding into three.

11 Rest in a cool place for 30 minutes.

12 Repeat points 10 and 11 twice more, giving three single turns in total.

13 Process according to the variety/varieties required (see recipes and methods below).

Place the butter on the dough, leaving one-third uncovered

Fold over the uncovered dough

Many different shapes and fillings can be used to make Danish pastries. Four examples are shown below.

Sultana roulade

Makes >	20
Almond cream	500 g
Sultanas	100 g

1 Carefully roll out the dough into a rectangle 18 cm wide and 3mm thick.

2 Spread the almond cream over the paste, leaving 2 cm of the long edge uncovered, eggwash before rolling up as for a Swiss roll.

3 Pinch the edge to seal and cut into 1–2 cm slices.

4 Lay on a papered baking sheet in staggered rows and prove at 22°C until nearly double in size.

5 Carefully eggwash and bake at 210°C for 25 minutes.

6 Allow to cool before brushing with boiled apricot glaze.

7 Finally brush over a thin coat of water icing.

8 Serve at room temperature.

Roll up the paste

Shape the roulade

Apple envelopes

Makes (approximately) >	20
Almond cream	500 g
Apple slices	From 2 large apples
Cinnamon sugar	100 g

1 Roll out the dough into a large square 3 mm thick.

2 Cut into smaller 8 cm squares.

3 Pipe a line of almond cream diagonally on the paste.

4 Lay on two slices of apple which have been dipped in cinnamon sugar.

5 Fold over the corners diagonally and seal with eggwash.

6 Lay on a papered baking sheet in staggered rows and prove at 22°C until nearly double in size.

7 Carefully eggwash and bake at 210°C for 25 minutes.

8 Allow to cool before brushing with boiled apricot glaze.

9 Finally brush over a thin coat of water icing.

10 Serve at room temperature.

Folding the envelopes

Windmills

Makes (approximately) >	20
Pastry cream	500 g
Apricot halves or plum halves	20

1. Roll out the dough into a large square 3 mm thick.
2. Cut into smaller 8 cm squares.
3. Make a cut on each corner and pipe a bulb of pastry cream in the centre.
4. Place the fruit on the pastry cream.
5. Fold alternate corners to meet in the centre, sealing with eggwash.
6. Repeat points 4–8 of the sultana roulade process described above.

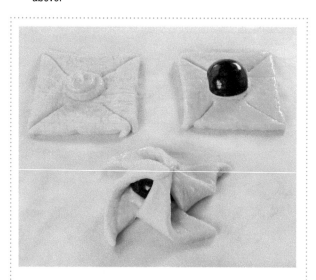

Folding the windmills

Cockscombs

Makes (approximately) >	20 pieces
Pastry cream	500 g
Stoned cherries	250 g

1. Roll out the dough into a long strip 10 cm wide and 3 mm thick.
2. Pipe the pastry cream down the length of the paste using a large plain piping tube.
3. Cover with halves of cherry, eggwash one edge and fold the other over to enclose the filling.
4. Make a series of 1 cm cuts down the length where the paste meets.
5. Cut into 8 cm lengths and bend outwards, causing the cuts to open up.
6. Repeat points 4–8 of the sultana roulade process described above.

Shaping the cockscombs

Note

Good-quality Danish pastries will have:

- uniform shapes and sizes
- soft light texture with no dryness
- uniform colour and even glaze
- thin water icing crust with a translucent appearance.

Test yourself

Level 2

1 Describe the following products in terms of shape, filling (if any) and finish:

   a Swiss buns

   b Chelsea buns

   c Hot cross buns.

2 How does dried yeast differ from fresh in terms of appearance and use?

Level 3

3 When making bread what are the advantages of using the ADD method compared with traditional techniques using BFT?

4 Describe the production methods used to make the following products:

   a Bread rolls

   b Brioche à tête

   c Danish pastries.

6 Hot desserts

This chapter covers the hot desserts component of:
→ **NVQ level 2 Prepare, cook and finish basic hot and cold desserts**
→ **VRQ level 2 Produce hot and cold desserts and puddings**
→ **NVQ level 3 Prepare, cook and finish complex hot desserts**
→ **VRQ level 3 Produce hot, cold and frozen desserts.**

In this chapter you will:
→ **Prepare and cook hot desserts and puddings using correct tools and equipment, and safe and hygienic practices (level 2)**
→ **Prepare and finish hot desserts and puddings to the recipe specifications, in line with current professional practice (level 3).**

Recipes in this chapter · · · · · · · · · · · · · · · · ·

Recipe	Page	Level 2	Level 3
1 Steamed sponge puddings	179	✓	
2 Sticky toffee pudding	180	✓	
3 Bramley apple spotted dick	181	✓	
4 Soufflé pudding	182	✓	
5 Baked apple	183	✓	
6 Orange pancakes	184	✓	
7 Apple Charlotte	186	✓	✓
8 Fruit fritters	188	✓	
9 Griottines clafoutis	190	✓	
10 Christmas pudding	191	✓	
11 Tarte tatin	192		✓
12 Apple crumble tartlet	194		✓
13 Rice pudding	195	✓	
14 Chocolate fondant	196		✓
15 Baked Alaska	198		✓
16 Vanilla soufflé	199		✓
17 Lemon curd flourless soufflé	202		✓
18 Crêpe soufflé	203		✓
19 Apple strudel	204		✓
20 Almond and apricot samosas	206		✓
21 Bread and butter pudding	207	✓	

Introduction

This chapter is where the word 'pudding' really comes into its own. It is often used in place of the word dessert but the original puddings were derived from a medieval mix of cereal, honey, wild fruits and meat, which was stuffed inside an animal stomach and boiled.

Later in the seventeenth century the meat was removed (suet now being the only reminder of its origins). The pudding cloth was first referenced in a recipe in 1617, replacing the animal stomach. One advantage of wrapping the pudding in a cloth was so that the poorer classes could cook 'pudding' in the same pot as the meat (the principal method of cookery at this time was a cauldron suspended over a fire). Later on, ground almonds, spices and dried fruits started to appear in puddings.

Mincemeat did at one time contain meat, with the other ingredients acting as a form of preservation. These would be saved and made into a pie around Christmas time (Christmas Pye). In America mincemeat is known as 'fruit mince'.

Suet puddings reached their peak in popularity in Victorian times – Prince Albert is credited with introducing the 'plum pudding' as part of the traditional Christmas dinner.

Today puddings are steamed in bowls rather than boiled in bags, and suet and steamed sponge puddings have enjoyed a revival in recent times, despite not being the healthiest of dishes.

The list of recipes here is by no means exhaustive – there are many more hot desserts than those included. There is, however, an example from each category, and variations and examples of alternatives have been given. This chapter includes traditional puddings and a variety of hot desserts, which are perhaps more in tune with a modern lifestyle.

Techniques and ingredients

There are no techniques or ingredients specific to hot desserts that are not found and explained elsewhere in this book.

Fresh ingredients

The principal fresh ingredient when making hot desserts is egg. When using eggs, obviously, they should be as fresh as possible, and if you use free range you will gain a deeper coloured yolk and a stronger white – this is particularly important when making a soufflé or desserts which contain meringue.

All dry ingredients must be stored correctly at around 20°C in a well-ventilated environment protected from contamination by pests or moisture.

Cooking methods

Boiling, simmering, baking and deep fat frying are all covered elsewhere in this book, but the following three methods deserve a special mention in relation to this chapter.

Steaming

This involves cooking in steam produced by boiling water. It is a more gentle method of cookery that suits dishes such as suet puddings, allowing the hard fat to melt more easily over a longer cooking time. Because of the low temperature, steaming does not give any colour to the food, hence the old fashioned named for a steamed suet roll is 'dead man's leg'. When steaming, make sure the products being steamed are properly sealed so they don't become waterlogged.

Combination ovens can provide a facility which allows baking (dry heat) with steam or steaming at a temperature higher than 100°C. This, unlike steam on its own, can add some colour to the product.

Health and safety !

Be careful if using a high pressure commercial steamer – stand aside and open the door carefully to avoid being burnt by the escaping steam.

Bain-marie

This is often used as a cookery method when baking desserts which contain custard. It consists of placing the products in a tray, half filling with hot water and baking in the oven. This allows the products to cook without boiling.

> ### Health and safety !
>
> Always place the tray in the oven before adding the hot water, and be extremely careful when removing after the cooking is complete. Have somewhere cleared to put the tray down and do not attempt to carry it across the kitchen.

Stewing/poaching/roasting

Fruit can be stewed in syrup or fruit juice, although nowadays this will rarely be seen on a menu. More often than not fruit is either poached or roasted. For a stew the fruit would usually be cut into pieces and the liquid thickened. Poached fruit is often cooked whole, with the liquid un-thickened and served cold. The term 'roasted', more commonly associated with meat, is now often used to describe vegetables and fruit which are cooked in a dry heat with the aid of fat.

Food safety and hygienic practice

It is recommended where a product or dish is cooked at a low temperature (bread and butter pudding, for example) or where the dish is required to have an undercooked centre (soufflé) that pasteurised eggs are used.

Allergies

It is estimated that up to one in three people will at some time in their lives suffer an allergy. Food is one of the most common allergens. Nuts, dairy produce and wheat products can all cause an allergic reaction. Anaphylaxis, usually associated with nuts, is one of the most serious allergic reactions, and if not treated quickly may cause death.

It is therefore vital that dishes are correctly labelled and staff are thoroughly briefed regarding the ingredients they contain. If a dish is sold as suitable for a particular dietary requirement, then care should be taken to ensure it is not contaminated at any time during or after its production.

1   Steamed sponge pudding

Makes >	10 individual puddings
Basic recipe	
Butter	250 g
Caster sugar	250 g
Self-raising flour	250 g
Baking powder	15 g
Eggs	6
Milk	20–40 g

Flavours for sponge puddings

- **Vanilla:** Scrape out the seeds from a vanilla pod and add to the mixture at the creaming stage. Use the pod to flavour the accompanying sauce à l'anglaise.
- **Chocolate:** Replace 50 g of the flour in the basic recipe with an equal amount of cocoa powder, add 150 g ground almonds, a pinch of sea salt, a splash of coffee essence and two more eggs. Add extra butter to the base of the mould and cover with muscavado sugar before filling. Serve with hot chocolate sauce.
- **Lemon or orange:** Add the grated zest of a large lemon or an orange to the mixture at the creaming stage. Serve with lemon or orange sauce as appropriate.
- **Sultana:** Add 100 g of washed sultanas and a few drops of vanilla compound. Serve with sauce à l'anglaise. The sultanas may be replaced by any dried fruit.
- **Ginger:** Add 10 g of ground ginger (sieved in and added with the dry ingredients) and 50 g of finely diced stem ginger folded in at the end. Serve with sauce à l'anglaise.
- **Golden syrup:** After buttering the mould pour into the base a generous layer of warmed golden syrup before adding the sponge mixture. Serve with sauce à l'anglaise sweetened with golden syrup.
- **Chocolate and fig/date:** Add six diced figs or dates to the chocolate recipe above.

1   Cream the butter and sugar until lighter in colour and approximately 30 per cent increase in volume is achieved.

2   Sieve the flour and baking powder twice to ensure even dispersion.

3   Beat the eggs and gradually add to the butter and sugar, beating in each addition before adding the next.

4   Finally fold in the dry ingredients.

5   Add enough milk to achieve a dropping consistency.

6   Fill buttered dariole moulds three-quarters full.

7   Place on a disc of silicone paper, and cover with foil, create a pleat to allow for expansion and crimp the edges around the lip of the mould to seal.

8   Steam for 45–50 minutes.

Note

A good-quality steamed sponge pudding will be light, moist and slightly spongy in texture. As steaming at 100°C imparts no colour, it should be as light in colour as the ingredients allow.

Sticky toffee pudding

Makes >	10 puddings
Dates	375 g
Water	625 g
Butter	125 g
Caster sugar	375 g
Eggs	5
Soft flour	375 g
Baking powder	10 g
Vanilla compound	1 tsp

1 Remove stones and chop the dates, place in the water and simmer for about 5 minutes until soft, set aside to cool.

2 Butter and sugar individual dariole moulds.

3 Cream the butter and sugar until light in colour and aerated.

4 Gradually add the beaten eggs, beating continuously.

5 Sieve the flour and baking powder twice and fold in.

6 Finally add the dates and water, and vanilla compound.

7 Fill the moulds three-quarters full and bake at 180°C for 30–35 minutes.

For the sticky toffee sauce

Granulated sugar	600 g
Unsalted butter	300 g
Double cream	450 g
Brandy	30 ml

1 To make a dry caramel, heat an empty pan. Add 1 tablespoon of sugar, then gradually add more as it melts, until a deep golden colour is achieved.

2 At the same time, cut the butter into small cubes, add to the cream and heat.

3 Gradually add the hot cream and butter to the caramel a little at a time. (Starting to add this arrests the cooking of the caramel.)

4 Finally stir in the brandy.

5 To serve, coat the pudding with the sauce and serve with vanilla or milk ice cream.

Professional tip

Sticky toffee pudding can be steamed instead of baked. If steaming, remember to cover with a disc of silicone paper and seal with a pleated square of foil, as for steamed puddings (Recipe 1).

3 Bramley apple spotted dick

Makes >	10 individual puddings
Soft flour	350 g
Salt	pinch
Baking powder	20 g
Suet	150 g
Light brown sugar	100 g
Currants	150 g
Lemon zest	1
Bramley apples	2 medium sized
Milk	250 ml

1 Sieve the flour, salt and baking powder into a bowl.

2 Stir in the suet, sugar, currants and grated lemon zest.

3 Peel and dice the apple into small cubes and add to the ingredients above.

4 Stir in the milk to form a sticky dough.

5 Divide the mixture between buttered dariole moulds (or similar).

6 Cover and seal the tops with foil and steam for 1½ hours.

7 Serve with crème anglaise or custard sauce.

Professional tip

Vegetarian suet can be used in this recipe.

Alternatively, this recipe can be cooked in the oven in a bain-marie.

Try something different

Serve with an apple and vanilla compote.

Vanilla pod	1
Sugar	300 g
Water	75 ml
Apples	4–5
Sultanas	50 g
White wine	50 ml

1 Split the vanilla pod and bring to the boil with the sugar and water.

2 Prepare the apples as for the spotted dick above.

3 Add the apples, sultanas and wine to the boiling syrup, remove from the heat and leave to stand before serving.

4 Soufflé pudding

Makes >	10 puddings
Milk	375 ml
Butter	50 g
Flour	50 g
Eggs	6
Caster sugar	50 g

1 Slowly bring the milk to boil in a sauteuse or heavy-bottomed pan.

2 Cream together the butter and flour (beurre manié).

3 Add small pieces of the beurre manié to the boiling milk, one at a time, whisking in each addition before adding the next.

4 Bring to the boil and simmer for a couple of minutes before removing from the heat.

5 Separate the eggs, put the whites to one side and whisk the yolks into the hot mixture. This forms the panada.

6 Cover with cling film and set aside to cool.

7 At this stage add any flavouring (see variations below).

8 Whisk the egg whites and sugar to a creamy but firm consistency, and fold into the base mixture.

9 Three-quarters fill buttered and sugared dariole moulds and place in a tray of boiling water (bain-marie) on top of the stove. Allow to gently simmer. When the mixture has risen to the top of the mould, place the whole tray in the oven at 210°C and bake for 15–20 minutes.

10 Unmould and serve with an appropriate sauce.

Professional tip

When boiling milk, rinse out the saucepan with cold water first but don't dry before adding the milk, this will help prevent the milk catching on the bottom.

Variations

- **Chocolate:** Add to the milk 60 g grated couverture.
- **Coffee:** Add to the milk 15 g instant coffee or coffee essence.
- **Lemon, orange or both (St Clements):** Add to the milk the grated zest of lemon or orange or both.
- **À l'indienne:** Add 60 g finely diced stem ginger to the mixture after adding the egg yolks. Serve with sauce à l'anglaise after infusing the milk with grated fresh ginger.
- **Grand Marnier:** Add a small dice of sponge fingers which have been sprinkled with Grand Marnier. Fold in just before the egg whites.
- **Saxon:** Flavour with vanilla.
- **Sans souci:** Add diced cooked apples and currants.

Note

A pudding soufflé is the only type of soufflé that is traditionally unmoulded. All other types of soufflé are served in the mould in which they are cooked or prepared.

By leaving them to stand in the bain-marie after cooking, pudding soufflés can be held for up to 15–20 minutes before serving.

5 Baked apple

Makes >	10 portions
Large dessert apples	10
Mincemeat filling	
Mincemeat	200 g
Grated orange zest	1
Grated apple	1
Rum	20 ml
Soft butter	200 g
Demerara sugar	approx. 300 g
Apple juice	500 ml

1 Mix together the mincemeat, zest, grated apple and rum.

2 Peel and core the apples, stuff with the filling.

3 Brush the outside of the apples generously with soft butter and roll in the demerara sugar to coat.

4 Place in a deep baking tray slightly spaced apart.

5 Pour in the apple juice.

6 Bake in the oven at 190°C for around 30–40 minutes basting frequently.

7 Allow to rest for 5 minutes

8 Strain the juices into a pan and reduce slightly

9 Spoon the juices over the apple and serve with either fresh cream, ice cream, crème fraiche or crème anglaise.

Note

Traditionally this dessert has been made using cooking apples with the skin left on and scored horizontally to allow for a controlled expansion.

This recipe recommends a dessert apple like a Braeburn or Granny Smith that is peeled but will cook through without disintegrating.

Try something different

Caramelised baked apple: Instead of brushing with butter and rolling in sugar, make a caramel and pour over each apple before cooking.

Pommes en cage: Cover the apple with a trellis of puff pastry before baking.

6   Orange pancakes (crêpes Suzette)

Said to have been the result of an accident when the pan caught fire, crêpes Suzette was named after a dining companion of Edward VII at his request.

Makes >	10 portions
Strong flour	250 g
Salt	pinch
Eggs	2
Milk	625 ml
Melted butter	25 g
Oil for cooking	

Note

Crêpe pans are heavy cast iron, flat-bottomed, low-lipped pans specifically for making crêpes, and should not be used for any other purpose, or they will start to stick. It is necessary to 'temper' the pans occasionally, by filling with salt and heating to a high temperature, so any moisture is drawn out of the metal. The salt is tipped out and the pan wiped with kitchen paper, after which it is ready to use.

Preparing and cooking the pancakes

1 Sieve the flour and salt into a bowl.
2 Make a well, add the eggs and half the milk, whisk to a thick batter and gradually add the rest of the milk and the melted butter.
3 Pass through a conical strainer. Allow to rest before using.
4 Heat the pancake pan and wipe thoroughly with kitchen paper, add a little oil and heat until it starts to smoke.
5 Add some of the batter, swirling the pan until covered in a thin layer, cook for a few seconds until lightly coloured.
6 Turn or 'toss' and cook on the other side.
7 Turn out onto a plate and sprinkle with sugar.
8 Repeat the process until all the mixture has been used, stacking the pancakes one on top of the other.

Professional tips

● Pancakes can be made in advance, separated by layers of paper, stacked in units of about ten portions. They can then be wrapped in cling film, labeled, dated and stored in the fridge overnight or the freezer for a longer period.
● This is not a dish suitable for large numbers of people – up to four servings is manageable.

For the Suzette

Makes >	2 portions
Butter	60 g
Oranges (grated zest and juice)	1
Caster sugar	60 g
Grand Marnier	30 ml
Cognac	30 ml
Pancakes	4

1 Melt the butter in a shallow pan, add the zest, sugar, orange juice and Grand Marnier, boil rapidly to reduce until thickened.

2 One by one add the pancakes to the pan, fold into four, lift out and keep hot on the plate.

3 Add the cognac and flambé.

4 Pour over the crêpes and serve.

5 Some orange segments may be added at the flambé stage.

Note

Traditionally this dish was finished and flambéed at the table, which not only provided the customers with visual entertainment, they could also enjoy the aromas provided by oranges and alcohol.

Variations

Other fruits can be used with the appropriate spirit or liqueur, such as apples and calvados, or pears with poire William. These fruits would need to be cooked in the liquor before the dish is flambéed.

Apple Charlotte

Makes >	8 individual or 2 small charlotte moulds
Dessert apples (Cox's)	1 kg
Butter	30 g
Caster sugar	100 g
Lemon zest	1
Breadcrumbs	
Large thin sliced bread loaf	1
Clarified butter	250 g

1 Peel, core and cut the apples into thick slices.

2 Melt the butter in a pan, add the sugar and finely grated lemon zest.

3 Add the apples and simmer until barely cooked, stir in some breadcrumbs to absorb any liquid.

4 Cut out circles of bread for the top and base of the moulds.

5 Cut the crusts and the rest of the bread into fingers 2–3cm wide, depending on whether you are making individual or larger charlottes.

6 Butter the moulds, dip half the circles in the clarified butter and place them butter side down in the base of each mould.

7 Next dip the bread fingers and line around the outside of the moulds, slightly overlapping.

8 Fill the centre with the apple filling, pressing in carefully.

9 Dip the rest of the circles in the butter and place on top, press firmly.

10 Bake at 230°C for 30–40 minutes until the bread is coloured and crisp.

11 Allow to cool slightly before unmoulding, serve with hot apricot sauce or crème anglaise.

Faults

A common fault is that the Charlotte collapses when unmoulded (the larger versions are more prone to this). The main reasons for this are:

- The filling is too wet. To avoid this ensure that cooking apples are never used, and do not overcook.
- The bread is not baked crisp enough to withstand the pressure of supporting the filling.

If making individual Charlottes, be careful the ratio of bread to filling is not compromised by using bread that is cut too thick or overlapping too much.

Try something different

Replace the apples with pears or use a mixture of both.

Professional tip

As this dessert will most likely be plated and sent from the kitchen, it will more often than not be made in individual portions. Larger versions (such as that in the first photograph) would be suitable for family or silver service, but could not be cut and made to look presentable if served from the kitchen.

8 Fruit fritters (beignets aux fruits)

For the batter

	Makes >	10 portions
Flour		500 g
Salt		pinch
Eggs		3
Milk		625 ml
Sunflower oil		75 ml
Caster sugar		25 g

1 Sieve the flour and salt.

2 Separate the eggs, add the egg yolks and half the milk, whisk until smooth.

3 Add the rest of the milk and the oil, pass through a conical strainer, cover and allow to rest for 30 minutes before using.

4 Just before using, whisk the egg whites with 25 g sugar until soft peaks form and fold into the batter.

To make the fritters

The fruit you use should be firm but ripe. Use approximately 1 kg in total.

Apples	peeled, cored and cut into thick slices
Banana	peeled and cut into three pieces on the slant
Pineapple	peeled, cored, cut into slices and halved
Plums	halved and stoned
Figs	cut into quarters
Apricots	halved and stoned
Kirsch	60 ml
Caster sugar	50 g
Lemon juice	20 ml

1 Place the prepared fruit in a bowl and add the kirsch, sugar and lemon juice, mix carefully (do not use your hands) and leave to marinate for 30 minutes.

2 Set a deep fat fryer at 180°C (make sure the oil is clean).

3 Drain the fruit (do not throw away the liquid).

4 Dip the individual pieces of fruit in flour, then batter, transfer to the fryer and cook for 2–3 minutes until a pale golden colour, lift out and drain on absorbent paper.

5 Sprinkle with icing sugar and serve hot on a paper napkin, with a fruit coulis or a sabayon flavoured with the saved liquid from the fruit.

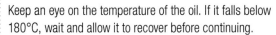

Professional tip

Keep an eye on the temperature of the oil. If it falls below 180°C, wait and allow it to recover before continuing.

Pre-dessert

It is fashionable, for example on taster menus, to serve a pre-dessert. This is a small portion of dessert served before the main dessert. Beignets could easily be adapted for this (for example, strawberries in tempura batter served with black pepper ice cream).

Variation: tempura batter

Egg yolks	1
Ice cold sparkling water	450 ml
Self-raising flour	125 g
Cornflour	125 g

1 Beat the egg yolk, add the water.

2 Add half the sieved flours, lightly mix, add the rest of the flours and mix lightly with chopsticks or similar. (This batter will be lumpy but it should never be over mixed. Throw in a few ice cubes to keep the mixture really cold.)

3 Use this batter to prepare the fritters, as described above.

9 Griottines clafoutis

This dish originates from the Limousin region of France and is traditionally made with black cherries.

	Makes >	10 portions
Eggs		5
Caster sugar		100 g
Flour		100 g
Milk		450 ml
Kirsch		20 ml
Griottine cherries		70
Neige décor to dust		

1 Whisk the eggs and sugar together.
2 Add the flour and whisk until smooth.
3 Add the milk and kirsch, mix well and pass through a conical strainer.
4 Brush 10 sur-le-plat dishes with soft butter.
5 Add 7 griottines to each dish and pour over the batter.
6 Bake at 200°C for 12–15 minutes until the batter has risen and set.
7 Serve warm, dusted with neige décor or icing sugar and a kirsch sabayon.

Variations

● Griottines work particularly well in this dish, but could be substituted for any other hard or fleshy fruit.
● Substitute 25 per cent of the flour with ground almonds or hazelnuts, or to make a richer mixture substitute up to half the milk for cream.

Note

Good-quality clafoutis should be a golden brown and will rise around the edges but stay flatter in the centre. Check to make sure they are set – if not, bake for a little longer.

Professional tip

● Like most batters, this one benefits from resting before cooking and can be made the day before.
● Clafoutis can also be made in Yorkshire pudding tins and served unmoulded.

10 Christmas pudding

	Makes >	2 × 1 kg puddings
Currants		175 g
Sultanas		175 g
Raisins		350 g
Guinness		500 ml
Cognac		50 ml
Strong flour		175 g
Mixed spice		4 g
Nutmeg		4 g
Cinnamon		4 g
Breadcrumbs		175 g
Suet		350 g
Mixed peel		80 g
Ground almonds		80 g
Eggs		4
Soft dark brown sugar		175 g
Salt		3 g
Lemon juice and zest		1

1 Place all the dried fruit in a bowl, pour over the Guinness and cognac and leave to soak overnight.

2 Sieve the flour with the mixed spice, grate over the nutmeg and cinnamon and place in a large bowl.

3 Add the breadcrumbs, suet, peel and ground almonds and mix well.

4 Make a well in the centre.

5 Whisk the eggs, sugar and salt, add the lemon juice and zest.

6 Pour the wet ingredients into the well and add the soaked fruit.

7 Mix well.

8 Place mixture into two buttered pudding basins, cover with a disc of silicone paper and seal with foil, crimping around the edges.

9 Steam for 7 hours, cool and store in the fridge.

10 Re-heat in the steamer for a couple of hours.

11 Serve with brandy or rum sauce and/or brandy butter.

Professional tip

Like a Christmas fruit cake, Christmas pudding is best made in September and allowed to mature.

11 Tarte tatin

1 First make a dry caramel by placing the granulated sugar in a hot heavy-based saucepan. When the sugar reaches a deep amber colour pour out onto a silicone mat and leave to cool completely.

2 When cold, crush the caramel into small pieces.

3 Take a medium-sized sauteuse and spread the butter thickly around the base and sides, sprinkle over the caster sugar and sprinkle over the caramel. (Any spare caramel can be stored in an airtight container for later use.) Place on the heat to melt.

4 Peel, core and halve the apples (if large cut into quarters) and pack into the sauteuse core side up and with the cores running horizontally, not facing outside.

5 Roll out the pastry 2 mm thick and leave to rest.

6 Place the pan on a medium heat for 10–12 minutes to allow the caramel to melt and infuse.

7 Quickly lay over the pastry and trim, tucking the edges down the side of the pan.

8 Bake at 220°C for 15 minutes until the pastry is crisp and the apples cooked through.

9 Invert onto a hot plate – please be aware this procedure can be tricky and requires two dry cloths and very careful handling.

10 Serve with cream, crème fraiche or ice cream.

Makes >	10 portions
Soft butter	120 g
Caster sugar	120 g
Dessert apples	5
Crushed caramel	120 g
Puff paste (Chapter 3, Recipe 6)	150 g
For the caramel	
Granulated sugar	500 g

Cook the melted butter, sugar and caramel

Lay in the apple pieces

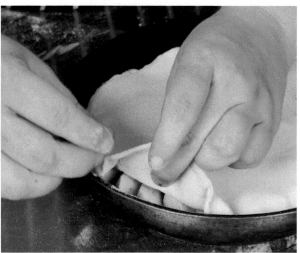

Lay the pastry over the top once the apples are half-cooked

Tuck in the edges

Turn out the tart carefully

Not all apples are suitable for this dish. Never use cooking apples, as if they contain a high water content they are likely to collapse and turn to purée.

Variations

- This dish can be made using pears instead of apples.
- Tarte tatin can also be made in individual moulds (as shown) if preferred.

The tarte tatin was created in 1888 by French sisters Caroline and Stéphanie Tatin, who owned and ran the Hotel Tatin in the Loire valley.

Health and safety

Extreme care should be taken when preparing the caramel as the sugar reaches very high temperatures and can cause serious burns.

12 Apple crumble tartlet

Makes >	10 individual tarts
Sweet paste (Chapter 3, Recipe 2)	500 g
Dessert apples	5
For the filling	
Soured cream	500 ml
Caster sugar	70 g
Plain flour	75 g
Egg	1
Vanilla extract	
Crumble	
Plain flour	80 g
Chopped walnuts	60 g
Brown sugar	65 g
Ground cinnamon	pinch
Salt	pinch
Unsalted butter (melted)	65 g
Icing sugar, to garnish	

1 Line individual tartlet moulds with the sweet paste.
2 Peel, core and finely slice the apples, and divide between the tartlets.
3 Whisk together the soured cream, sugar, flour, egg and a few drops of vanilla, pass through a conical strainer.
4 Pour over the apples and bake at 190°C for 10 minutes.
5 Combine the dry crumble ingredients and mix with the melted butter.
6 Divide the crumble mixture between the tartlets and bake for a further 10 minutes.
7 Allow to cool slightly before unmoulding, dust with icing sugar and serve with sauce à l'anglaise.

Try something different
This dish could be made with pears or plums instead of apples.

13 Rice pudding

Makes >	10 portions
Milk	650 ml
Vanilla pod	1
Short grain rice	60 g
Liaison	
Pasteurised egg	60 g
Caster sugar	60 g
Butter (diced)	30 g

1 Rinse a heavy pan with cold water and add the milk.
2 Split the vanilla pod, scrape out the seeds and add along with the pod.
3 Slowly bring to the boil.
4 Wash the rice and sprinkle into the boiling milk, stir, cover with a lid and allow to simmer until the rice is tender.
5 In a bowl whisk the eggs and sugar and drop in the butter.
6 Ladle a quarter of the boiling milk and rice onto the liaison, mix well and return all to the pan, carefully cook out until the mixture thickens, before removing from the heat (it must not be allowed to boil).
7 Place into suitable individual or large (usually china) dishes.
8 Grate with nutmeg and glaze under the salamander.
9 Serve with a warm seasonal fruit compote.

Try something different
- Serve with the apple compote suggested in Recipe 3.
- Place some good-quality jam in the base of the dish.
- Place in a serving dish before piping meringue on top and baking in a hot oven until coloured.
- Leftover rice pudding can be used as an alternative to crème pâtissière or frangipane as a filling for a baked flan.

Health and safety ⚠

- As this recipe requires that the ingredients are not boiled, ensure all work surfaces and equipment are kept scrupulously clean.
- It is recommended that pasteurised eggs are used.
- Rice pudding can be held over service at a temperature of no less than 75°C for 2 hours.
- Any leftover rice pudding must be cooled to below 5°C within 20 minutes, labelled and stored in a refrigerator.

Professional tip
Rinsing out the saucepan with cold water and adding the milk to the wet pan will help prevent the milk from catching on the bottom.

14 Chocolate fondant

	Makes >	10 portions
Unsalted butter		260 g
Dark couverture		260 g
Eggs, pasteurised		120 g
Egg yolks, pasteurised		40 g
Caster sugar		150 g
Instant coffee		5 g
Plain flour		110 g
Baking powder		5 g
Cocoa powder		75 g
Salt		pinch

1. Melt the butter and couverture together.
2. Warm the eggs, egg yolks, sugar and coffee and whisk to the ribbon stage.
3. Sieve all the dry ingredients twice.
4. Fold the chocolate and butter into the eggs.
5. Fold in the dry ingredients.
6. Pipe into individual stainless steel rings lined with silicone paper and placed on a silicone-paper-lined baking sheet.
7. Bake at 190°C for 5 minutes.
8. Carefully slide off the rings and serve with vanilla ice cream.

Melt the chocolate and butter in small pieces

Fold the melted chocolate into the egg mixture

Add the dry ingredients

To make a contrasting centre, add white chocolate pieces on a base of the chocolate mixture

Pipe in more of the chocolate mixture until the mould is full

Professional tips

These fondants can be kept in the refrigerator and cooked to order. If they are chilled, then extend the cooking time by 2 minutes.

Because they have a liquid centre, they are very delicate; if piped inside a ring they will be much easier and quicker to serve, rather than trying to turn them out of a mould.

Precise timing is essential or the centre of the fondant will not be liquid.

Try something different

- Try adding salted caramel to the centre by making and freezing it in ice cube trays.
- Alternatively, serve with malt ice cream (just add malt powder instead of vanilla and mix in some crushed chocolates) or replace the cream with crème fraiche to give a less rich ice cream.
- Prepare fondants in moulds lined with melted butter and roasted sesame seeds.

Note

Chocolate fondant should have a liquid centre with a rich buttery, chocolate taste. Like most recipes, the quality of the finished product relies on the quality of the ingredients. Always use good-quality chocolate (couverture) which contains a high percentage cocoa butter and solids.

15 Baked Alaska (omelette soufflé surprise)

Note

This is one of those desserts that does not clearly fit into a category, and it is debatable as to whether this should be classified as a hot dessert as the filling will remain frozen. However, as it is baked and coloured in a very hot oven, this is probably the most appropriate chapter for it.

Professional tips

- These are best made in advance and held in the freezer, then flashed through the oven just before serving. It is now common practice to colour these with a blowtorch, but the meringue will have a much better texture and more even colouring if they are finished in the oven.
- If making individual baked Alaskas, as in the photographs, take care not to upset the balance between filling and meringue – when scaled down it is easy to pipe on too much meringue.

Makes >	10 individual portions
Vanilla ice cream or parfait	10 × 5 cm diameter rings
Roulade sponge (Chapter 8, Recipe 2)	1 sheet
Stock syrup flavoured with rum or kirsch	50 ml
Italian meringue (Chapter 3, Recipe 20)	500 g

1 Sit the ice cream or parfait on a base of sponge.

2 Cut more sponge to fit and completely cover (as in the photo).

3 Brush all over with the syrup.

4 Set on squares of silicone paper, coat with the meringue and decorate by piping on a design with a small plain tube.

5 Dust with icing sugar and place in a very hot oven 230°C for 2–3 minutes until the meringue is coloured.

6 Serve immediately with crème anglaise or a fruit coulis.

Health and safety

Under no circumstances should this dessert be re-frozen once it has been removed from the freezer and baked. Ice cream is highly susceptible to contamination by bacteria which can cause food poisoning.

Variations

Classic variations would be **omelette soufflé milady**, which contains poached sliced peaches with vanilla or raspberry ice cream, and **omelette soufflé milord**, which contains poached sliced pear with vanilla ice cream.

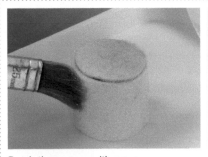

Brush the sponge with syrup

Pipe meringue to cover

Pipe swirls of meringue to decorate

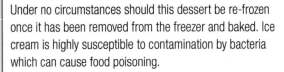

16 | Vanilla soufflé

Makes >	10 individual soufflés
Milk	500 ml
Vanilla pods	2
Butter	75 g
Strong flour	60 g
Egg yolks	10
Caster sugar	50 g

For every 400 g of the above mixture, use the following quantity of egg whites, sugar, cornflour and lemon juice.

Egg whites	150 g
Caster sugar	60 g
Cornflour	12 g
Lemon juice	2–3 drops

1 Rinse out a heavy saucepan with cold water and add the milk, split the vanilla pod, scrape out the seeds and add both to the milk and put on the heat to boil.

2 Melt the butter in another heavy pan, add the flour and cook out to form a white roux, gradually start adding the boiling milk, mixing in each addition before adding the next.

3 When all the milk has been added, allow to simmer for a few minutes.

4 Whisk the egg yolks and sugar, add to the mixture in the saucepan and keep stirring over the heat until the mixture starts to bubble around the edges. This forms the panada.

5 Pour onto a clean tray and cover with cling film to prevent a skin forming, allow to cool. (This can be kept in the fridge until needed as soufflés must be cooked to order.)

6 Take 400 g of the base and beat until smooth.

7 Whisk the whites, sugar, cornflour and lemon juice to form firm peaks.

8 Beat the base mixture in a clean bowl until smooth, add one third of the whites and mix in, very carefully fold in the remaining whites.

9 Carefully fill prepared individual china ramekins (see Professional tips), level the top and run your thumb around the edge moving the mixture away from the lip of the mould.

10 Space well apart on a solid baking sheet (if they are close together they will not rise evenly and will bake stuck together).

11 Place immediately in the oven at 215°C for 12–14 minutes.

12 The soufflés should rise out of the moulds by around 5–6 cm and have a flat top with no cracks.

13 Dust with icing sugar and serve immediately with fruit coulis and/or ice cream (chocolate or vanilla).

Prepare the mould (see Professional tips)

Knock one third of the meringue into the panada

Fold in the remaining meringue

Thumb the edge

Professional tips

Hot soufflés have a largely undeserved reputation for being problematic, but if the following points are observed they should pose no problems for a competent chef.

- Mould preparation is very important. The rule is to butter the sides twice and the bottom once. Always use soft, not melted, butter (then it stays where you put it). Give the moulds one coat all over to start with, then place in the fridge to set before giving the sides only a second coat (giving the base two coats results in a puddle of butter left in the bottom). After giving the sides a second coat of butter the moulds are usually coated with sugar.
- When beating the whites, the addition of sugar, lemon juice and cornflour helps to strengthen the gluten and stabilise the mixture. When whisked they should be firm but creamy. Over-whisking ruins them. Scald the bowl first to remove any trace of fat, which will prevent the whites from reaching their full potential. The above recipe uses more egg white than is needed – always take from the centre and leave behind the egg white from the edge, which is never a good egg white. The quantity of whites added is approximately the same in volume as the base, but whatever the method used the key points are always the same.
- Be organised with your timings. Make sure you have the moulds fully prepared and ready and make sure the oven is up to heat well before the whites are whisked. Once mixed the soufflé should be cooked (and served) immediately, although some recipes, including this one, may be held before baking for up to 2 hours.

Note

The following points identify the features of a good-quality soufflé.

- The point of a soufflé is that it will rise out of the dish in which it is cooked – it should rise evenly, at least 3–4 cm.
- The term used to describe the centre of a cooked soufflé is 'baveuse', which means slightly undercooked.
- A soufflé should have flavour, which can sometimes almost get overlooked with the emphasis on getting an impressive 'rise' – too much egg white can dilute the flavour.
- Finally, a soufflé should look like it tastes – if the flavour is raspberry then that should be obvious from the colour.

Faults

Soufflé does not rise:
- under- or overbeaten whites
- wrong proportion of whites to base
- mixture left to stand before cooking
- moulds not buttered correctly.

Soufflé does not rise evenly:
- moulds not prepared correctly (mixture has stuck to the mould on one side)
- uneven heat in the oven.

Soufflé rises but drops back:
- too much egg white used.

Soufflé has a cracked top:
- too much egg white used
- egg white not mixed in properly
- egg white is overbeaten.

Variations

The recipe above is a basic recipe and can easily be adapted. For example, a **chocolate soufflé** can be made by adding melted chocolate and cocoa powder to the base. This will firm up the base mixture so the whites mixture will need to be increased by 25–30 per cent to compensate.

Soufflés can be made in many different flavours and combinations. Recipes can vary considerably. For example, **fruit soufflés** can be made using a sabayon or a boiled sugar base. **Liqueur soufflés** can be made using a béchamel (like the one above) or a crème pâtissière base.

As an alternative to coating the moulds with sugar, if compatible, dust them with cocoa powder, grated chocolate or try adding some cinnamon to the sugar.

Professional tip

For liqueur soufflés, sprinkle the liquor over a small dice of sponge fingers and fold into the mixture just before the whites. This will trap the liquor in small pockets concentrating the flavour, and will not evaporate during cooking.

17 | Lemon curd flourless soufflé

Makes >	10 individual soufflés
Lemon curd	
Eggs	2
Caster sugar	100 g
Lemons (juice of)	5
Unsalted butter	60 g
Cornflour	15 g
Soufflé	
Eggs	9
Caster sugar	190 g
Lemons (zest and juice)	5
Cream of tartar	pinch
Egg-white powder	pinch

For the lemon curd

1 Prepare by whisking all the ingredients over a pan of simmering water until the mixture thickens.

2 After preparing the soufflé moulds as described in the previous recipe, divide the lemon curd mixture between the prepared ramekins.

For the soufflé

1 Separate the eggs, place together the yolks, sugar, lemon zest and juice and whisk together well.

2 In a separate bowl whisk the whites with the cream of tartar and egg-white powder until soft but in strong peaks.

3 Carefully fold the two mixtures together.

4 Carefully fill the prepared moulds, level the top and run your thumb around the edge moving the mixture away from the lip of the mould.

5 Place well apart on a heavy baking sheet and bake at 200°C for 16–18 minutes.

6 Dust with icing sugar and serve immediately.

Note

As this dessert is gluten-free it would be suitable for coeliacs or those with a wheat intolerance.

18 Crêpe soufflé

This recipe makes 10 portions.

For the crêpes

Use crêpes from Recipe 6, which should be undercooked.

For the soufflé filling

Use Recipe 16 above, appropriately flavoured.

1 Fold the pancakes into four and fill the two pockets with soufflé mixture, this is best done using a piping bag and a large plain tube, or simply fold in half and fill.

2 Place well apart on a silicone-paper-covered baking sheet and bake immediately at 210°C for 6–7 minutes.

3 Dust with icing sugar and serve with a fruit coulis and/or ice cream, or a flavoured syrup.

Try something different

An alternative to this dish is to take two pancakes per portion, place some soufflé mixture in the centre, eggwash the edge and place the second pancake on top and seal. After baking as above the result should be a pancake ball.

19 Apple strudel

Makes >	10 portions
Strudel paste (Chapter 3, Recipe 8)	500 g
Melted butter	120 g
Fresh bread crumbs	Approx. 100 g
Filling	
Lemons	1
Large dessert apples	5
Caster sugar	to taste
Ground cinnamon	to taste
Nibbed almonds	to taste
Sultanas	to taste

Make the paste as specified in Chapter 3, Recipe 8 and leave to rest between two oiled plates for an hour.

For the filling

1 Grate the zest and juice the lemon, place in a large bowl.
2 Peel and core the apples before cutting into 2 mm-wide batons, add to the lemon juice and zest.
3 Add the other ingredients to taste (the quantity of sugar will be influenced by the sweetness of the apples and the rest, other than the cinnamon, will be around a couple of handfuls each).
4 Mix well with a spoon, cover and leave to stand – about half an hour before the paste is rolled tip the filling into a colander to drain.

Processing the paste

This must be done on a large cloth (an old, but without holes, tablecloth or similar).

1 Lay the cloth over a table you can walk all the way around (i.e not one against a wall).
2 Lightly flour the cloth and ease the paste off the plate using a plastic scraper.
3 Do not knead or work at all – just lightly flour the top and start rolling carefully, lifting the paste up and flouring underneath until it reaches a size where it is too big to roll.
4 Reach underneath the paste and using the backs of your hands, carefully stretch, until the paste is gossamer thin and covers the table. It should be thin enough to read a newspaper through, literally.
5 You must now work quickly – brush the paste with melted butter, and sprinkle a good layer of crumbs over half the paste only.
6 Place a layer of apples on the crumbs only and start to roll up using the cloth as shown in the final photo.
7 Twist the ends, lift onto a silicone-paper-covered baking sheet, brush the outside with melted butter.
8 Bake at 190°C for 30–35 minutes.
9 Allow to settle, dust with icing sugar, cut diagonally with a serrated sponge knife and serve with sour cream.

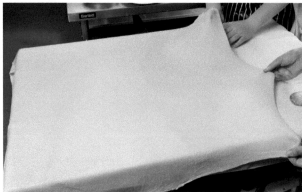

Fully stretched pastry

Fold the side in

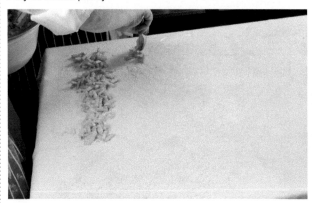

After dusting, add the filling

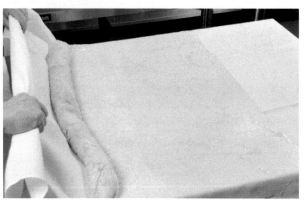

Roll up in a cloth

Faults

The paste develops holes during stretching:

● Rough handling, or taking too long to process the paste so it starts to dry out. This is a difficult and an awkward process for one person – ask a friend and get them to work opposite you, this will make the work easier and speed up the process.

The strudel splits during cooking:

● the filling is too wet
● not enough breadcrumbs used – these act as insulation and are there to absorb moisture
● apples have been cut too large and/or in irregular shapes so they puncture the paste.

Tastes like cardboard with a hint of apple:

● Paste too thick, under-filled and overbaked.

Health and safety

When mixing fruit always use a spoon, never your hands, which can cause the fruit to ferment.

Try something different

● Cake crumbs, ground almonds, brioche crumbs or any combination of these can be used in place of breadcrumbs.
● Instead of breadcrumbs, make a chaplure by shallow-frying breadcrumbs to a golden brown in butter.
● Replace the apples with pears or stoned cherries soaked in kirsch and replace the almonds with pine nuts.
● Use dried cranberries or dried apricots in place of sultanas, and try adding diced stem ginger if using pears.
● The almonds can be replaced by hazelnuts, and mixed spice instead of cinnamon.
● It is traditional to serve strudel with soured or acidulated cream but crème fraiche, sauce à l'anglaise and/or ice cream will also complement the dish well.

Professional tip

As with so many processes in the pastry kitchen, it is vital to have all ingredients prepared and ready before the paste is processed.

Almond and apricot samosas

Pipe the filling on to the paste

Makes >	10 portions
Spring roll paste	1 pkt
Beaten egg white or flour paste to seal	
Filling	
Almond cream (Chapter 3, Recipe 18)	250 g
Diced dried apricots soaked in rum	50 g

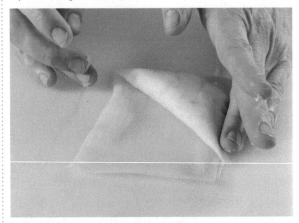

Fold diagonally into a triangular parcel

1 Open the spring roll paste and keep covered with a damp cloth so it does not dry out.

2 Cut paste in half, which should give two rectangles 15 × 10 cm.

3 Pipe a little of the filling on to the paste and fold diagonally three times to form triangular parcels, brush the exposed edge with beaten egg white and seal.

4 Allow two per portion, deep fry at 180°C for 5 minutes until coloured and an even golden brown, drain on absorbent paper.

5 Dust with icing sugar and serve with caramel ice cream (beurre de Paris).

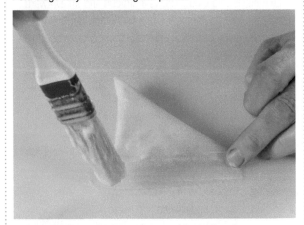

Brush the exposed edge with egg white and seal

Try something different

● Fill with a mixture of diced poached pears, stem ginger and mascarpone cheese.

● Fill with semi-poached and dried rhubarb with stem ginger, and serve with a white chocolate sauce.

● Instead of making into triangles, the pastry could alternatively be made into spring rolls.

All these recipes would be suitable to serve as a pre-dessert if they were made smaller.

21 Bread and butter pudding

Makes >	10 portions
Washed sultanas	100 g
Thin slices of white bread	approx. 5
Melted butter	200 g
Custard	
Vanilla pod	1
Milk	300 ml
Cream	300 ml
Eggs	5
Caster sugar	100 g
Nutmeg	
Apricot jam	100 g

1 Butter an earthenware or other suitable dish and sprinkle with the sultanas.

2 Cut the crusts off the bread, dip in melted butter on both sides and cut in half diagonally.

3 Arrange overlapping bread slices neatly in the dish.

4 Sprinkle with more sultanas and cover with another layer of bread.

5 To make the custard, split the vanilla pod, add to the milk and cream and slowly bring to the boil.

6 Whisk the eggs and sugar together and add the boiling liquid, leave to infuse for 5 minutes before passing through a conical strainer.

7 Pour the custard over the bread and grate on some fresh nutmeg.

8 Place in a bain-marie and place in a moderate oven at 160°C, pour hot water into the bain-marie until it comes half way up the dish.

9 Bake for around 45 minutes until the custard is just set.

10 Once removed from the oven, sprinkle with sugar and place under the salamander to crisp up and colour the top.

11 Finally, brush with boiled apricot glaze and serve with pouring cream or crème fraiche.

Professional tip

Add the custard in two or three lots, allowing it to soak in before adding the next. This will prevent the bread floating to the top.

Try something different

- This pudding can be made in individual dishes or baked in a tray and cut and plated.
- Try using alternatives to bread such as fruit loaf, brioche, baguette slices or panettone.
- Soak the sultanas in rum the day before.
- Try adding a layer of caramelised apple slices.
- A chocolate version can be made by adding couverture (good-quality chocolate) to the custard.
- There are several ways to make this dessert healthier. Reduce the sugar content and add more dried fruit, apricots or cranberries. This dessert can also be made using milk only, semi-skimmed or skimmed. You can also dip the bread in the butter on just one side to reduce the fat content.

Video: bread and butter pudding, http://bit.ly/18ZKes8

Test yourself

Level 2

1 What is the purpose of a bain-marie?

2 Why is steaming the preferred method when cooking puddings made with suet?

Level 3

3 Describe the correct procedure when scaling up (increasing) or scaling down (decreasing) a recipe.

4 List the differences between a hot soufflé and a pudding soufflé.

5 Name two ingredients that can be added to a meringue as stabilisers.

This chapter covers the cold and frozen dessert components of:
→ **NVQ level 2 Prepare, cook and finish basic hot and cold desserts**
→ **VRQ level 2 Produce hot and cold desserts and puddings**
→ **NVQ level 3 Prepare, cook and finish complex cold desserts**
→ **VRQ level 3 Produce hot, cold and frozen desserts.**

In this chapter you will:
→ **Work safely and hygienically, using the correct equipment**
→ **Apply quality points and evaluate the finished cold dessert.**

Recipes in this chapter ·

Recipe	Page	Level 2	Level 3
1 Tropical fruit plate	220	✓	✓
2 Fresh fruit salad	221	✓	
3 Poached fruits	222	✓	
4 Fruit mousse	223	✓	
5 Bavarois	224	✓	✓
6 Vanilla panna cotta	226	✓	
7 Fruit fool	227	✓	
8 Meringue	228	✓	✓
9 Vacherin	229	✓	
10 Black Forest vacherin	230		✓
11 Lime soufflé frappé	231		✓
12 Trifle	232	✓	
13 Lime and mascarpone cheesecake	233		✓
14 Baked blueberry cheesecake	234		✓
15 Baked apple cheesecake	235		✓
16 Vanilla ice cream	236	✓	✓
17 Rich vanilla ice cream	237		✓
18 Lemon curd ice cream	237		✓
19 Caramel ice cream	238		✓
20 Peach ice cream	239		✓
21 Chocolate ice cream	239		✓
22 Malt ice cream and caramel sauce	240		✓
23 American orange ice cream	240		✓
24 Chocolate sorbet	240		✓
25 Apple sorbet	241		✓
26 Lemon sorbet	242		✓
27 Grapefruit water ice	242		✓
28 Vanilla water ice	243		✓
29 Lemon water ice	243		✓
30 Champagne water ice	243		✓
31 Pistachio and chocolate crackle ice cream (using a Pacojet)	244		✓

32 Savoury ice cream using maltodextrin and a Pacojet	245		✓
33 Lemon and ginger sorbet (using a Pacojet)	245		✓
34 Blackberry sorbet (using a Pacojet)	245		✓
35 Bombe glacée	246		✓
36 Orange brandy granita	248		✓
37 Orange and Cointreau iced soufflé	248		✓
38 Raspberry parfait	249		✓
39 Peach Melba	250	✓	
40 Pear belle Hélène	250	✓	
41 Crème caramel	251	✓	
42 Chocolate mousse	252	✓	✓
43 Raspberry or strawberry mousse	253		✓
44 Orange mousse with biscuit jaconde	254		✓
45 Blackcurrant delice	255		✓
46 Empress rice	256		✓
47 Strawberry Charlotte	257		✓
48 Tiramisu torte	258		✓
49 Raspberry and chocolate truffle cake	259		✓
50 Floating islands	260		✓
51 Crème brûlée	261	✓	✓
52 Vanilla crème brûlée (gastro tray method for large-scale production)	261		✓
53 Unmoulded crème brûlée with spiced fruit compote	262		✓
54 Crème beau rivage	263		✓
55 Gateau MacMahon	264		✓
Pre-desserts			
56 Petits pots au chocolat	265	✓	✓
57 Passion fruit and white chocolate posset with Champagne jelly	266		✓
58 Honey chocolate panna cotta, banana passion fruit jam with almond streusel	267		✓
59 Snow egg with Cointreau cream	268		✓

Iced confections

Traditional ice cream is made from a basic egg custard sauce (sauce à l'anglaise). The sauce is cooled and mixed with fresh cream. It is then frozen by a rotating machine where the water content forms ice crystals and the mixture is aerated.

Ice cream should be served at around −13°C; this is the correct eating temperature – it is too hard if it is any colder and too soft if it is any warmer. Long-term storage should be between −18°C and −20°C.

The traditional method of making ice cream uses only egg yolks, sugar and milk/cream in the form of a sauce à l'anglaise base. Modern approaches to making ice cream use stabilisers (see Appendix) and different sugars as well as egg whites. This can help to reduce the sometimes high wastage of egg whites that might otherwise occur in the pastry section.

Classification of iced confections

Water ices, sorbets and granita

Water ices are a simple syrup flavoured with a fruit purée and lemon juice. They contain no milk, eggs or cream, so therefore are not subject to the ice cream regulations. They are frozen in the same manner as normal ice cream. Also included in this type of ice is a sorbet. This is a very light, slightly less sweet water ice, usually lemon, to which Italian meringue is added during the churning process after the base mixture has partially frozen. It is sometimes flavoured with champagne, liqueurs or wine. Granita is similar to a water ice, less sweet than a sorbet and containing no meringue, thus producing a slightly granular texture normally achieved by pouring into a shallow tray, freezing, then scraping the surface into frozen granules and serving in frozen glasses.

For many pastry dishes (for example, sorbets) sugar syrups of a specific density are required. The density is measured by a hydrometer known as a saccharometer – a device that measures the thickness of stock syrups. This is measured in degrees Baumé. The instrument is a hollow glass tube sealed at each end. One end is weighted with lead shot so that when it is placed in the solution it floats upright. The scale marked on the side of the saccharometer is calibrated in the Baumé scale from 0–45, where 0 represents the density of water and 45 represents the density of a heavy-saturated sugar solution. The instrument thus measures the amount of sugar in the solution. To increase the density add more sugar to the warm syrup, to decrease it add water.

The following Baumé scale would apply to the following water ices:

Water ice	=	17° Baumé
Sorbet	=	15° Baumé
Granita	=	14° Baumé
Sorbet syrup	=	700g sugar to 1 litre boiling water (17° Baumé)

Another instrument that measures the amount of sugar in a syrup is a refractometer. This piece of equipment measures the percentage size of the sugar crystal and is calibrated according to the Brix scale. 17° Baumé equates to 30 per cent Brix. Photos of both pieces of equipment can be seen in the Appendix.

Table 7.1 Conversion chart for Baumé to Brix scale and common uses in the patisserie

Baumé	Brix	Product
14°	26%	Granita
15°	28%	Sorbet
17°	31%	Water ice
20°	37%	Compote of fruits
22°	40%	Savarin syrup
28°	52%	Syrup for pâte à bombe
34°	63%	Confiture of fruits
17°	31%	Sorbet syrup

Cream ice

This is a mixture of milk, sugar, eggs, cream and flavouring. The base is made from a crème anglaise using these ingredients. The soft creamy texture is achieved by churning and aerating as the mixture is freezing, this can only be achieved by freezing the mixture in an ice cream machine known as a sorbetiere. Modern methods of processing cream ices use Pacojet machines. These machines give a very smooth texture and allow the ice cream to be churned prior to each service ensuring consistency in texture (see Appendix).

Biscuit glacé

Biscuit glacé is a light solid-type ice cream that can be produced without the aid of a sorbetiere. The aeration is achieved from a sabayon made from egg yolks and stock syrup whisked over a bain-marie of hot water.

Once cold, whipped cream, Italian meringue and flavourings are folded through. Traditionally the mixture is then deposited into a 'biscuit glacé mould' and then frozen. The glace may be made in two or more flavours arranged in layers and could also contain various garnishes such as nuts and crystallised fruits. The moulds used are rectangular in shape and for service the biscuit is sliced into portions.

Bombe glacée

These are also frozen in moulds. Bombe moulds are dome-shaped and are fitted with a screw plug in the bottom to facilitate removal. The moulds are first lined 'chemise' with cream or water ice and then the centre is filled with a pâte à bombe mixture, so producing two or more different flavours, colours and textures. They are decorated and served whole. Bombes are made of various combinations of ices depending on the title of the bombe.

Parfait glacé

Parfait gets its name from the mould it is shaped in. Modern interpretations make the parfait and deposit it into ring moulds lined with jaconde sponge, then freeze it, remove it from the moulds and plate with a suitable decoration.

Soufflé glacé

There are two different types of soufflé glacé – cream base and fruit purée base. Cream-based iced soufflé is made by preparing a pâte à bombe and folding through whipped cream, Italian meringue and flavourings. Purée-based iced soufflé is prepared with fruit purée and folded through with whipped cream and Italian meringue. Both types are frozen in a soufflé mould with a paper collar. Once frozen the paper collar is removed and the exposed sides can be decorated with chopped pistachio nuts, Bres (caramelised nuts), toasted coconut, chocolate shavings, etc.

Table 7.2 Iced confection identification chart

Name of mixture	Egg yolk	Water	Sugar	Syrup	Cream	Fruit purée	Egg white	Milk	Other
Vanilla ice cream	✓		✓		✓			✓	Vanilla, sometimes inverted sugar, ice cream stabiliser, glucose
Pâte à bombe	✓	✓	✓		✓				
Pâte à biscuit glacé	✓	✓	✓		✓		Italian meringue		Flavour
Soufflé glacé cream base	✓	✓	✓		✓		Italian meringue		Flavour
Soufflé glacé fruit base					✓	✓	Italian meringue		
Parfait glacé	✓	✓	✓		✓				Flavour
Sorbet		✓		15° Baumé			Italian meringue		Flavour
Water ice				17° Baumé					Flavour
Granita				14° Baumé					Flavour
Spoons				20° Baumé			Same as sorbets with double quantity of Italian meringue		Flavour, wine
Marquise				17° Baumé	✓		Italian meringue		Diced strawberries and pineapple macerated in kirsch
Cassata					✓		Italian meringue		Glacé fruits, mould is lined with three flavours of ice cream

Video: equipment for ice cream, http://bit.ly/18ZKhUO

The ice cream regulations

The Dairy Products (Hygiene) Regulations 1995 apply to the handling of milk-based ice cream and the Ice Cream Heat Treatment Regulations 1959 and 1963 apply to non-milk-based ice cream in any catering business or shop premises. The production process must also take into consideration the Food Hygiene Regulations of 2006.

The regulations state that:
- Ice cream must be obtained from a mixture which has been heated to any of the temperatures in Table 7.3 for the times specified.
- The mix must be reduced to a temperature of not more than 7.2°C within 1½ hours. This temperature must not be exceeded until freezing begins.
- If the ice cream becomes warm (above −2.2°C) it cannot be sold/used until it has been heated again as described above.
- A complete cold mix which is reconstituted with water does not need to be pasteurised first to comply with these regulations.
- A complete cold mix reconstituted with water must be kept below −2.2°C once it has been frozen.

Table 7.3

Temperature	Time (not less than)
65.5°C	30 minutes
71.1°C	10 minutes
79.4°C	15 seconds

Ice cream needs this treatment to kill harmful bacteria. Freezing without the correct heat treatment does not kill bacteria; it allows them to remain dormant. The storage temperature for ice cream should not exceed −20°C ideally, although standard freezers operate between −18°C and −22°C.

The rules for sterilised ice cream are the same except that:
- The temperature for the heat treatment must not be less than 149.9°C for at least 2 seconds.
- If the sterilised mix is kept in unopened, sterile and air-tight containers, there is no requirement to refrigerate the mixture before it is frozen.

- In the case of non-milk based products, the temperature of opened containers must not exceed 7.2°C, except where food mixtures are added that have a pH of 4.5 or less to make water ice or similar products and the combined product is frozen within 1 hour of combination.

Any ice cream sold must comply with the following compositional standards:
- It must contain not less than 5 per cent fat and not less than 2.5 per cent milk protein (not necessary in natural proportions).
- It must conform to the Dairy Product Regulations 1995.

For further information contact the Ice Cream Alliance (see www.ice-cream.org).

The ice-cream making process

1 Weighing: ingredients should be weighed precisely in order to ensure the best results and, what is more difficult, regularity and consistency.
2 Pasteurisation: this is a vital stage in making ice cream. Its primary function is to minimise bacterial contamination by heating the mixture of ingredients to 85°C, then quickly cooling it to 4°C.
3 Homogenisation: high pressure is applied to cause the explosion of fats, which makes ice cream more homogenous, creamier, smoother and much lighter. This is not usually done for homemade ice cream.
4 Ripening: this basic but optional stage refines flavour, further develops aromas and improves texture. This occurs during a rest period (4–24 hours) at 3°C, which gives the stabilisers and proteins time to act, improving the overall structure of the ice cream. This has the same effect on a crème anglaise, which is much better the day after it is made than it is on the same day.
5 Churning: the mixture is frozen while at the same time air is incorporated. The ice cream is removed from the machine at about −10°C.

Refer to the Appendix for further details on the specialist equipment used.

The main components of ice cream

- **Sucrose** (common sugar) not only sweetens ice cream, but also gives it body. An ice cream that contains only sucrose (not recommended) has a higher freezing point.

  The optimum sugar percentage of ice cream is between 15 and 20 per cent.

  As much as 50 per cent of the sucrose can be substituted with other sweeteners, but the recommended amount is 25 per cent.

- Ice cream that contains **dextrose** (another type of sugar) has a lower freezing point, and better taste and texture.

  The quantity of dextrose used should be between 6 and 25 per cent of the substituted sucrose (by weight).

- **Glucose** (another type of sugar) improves smoothness and prevents the crystallisation of sucrose.

  The quantity of glucose used should be between 25 and 30 per cent of the sucrose by weight.

- **Atomised glucose** (glucose powder) is more water absorbent, so helps to reduce the formation of ice crystals.

- **Inverted sugar** is a paste or liquid obtained from heating sucrose with water and an acid (such as lemon juice). Using inverted sugar in ice cream lowers the freezing point.

  Inverted sugar also improves the texture of ice cream and delays crystallisation.

  The quantity of inverted sugar used should be a maximum of 33 per cent of the sucrose by weight. It is very efficient at sweetening and gives the ice cream a low freezing point.

- **Honey** has very similar properties as those of inverted sugar.

- The purpose of **cream** in ice cream is to improve creaminess and taste.

- **Egg yolks** act as stabilisers for ice cream due to the lecithin they contain – they help to prevent the fats and water in the ice cream from separating.

  Egg yolks improve the texture and viscosity of ice cream.

- The purpose of other **stabilisers** (e.g. gum Arabic, gelatine, pectin) is to prevent crystal formation by absorbing the water contained in ice cream and making a stable gel.

  The quantity of stabilisers in ice cream should be between 3 g and 5 g per kg of mix, with a maximum of 10 g.

  Stabilisers promote air absorption, making products lighter to eat and also less costly to produce, as air makes the product go further.

Note

What you need to know about ice cream and water ices:

- Hygienic conditions are essential while making ice cream – personal hygiene and high levels of cleanliness in the equipment and the kitchen environment must be maintained.
- An excess of stabilisers in ice cream will make it sticky.
- Stabilisers should always be mixed with sugar before adding, to avoid lumps.
- Stabilisers should be added at 45°C, which is when they begin to act.
- Cold stabilisers have no effect on the mixture, so the temperature must be raised to 85°C.
- Ice cream should be allowed to 'ripen' for 4–24 hours. This is a vital step that helps improve its properties.
- Ice cream should be cooled quickly to 4°C, because micro-organisms reproduce rapidly, particularly between 20°C and 55°C.
- Water ices and sorbet are generally more refreshing and easier to digest than ice cream.
- Fruit for water ices and sorbets must always be of a high quality and perfectly ripe.
- The percentage of fruit used in water ices and sorbets varies according to the type of fruit, its acidity and the properties desired.
- The percentage of sugar will depend on the type of fruit used.
- The minimum sugar content in water ice and sorbets is about 13 per cent.
- As far as ripening is concerned, the syrup should be left to rest for 4–24 hours and never mixed with the fruit because its acidity would damage the stabiliser.
- Stabiliser is added in the same way as for ice cream.

Stabilisers

Gelling substances, thickeners and emulsifiers are all stabilisers. They are products we use regularly, each with its own specific function; but their main purpose is to retain water to make a gel. The case of ice cream is the most obvious, in which they are used to prevent ice crystal formation. They are also used to stabilise the emulsion, increase the viscosity of the mix and give a smoother product that is more resistant to melting. There are many stabilising substances, both natural and artificial.

Edible gelatine

Edible gelatine is extracted from animals' bones (for example, pork and veal) and, more recently, fish skin. Sold in sheets of 2 g, it is easy to precisely control the amount used and to manipulate it. Gelatine sheets must always be washed thoroughly with lots of cold water to remove impurities and any remaining odours. They must then be drained before use.

Gelatine sheets melt at 40°C and should be melted in a little of the liquid from the recipe before adding it to the base.

Pectin

Pectin is another commonly used gelling substance because of its great absorption capacity. It comes from citrus peel (orange, lemon, etc.), though all fruits contain some pectin in their peel.

It is a good idea always to mix pectin with sugar before adding it to the rest of the ingredients.

Agar-agar

Agar-agar is a gelatinous marine algae found in Asia. It is sold in whole or powdered form and has a great absorption capacity. It dissolves very easily and, in addition to gelling, adds elasticity and resists heat (this is classified as a non-reversible gel).

Other stabilisers

- **Carob gum**, which comes from the seeds of the carob tree, makes sorbets creamier and improves heat resistance.
- **Guar gum** and **carrageenan** are, like agar-agar, extracted from marine algae and are some of many other existing gelling substances available but they are used less.

For details of other gels and stabilisers see the Appendix.

Milk

Full-cream, skimmed or semi-skimmed milk can be used for the desserts in this chapter.

Milk is a basic and fundamental element of Western diets. It is composed of water, sugar and fat (with a minimum fat content of 3.5 per cent). It is essential in an infinite number of products, from creams, ice creams, yeast doughs, mousses and custards to certain ganaches, cookies, tuiles and muffins.

Milk has a slightly sweet taste and little odour. Two distinct processes are used to conserve it:

- **Pasteurisation** – the milk is heated to between 73°C and 85° for a few seconds, then cooled quickly to 4°C.
- **Sterilisation (UHT)** – the milk is heated to between 140°C and 150°C for 2 seconds, then cooled quickly.

Milk is homogenised to disperse the fat evenly, since the fat has a tendency to rise to the surface (see 'Cream', below).

Here are some useful facts about milk:

- Pasteurised milk has a better taste and aroma than UHT milk.
- Milk is a useful for developing flavour in sauces and creams, due to its lactic fermentation.
- There are other types of milk, such as sheep's milk, that are very interesting to use in many restaurant desserts.
- Milk is much more fragile than cream. In recipes, adding it in certain proportions is advisable for a much more subtle and delicate final product.

Cream

Cream is used in many recipes because of its great versatility and capabilities.

Cream is the concentrated milk fat that is skimmed off the top of the milk when it has been left to sit. A film forms on the surface because of the difference in density between the fat and liquid. This process is speeded up mechanically in large industries by heating and using centrifuges.

Cream should contain at least 18 per cent butter fat. Cream for whipping must contain more than 30 per cent butterfat. Commercially frozen cream is available in 2 kg and 10 kg slabs. Types, packaging, storage and uses of cream are listed in Table 7.4.

Whipping and double cream may be whipped to make them lighter and to increase volume. Cream will whip more easily if it is kept at refrigeration temperature. Indeed, all cream products must be kept in the refrigerator for health and safety reasons. They should be handled with care and, as they will absorb odour, they should never be stored near onions or other strong-smelling foods.

As with milk, there are two main methods for conserving cream:
- **Pasteurisation** – the cream is heated to between 85°C and 90°C for a few seconds and then cooled quickly; this cream retains all its flavour properties.
- **Sterilisation (UHT)** – this consists of heating the cream to between 140°C and 150°C for 2 seconds; cream treated this way loses some of its flavour properties, but it keeps for longer.

Always use pasteurised cream whenever possible; for example, in the restaurant when specialities are made for immediate consumption, such as 'ephemeral', patisserie (dessert cuisine) with a short life (for example, a chocolate bonbon or a soufflé) that will be consumed immediately.

Here are some useful facts about cream:
- Cream whips with the addition of air, thanks to its fat content. This retains air bubbles formed during beating.
- Understand how to use fresh cream; remember that it is easily over-whipped.
- Cream adds texture and enriches.
- Once cream is boiled and mixed or infused with other ingredients to add flavour, it will whip again if first left to cool; when preparing a chocolate Chantilly, for example.
- To whip cream well, it must be cold (around 4°C).
- Cream can be infused with other flavours when it is hot or cold. The flavour of cream will develop if left to infuse prior to preparation.
- Always ensure that the bowl being used to whip the cream is cold.

Piping fresh cream
- The piping of fresh cream is a skill; like all other skills it takes practice to become proficient. The finished item should look attractive, simple, clean and tidy, with neat piping.
- All piping bags should be sterilised after each use, as these may well be a source of contamination; alternatively use a disposable piping bag.
- Make sure that all the equipment you need for piping is hygienically cleaned before and after use to avoid cross-contamination.
- Stabilisers may be added to whipped cream which helps to prevent water leakage and gives a firmer piped cream

Eggs

Eggs are one of the principal ingredients in cooking, and essential for many desserts. Their great versatility and extraordinary properties as a thickener, emulsifier and stabiliser make their presence important in various creations in patisserie: sauces, creams, sponge cakes, custards and ice creams. Although it is not often the main ingredient, it plays specific and determining roles in terms of texture, taste and aroma. The egg is fundamental in products such as brioches, crème anglaise, sponge cakes and crème pâtissière. The extent to which eggs are used (or not) makes an enormous difference to the quality of the product.

A good custard cannot be made without eggs, as they cause the required coagulation and give it the desired consistency and finesse.

Eggs are also an important ingredient in ice cream, where their yolks act as an emulsifier, due to the lecithin they contain, which aids the emulsion of fats.

Table 7.4

Type of cream	Legal minimum fat (%)	Processing and packaging	Storage	Characteristics and uses
Half cream	12	Homogenised; may be pasteurised or ultra-heat treated	2–3 days	Does not whip; used for pouring; suitable for low-fat diets
Cream (single cream)	18	Homogenised; pasteurised by heating to 79.5°C for 15 seconds, then cooling to 4.5°C; packaged in bottles and cartons, sealed with foil caps; may be available in bulk	2–3 days in summer; 3–4 days in winter under refrigeration	A pouring cream suitable for coffee, cereals, soup or fruit; added to cooked dishes and sauces; does not whip
Whipping cream	35	Not homogenised; pasteurised and packaged like single cream		Ideal for whipping; suitable for piping, cake and dessert decoration; used in ice cream, cake and pastry fillings
Double cream	48	Slightly homogenised; pasteurised and packaged like single cream		A rich pouring cream; will whip; floats on coffee or soup
'Thick' double cream	48	Heavily homogenised; pasteurised and packaged like single cream; usually only sold in domestic quantities		A rich, spoonable cream; cannot be poured
Clotted cream	55	Heated to 82°C then cooled for about 4½ hours; the cream crust is then skimmed off; packed in cartons, usually by hand; may be available in bulk		Very thick; has its own special flavour and colour; used with scones, fruit and fruit pies
Ultra-heat treated (UHT) cream	12 (half), 18 (single) or 35 (whipping)	Homogenised; heated to 132°C for 1 second, then cooled immediately; aseptically packaged in polythene and foil-lined containers; available in catering-size packs	6 weeks if unopened; does not need refrigeration; usually date stamped	A pouring cream; 35% UHT will whip

Eggs are used for several reasons:

- They act as a texture agent in, for example, crème pâtissière and ice creams.
- They enhance flavours.
- They give volume to whisked sponges and batters.
- They act as a thickening agent, e.g. in crème anglaise.
- They act as an emulsifier in products such as mayonnaise and ice cream.
- They aerate and give lightness to pastry products, for example mousses and parfaits.

Quality and nutritional value

Important facts about eggs:

- A fresh egg (in shell) should have a small, shallow air pocket inside it.
- The yolk of fresh egg should be bulbous, firm and bright.
- The fresher the egg, the more viscous (thick and not runny) the egg white.
- Eggs should be stored away from strong odours as their shells are porous and smells are easily absorbed.
- In a whole 60 g egg, the yolk weighs about 20 g, the white 30 g and the shell 10 g.

Egg yolk is high in saturated fat. The yolk is a good source of protein and also contains vitamins and iron. The egg white is made up of protein (albumen) and water. The egg yolk also contains lecithin, which acts as an emulsifier in dishes and commodities such as mayonnaise and chocolate – it helps to keep the ingredients mixed, so that the oils and water do not separate.

Working with egg whites

- To avoid the danger of salmonella, if the egg white is not going to be cooked or will not reach a temperature of 70°C, use pasteurised egg whites. Egg white is available chilled, frozen or dried.
- Equipment must be thoroughly clean and free from any traces of fat, as this prevents the whites from whipping; fat or grease prevents the albumen strands from bonding and trapping the air bubbles.
- Take care that there are no traces of yolk in the white, as yolk contains fat.
- A little acid (cream of tartar or lemon juice) strengthens the egg white, extends the foam and produces more meringue. The acid also has the effect of stabilising the meringue.
- If the foam is over-whipped, the albumen strands, which hold the water molecules with the sugar suspended on the outside of the bubble, are overstretched. The water and sugar make contact and the sugar dissolves, making the meringue heavy and wet. This can sometimes be rescued by further whisking until it foams up, but very often you will find that you may have to discard the mixture and start again.

Beaten egg white forms a foam that is used for aerating sweets and many other desserts, including meringues (Recipe 8).

Egg custard-based desserts

The essential ingredients for an egg custard are whole egg and milk.

Cream is often added to egg custard desserts to enrich them and to improve the feel in the mouth (mouth-feel) of the final product, sugar is also added to sweeten and vanilla to flavour.

Egg custard mixture provides the chef with a versatile basic set of ingredients that covers a wide range of sweets. Often the mixture is referred to as crème renversée. Although the ingredients are very similar, crème renversée is cooked in the oven, whereas crème anglaise is cooked on top of the stove.

Some examples of sweets produced using this mixture are:

- crème caramel
- bread and butter pudding
- diplomat pudding
- cabinet pudding
- queen of puddings
- baked egg custard.

Savoury egg custard is used to make:

- quiches
- tartlets
- flans.

When a starch such as flour is added to the ingredients for an egg custard mix, this changes the characteristic of the end product, as per crème pâtissière, for example.

Two other products based on an egg custard are **pastry cream** (also known as confectioner's custard or crème pâtissière – see Chapter 3, Recipe 12) and **sauce à l'anglaise** (see Chapter 3, Recipe 27 – it is also used as a base for some ice creams).

Basic egg custard sets by coagulation of the egg protein. Egg white coagulates at approximately 60°C, egg yolk at 70°C. Whites and yolks mixed together will coagulate at 66°C. If the egg protein is overheated or overcooked, it will shrink and water will be lost from the mixture, causing undesirable bubbles in the custard. This loss of water is called syneresis, commonly referred to as scrambling or curdling. This will occur at temperatures higher than 85°C. Therefore, a sauce à l'anglaise should be ideally cooked between 70°C and 85°C. The sauce will become thicker as it becomes closer to 85°C but is at risk of curdling (syneresis) beyond this temperature. Modern methods of cooking crème anglaise use a water bath and cook 'sous vide' set at 85°C for the required length of time, checking the temperature using a special probe designed for sous vide cooking. The word 'sous vide' literally translates to under vacuum.

Fruit

Fruit is used as an ingredient in many desserts.

Quality and purchasing

Fresh fruit should be:

- whole and fresh looking (for maximum flavour the fruit must be ripe but not overripe)
- firm, according to type and variety
- clean, and free from traces of pesticides and fungicides
- free from external moisture
- free from any unpleasant foreign smell or taste
- free from pests or disease
- sufficiently mature; it must be capable of being handled and travelling without being damaged
- free from any defects characteristic of the variety in shape, size and colour
- free of bruising and any other damage.

Soft fruits deteriorate quickly, especially if they are not sound. Take care to see that they are not damaged or overripe when purchased. Soft fruits should look fresh; there should be no signs of wilting, shrinking or mould. The colour of certain soft fruits is an indication of their ripeness (for example, strawberries or dessert gooseberries).

Food value

Fruit is rich in antioxidant minerals and vitamins. Antioxidants protect cells from damage by oxygen, which may lead to heart disease and cancer. The current recommendation is to eat five portions of fruit and vegetables each day.

Storage

Hard fruits, such as apples, should be left in boxes and kept in a cool store. Soft fruits, such as raspberries and strawberries, should be left in their punnets or baskets in a cold room. Stone fruits, such as apricots and plums, are best placed in trays so that any damaged fruit can be seen and discarded. Peaches and citrus fruits are left in their delivery trays or boxes. Bananas should not be stored in too cold a place because their skins will turn black.

Note: Healthy eating and desserts

Today desserts and puddings remain popular with the consumer, but there is now a demand for products with reduced fat and sugar content, as many people are keen to eat healthily. Chefs will continue to respond to this demand by modifying recipes to reduce the fat and sugar content; they may also use alternative ingredients, such as low-calorie sweeteners where possible and unsaturated fats. Although salt is an essential part of our diet, too much of it can be unhealthy, and this is something else that chefs should take into consideration.

Portion control with cold and frozen desserts

Portion control is extremely important to obtain the required portions from a set recipe and to calculate its food cost and selling price. It is also essential to calculate the number of portions for a set number of customers, especially large functions. There are a number of ways of achieving good portion control:

- use measured ladles
- use individual moulds, for example, ramekins, ring moulds
- use a torten divider which will divide the top of a mousse into uniform portions
- use ice-cream scoops
- use individual silicone moulds (these come in various individual shapes with 10, 15 or 20 units per mat)
- use piped crème Chantilly to indicate a portion from a multi-portion dessert (for example, on a cheesecake or trifle)
- follow the dish specification
- use cases made from meringue (nests), pastry (tartlets) or chocolate.

Faults in cold and frozen desserts

Water ice or sorbet once frozen is hard and difficult to scoop:
- There is not enough sugar in the recipe – test sugar content prior to freezing with saccharometer or refractometer and add more sugar if required.

Water ice or sorbet not setting firm enough:
- There is too much sugar in the recipe – test content prior to freezing with saccharometer or refractometer and dilute sugar content by adding water until the correct Baumé/Brix scale is achieved.

Curdled appearance to baked egg custard dessert such as crème caramel:
- It has been baked at too high a temperature causing the protein in the egg custard to over-coagulate.

Uneven surface on the top of a fruit delice:
- The base mixture has gelled too much prior to adding the cream and Italian meringue.
- The cream used for folding into the mixture is over whipped.

Curdled texture to crème Chantilly:
- The cream has been over-whipped.
- The cream is old.
- The cream is warm and not refrigerated prior to whipping.
- The whisking bowl is warm.

White granular texture in fruit mousse:
- The egg white has been over-whipped prior to adding the boiled sugar for Italian meringue.
- The cream has been over-whipped.

Heavy-textured chocolate mousse:
- The pâte à bombe is under-aerated.
- The cream is insufficiently whipped.
- The mix has been over-worked when folding in the cream and Italian meringue.

1 Tropical fruit plate

An assortment of fully ripe fruits, such as pineapple, papaya, mango, peeled, deseeded, cut into pieces and neatly dressed on a plate.

An optional accompaniment could be yoghurt, vanilla ice cream, crème fraiche, fresh or clotted cream.

2 Fresh fruit salad

Portions >	4 portions	10 portions
Orange	1	2–3
Dessert apple	1	2–3
Dessert pear	1	2–3
Cherries	50 g	125 g
Grapes	50 g	125 g
Banana	1	2–3
Stock syrup		
Caster sugar	50 g	125 g
Water	125 ml	375 ml
Lemon, juice of	½	1

1 For the syrup, boil the sugar with the water and place in a bowl.

2 Allow to cool, add the lemon juice.

3 Peel and cut the orange into segments.

4 Quarter the apple and pear, remove the core, peel and cut each quarter into two or three slices, place in the bowl and mix with the orange.

5 Stone the cherries, leave whole.

6 Cut the grapes in half, peel if required, and remove the pips.

7 Mix carefully and place in a glass bowl in the refrigerator to chill.

8 Just before serving, peel and slice the banana and mix in.

Try something different

- Any of the following fruits may be used: dessert apples, pears, pineapple, oranges, grapes, melon, strawberries, peaches, raspberries, apricots, bananas, cherries, kiwi fruit, plums, mangos, pawpaws and lychees. Allow about 150 g unprepared fruit per portion. All fruit must be ripe.
- Kirsch, Cointreau or Grand Marnier may be added to the syrup.
- A fruit juice (such as apple, orange, grape or passion fruit) can be used instead of syrup.

Poached fruits (compote des fruits)

Makes >	4 portions	10 portions
Stock syrup (Chapter 3, Recipe 24)	250 ml	625 ml
Fruit	400 g	1 kg
Sugar	100 g	250 g
Lemon, juice of	½	1

Apples, pears

1 Boil the syrup and sugar.
2 Quarter the fruit, remove the core and peel.
3 Place in a shallow pan in sugar syrup.
4 Add a few drops of lemon juice.
5 Cover with greaseproof paper.
6 Allow to simmer slowly, preferably in the oven, cool and serve.

Soft fruits (raspberries, strawberries)

1 Pick and wash the fruit. Place in a glass bowl.
2 Pour on the hot syrup. Allow to cool and serve.

Stone fruits (plums, damsons, greengages, cherries)

1 Wash the fruit, barely cover with sugar syrup and cover with greaseproof paper or a lid.
2 Cook gently in a moderate oven until tender.

Rhubarb

1 Trim off the stalk and leaf and wash.
2 Cut into 5 cm lengths and cook as above, adding extra sugar if necessary. A little ground ginger may also be added.

Gooseberries, blackcurrants, redcurrants

1 Top and tail the gooseberries, wash and cook as for stone fruit, adding extra sugar if necessary.
2 The currants should be carefully removed from the stalks, washed and cooked as for stone fruits.

Poached rhubarb and pear

Dried fruits (prunes, apricots, apples, pears)

1 Dried fruits should be washed and soaked in cold water overnight.
2 Gently cook in the liquor with sufficient sugar to taste.

Try something different

● A piece of cinnamon stick and a few slices of lemon may be added to the prunes or pears, one or two cloves to the dried or fresh apples.
● Any compote may be flavoured with lavender and/or mint.

4 Fruit mousse

Makes >	10 portions
Egg yolks	4
Sugar	50 g
Fruit purée	250 g
Gelatine	4 leaves
Lemon juice	½
Lightly whipped cream	250 g
Italian meringue	
Sugar	112 g
Egg whites	2
Cream of tartar	pinch
Glaze topping	
Stock syrup	150 ml
Fruit purée	150 ml
Gelatine	3 leaves, soaked in cold water

1 Mix the egg yolks and sugar together and slowly add the boiled fruit purée which has been flavoured with a squeeze of lemon juice.

2 Return to the stove and cook to 80°C until slightly thickened. Do not boil.

3 Add the previously softened gelatine to the warm purée and mix until fully dissolved. Chill down.

4 Prepare the Italian meringue by placing the sugar in a pan and saturating in water.

5 Boil the sugar to 115°C, then whisk the egg whites with a pinch of cream of tartar.

6 Once the egg whites are at full peak, gradually add the boiled sugar which now should have reached the temperature of 121°C. Whisk until cold.

7 Once the purée is cold, but not set, incorporate the Italian meringue and whipped cream.

8 Place into piping bag and pipe into the desired ring mould, normally lined with a suitable sponge such as a jaconde.

9 Level the surface using a palette knife and refrigerate.

10 Once set, glaze the surface, then refrigerate.

11 To remove from the mould, warm the outside of the mould with a blow torch and remove the ring mould.

For the glaze

Warm the syrup. Add the gelatine and stir until dissolved, then add the desired fruit purée. Apply to the surface of the chilled mousse whilst in a liquid state, but not hot.

5 | Bavarois (Bavarian cream)

Cream-based bavarois

	Makes >	10 portions
Milk		400 ml
Egg yolks		5
Caster sugar		75 g
Gelatine		7 leaves
Vanilla extract		5 drops
Lightly whipped cream		300 g
Pasteurised egg whites		150 g
Cream of tartar		pinch
Caster sugar		30 g

1 Prepare a crème anglaise with the milk, egg yolks and 75 g of the caster sugar.

2 Once cooked, add the previously soaked gelatine and mix until fully incorporated, followed by the kirsch.

3 Place the custard mixture over a bowl of iced cold water and chill, stirring occasionally with a spatula until the mixture starts to gel, but is still liquid.

4 Quickly fold in the whipped cream and cold meringue made from the pasteurised egg white, cream of tartar and 30 g of sugar.

5 Pour into mould.

6 Refrigerate until set.

7 To demould, dip the mould briefly in hot water and turn out onto a serving dish.

Flavours for bavarois

- **Raspberry or strawberry bavarois:** when the custard is almost cool, add 200 g of raspberry or strawberry purée. Increase the gelatine content by 2 leaves. Decorate with whole fruit and whipped cream.
- **Chocolate bavarois:** dissolve 50 g chocolate couverture in the milk. Reduce the gelatine content by 1 leaf. Decorate with whipped cream and grated chocolate.
- **Coffee bavarois:** proceed as for a basic bavarois, with the addition of coffee essence to taste.
- **Orange bavarois:** add grated zest and juice of 2 oranges and 1 or 2 drops orange colour to the mixture, and increase the gelatine by 2 leaves. Decorate with blanched, fine julienne of orange zest, orange segments and whipped cream.
- **Lemon or lime bavarois:** as orange bavarois, using lemons or limes in place of oranges.
- **Vanilla bavarois:** add a vanilla pod or a few drops of vanilla essence to the milk. Decorate with vanilla-flavoured sweetened cream (crème Chantilly).

Notes

Bavarois may be decorated with sweetened, flavoured whipped cream (crème Chantilly).

It is advisable to use pasteurised egg yolks and whites. Raw egg white is traditionally used to aerate and lighten a bavarois, but some chefs omit this ingredient.

Fruit-based bavarois

	Makes >	8 portions
Fruit purée		300 g
Lemon juice		1
Caster sugar		75 g
Gelatine		5 leaves
Lightly whipped cream		400 g

1 Heat the fruit purée, but do not boil. Add the lemon juice and sugar.
2 Add the previously soaked gelatine to the warm fruit purée.
3 Set the fruit purée over ice following the same guidelines as the cream-based bavarois.
4 Quickly fold in the lightly whipped cream.
5 Pour into mould.
6 Finish as for a cream-based bavarois.

> **Professional tip**
>
> Certain fruits, such as pineapple and kiwi, contain enzymes that break down the protein in gelatine, destroying its gelling properties. When using these fruits for a bavarois or gelatine mousse, first boil the purée to destroy the enzymes.

Video: mango mousse,
http://bit.ly/1fiivEH

6 Vanilla panna cotta

	Makes >	6 portions
Milk		125 ml
Double cream		375 ml
Aniseeds		2
Vanilla pod		½
Gelatine (soaked)		2 leaves
Caster sugar		50 g

1 For the panna cotta, boil the milk and cream, add aniseeds, infuse with the vanilla pod, remove after infusion.

2 Heat again and add the soaked gelatine and caster sugar. Strain through a fine strainer.

3 Place in a bowl set over ice and stir until it thickens slightly; this will allow the vanilla seeds to suspend throughout the mix (if desired) instead of sinking to the bottom.

4 Fill individual dariole moulds.

5 Place some fruit compote (see Recipe 3) on individual fruit plates, turn out the panna cotta, place on top of the fruit compote, and finish with a tuile biscuit.

7 Fruit fool

Method 1

Makes >	4 portions	10 portions
Fruit (apple, gooseberry, rhubarb, etc.)	400 g	1 kg
Water	60 ml	150 ml
Granulated or unrefined sugar	100 g	250 g
Cornflour	25 g	60 g
Milk, whole or skimmed	250 ml	625 ml
Caster or unrefined sugar	25 g	60 g

1 Cook the fruit in the water and granulated sugar to a purée. Pass through a sieve.

2 Dilute the cornflour in a little of the milk, add the caster sugar.

3 Boil the remainder of the milk.

4 Pour on the diluted cornflour, stir well.

5 Return to the pan on a low heat and stir to the boil.

6 Mix with the fruit purée. The quantity of mixture should not be less than 0.5 litres.

7 Pour into 4 (or 10) glass coupes or suitable dishes and allow to set.

8 Decorate with whipped sweetened cream or non-dairy cream. The colour may need to be adjusted slightly with food colour.

Method 2

Makes >	4 portions	10 portions
Fruit in purée (raspberries, strawberries, etc.)	400 g	1 kg
Caster sugar	100 g	250 g
Fresh whipped cream	250 ml	625 ml

Mix the ingredients and serve in coupes.

> **Professional tip**
>
> In methods 2 and 3 the fat content may be reduced by using equal quantities of cream and natural Greek-style yoghurt.

Method 3

Makes >	4 portions	10 portions
Cornflour	35 g	85 g
Water	375 ml	900 ml
Sugar	100 g	250 g
Fruit in purée (raspberries, strawberries, etc.)	400 g	1.25 kg
Cream	185 ml	500 ml

1 Dilute the cornflour in a little of the water.

2 Boil the remainder of the water with the sugar and prepared fruit until soft.

3 Pass through a fine sieve.

4 Return to a clean pan and reboil.

5 Stir in the diluted cornflour and reboil. Allow to cool.

6 Lightly whisk the cream and fold into the mixture.

7 Serve as for method 1.

Meringue

Unfilled meringues and vacherins

Makes >	4 portions	10 portions
Lemon juice or cream of tartar		
Egg whites, pasteurised	4	10
Caster sugar	200 g	500 g

1 Whip the egg whites stiffly with a squeeze of lemon juice or pinch of cream of tartar.

2 Sprinkle on the sugar and carefully mix in.

3 Place in a piping bag with a large plain tube and pipe on to silicone paper on a baking sheet.

4 Bake in the slowest oven possible or in a hot plate (110°C). The aim is to dry out the meringues without any colour whatsoever.

Whipping egg whites

The reason egg whites increase in volume when whipped is because they contain so much protein (11 per cent). The protein forms tiny filaments, which stretch on beating, incorporate air in minute bubbles then set to form a fairly stable puffed-up structure expanding to seven times its bulk. To gain maximum efficiency when whipping egg whites, the following points should be observed.

● Because of possible weakness in the egg-white protein it is advisable to strengthen it by adding a pinch of cream of tartar and a pinch of dried egg-white powder. If all dried egg-white powder is used no additions are necessary.

● Eggs should be fresh.

● When separating yolks from whites no speck of egg yolk must be allowed to remain in the white; egg yolk contains fat, the presence of which can prevent the white being correctly whipped.

● The bowl and whisk must be scrupulously clean, dry and free from any grease.

● When egg whites are whipped, the addition of a little sugar (15 g to 4 egg whites) will assist the efficient beating and reduce the chances of over-beating.

9 Vacherin with strawberries and cream

Makes >	4 portions	10 portions
Egg whites	4	10
Caster sugar	200 g	500 g
Cream (whipped and sweetened) or non-dairy cream	125 ml	300 ml
Strawberries, picked and washed)	100–300 g	250–750 g

1 Stiffly whip the egg whites. (Refer to the notes in Recipe 8 for more guidance.)

2 Carefully fold in the sugar.

3 Place the mixture into a piping bag with a 1 cm plain tube.

4 Pipe on to silicone paper on a baking sheet.

5 Start from the centre and pipe round in a circular fashion to form a base of 16 cm then pipe around the edge 2–3 cm high.

6 Bake in a cool oven at 100°C until the meringue case is completely dry. Do not allow to colour.

7 Allow the meringue case to cool then remove from the paper.

8 Spread a thin layer of cream on the base. Add the strawberries.

9 Decorate with the remainder of the cream.

Note

A vacherin is a round meringue shell piped into a suitable shape so that the centre may be filled with sufficient fruit (such as strawberries, stoned cherries, peaches and apricots) and whipped cream to form a rich sweet. The vacherin may be prepared in one-, two- or four-portion sizes, or larger.

Professional tip

Try 'diluting' the fat in the cream with some low-fat fromage frais.

Try something different

- Melba sauce (Chapter 3, Recipe 36) may be used to coat the strawberries before decorating with cream.
- Raspberries can be used instead of strawberries.

10 Black Forest vacherin

Makes >	12 portions
Egg whites	250 ml
Caster sugar	500 g
Lemon juice	5 ml
Vanilla essence	drop
Cornflour, sieved	30 g
Cocoa powder, sieved	50 g
Small discs of chocolate sponge	12
Kirsch syrup	100 ml
Cherries (fresh, tinned or griottines)	60–72
Pastry cream (Chapter 3, Recipe 12)	200 g
Kirsch	20 ml
Gelatine, soaked	2 leaves
Couverture, melted	200 g (approx.)
Double cream, whipped	400 ml
Chocolate shavings	
Icing sugar	
Cocoa powder (to dust)	

1 Whisk the egg white and one-quarter of the sugar until firm. Continue to whisk while streaming in half of the sugar.

2 Add the lemon juice and vanilla. Fold in the cornflour and cocoa powder, and the remaining quarter of the sugar.

3 Pipe this vacherin mixture into 12 rounds, 80 mm in diameter. Bake at 150°C for approx. 30 minutes.

4 Place a sponge disc on each vacherin. Moisten the sponge with Kirsch syrup and place 5 or 6 cherries on top.

5 Beat the pastry cream. Dissolve the gelatine in the warm Kirsch and then beat it into the pastry cream.

6 Beat in the melted couverture to taste. Fold in the cream.

7 Pipe the chocolate mixture onto the prepared bases in a spiral.

8 Cover with chocolate shavings. Dust with icing sugar.

11 Lime soufflé frappé

Makes >	10 portions	15 portions
Couverture	200 g	250 g
Sponge	10	15
Lime syrup	100 ml	150 ml
For the Swiss meringue		
Egg whites, pasteurised	190 ml	300 ml
Caster sugar	230 ml	340 ml
For the sabayon		
Whipping cream	600 ml	900 ml
Lime zest, finely grated and blanched, and juice	8	12
Egg yolks, pasteurised	200 g	300 g
Caster sugar	170 g	250 g
Leaf gelatine, soaked in iced water	9½	14
To decorate		
Confit of lime segments		
Moulded chocolate		

1 Use individual stainless steel ring moulds. Cut a strip of acetate, 8 cm wide, to fit inside each ring. Cut a 6 cm strip to fit inside the first, spread it with tempered couverture and place inside the first strip, in the mould.

2 Place a round of sponge in the base of each mould and moisten with lime syrup.

3 Make a Swiss meringue by whisking the pasteurised egg whites and sugar over a bain-marie of simmering water until a light and aerated meringue is achieved.

4 Whisk the cream until it is three-quarters whipped, then chill.

5 Whisk together the egg yolks, sugar and blanched lime zest. Boil the juice and pour it over the mixture to make the sabayon. Whisk over a bain-marie until it reaches 75°C, then continue whisking away from the heat until it is cold.

6 Drain and melt the gelatine. Fold it into the sabayon.

7 Fold in the Swiss meringue, and then the chilled whipped cream.

8 Fill the prepared moulds. Level the tops and chill until set.

9 To serve, carefully remove the mould, peel away the acetate, plate and decorate.

12 Trifle

Makes >	6–8 portions
Sponge	100 g
Jam	25 g
Tinned fruit (pears, peaches, pineapple)	1
Sherry (optional)	
Whipped sweetened cream or non-dairy cream	250 ml
Custard	
Custard powder	35 g
Milk, whole or skimmed	375 ml
Caster sugar	50 g
Cream (¾ whipped) or non-dairy cream	125 ml

1 Cut the sponge in half, sideways, and spread with jam. Dice it.

2 Place in a glass bowl or individual dishes and soak with fruit syrup drained from the tinned fruit; a few drops of sherry may be added.

3 Cut the fruit into small pieces and add to the sponge.

4 Dilute the custard powder in a basin with some of the milk, add the sugar.

5 Boil the remainder of the milk, pour a little on the custard powder, mix well, return to the saucepan and over a low heat and stir to the boil. Allow to cool, stirring occasionally to prevent a skin forming; fold in the three-quarters whipped cream.

6 Pour on to the sponge. Leave to cool.

7 Decorate with the whipped cream, and garnish as desired.

Try something different

- Other flavourings or liqueurs may be used in place of sherry (such as whisky, rum, brandy, Tia Maria).
- For raspberry or strawberry trifle use fully ripe fresh fruit in place of tinned, and decorate with fresh fruit in place of angelica and glacé cherries.
- A fresh egg custard may be used with fresh egg yolks (see Chapter 3, Recipe 27).

13 Lime and mascarpone cheesecake

Makes >	1 cheesecake
The base	
Packet of ginger biscuits	1
Butter, melted	200 g
The cheesecake	
Egg yolks, pasteurised	125 g
Caster sugar	75 g
Cream cheese	250 g
Mascarpone	250 g
Gelatine, softened in cold water	15 g
Limes, juice and finely grated zest of	2
Semi-whipped cream	275 ml
White chocolate, melted	225 g

Professional tip

Use a Microplane or micrograter for the lime zest.

1 Blitz the biscuits in a food processor. Mix in the melted butter. Line the cake ring with this mixture and chill until required.

2 Make a sabayon by whisking the egg yolks and sugar together over a pan of simmering water.

3 Stir the cream cheese and mascarpone into the sabayon until soft.

4 Meanwhile, warm the gelatine in the lime juice, and pass through a fine chinois. Also whip the cream.

5 Pour the gelatine and melted white chocolate into the cheese mixture.

6 Fold in the whipped cream with a spatula. Finally, whisk in the lime zest.

7 Pour over the prepared base. Chill for 4 hours.

14   Baked blueberry cheesecake

Makes >	1 cheesecake
The base	
Digestive biscuits	150 g
Butter, melted	50 g
The cheesecake	
Full-fat cream cheese	350 g
Caster sugar	150 g
Eggs	4
Lemon, zest and juice of	1
Vanilla essence	5 ml
Blueberries	125
Soured cream	350 ml

1　Blitz the biscuits in a food processor. Stir in the melted butter. Press the mixture into the bottom of a lightly greased cake tin with a removable collar.

2　Whisk together the cheese, sugar, eggs, vanilla and lemon zest and juice, until smooth.

3　Stir in the blueberries, then pour the mixture over the biscuit base.

4　Bake at 160°C for approx. 30 minutes.

5　Remove from the oven and leave to cool slightly for 10–15 minutes.

6　Spread soured cream over the top and return to the oven for 10 minutes.

7　Remove and allow to cool and set. Chill.

15 Baked apple cheesecake

Makes >	16 portions
Base	
Biscuit crumbs	225 g
Butter, melted	110 g
Caster sugar	30 g
Filling	
Apples, cooked, halved	Approx. 8
Cream cheese, full fat	800 g
Caster sugar	230 g
Cornflour	75 g
Eggs	2 (120 g)
Vanilla arome or essence	4 drops
Double cream	290 ml

1 Combine the ingredients for the base. Press the mixture into the bottom of two lined cake tins.

2 Place the cooked apple halves into the tins.

3 Cream the cheese and sugar together. Stir in the cornflour, eggs, vanilla and double cream.

4 Divide the filling between the two tins.

5 Bake at 160°C for approximately 40 minutes.

6 Allow to cool slightly, then remove from the mould and dust with icing sugar.

7 Decorate with dried apple slices.

16 | Vanilla ice cream

Makes >	8–10 portions
Egg yolks	4
Caster or unrefined sugar	100 g
Milk, whole or skimmed	375 ml
Vanilla pod or essence	1 pod/4 drops
Cream or non-dairy cream	125 ml

1 Whisk the yolks and sugar in a bowl until almost white.

2 Boil the milk with the vanilla pod or essence in a thick-based pan.

3 Whisk on to the eggs and sugar; mix well.

4 Return to the cleaned saucepan, place on a low heat.

5 Stir continuously with a heat-resistant spatula until the mixture coats the back of the spoon. The temperature should reach 82°C.

Try something different

- Coffee ice cream: add coffee essence, to taste, to the custard after it is cooked.
- Chocolate ice cream: add 50–100 g of chopped couverture to the milk before boiling.
- Strawberry ice cream: add 125 ml of strawberry pulp in place of 125 ml of milk. The pulp is added after the custard is cooked.
- Rum and raisin ice cream: soak 50 g raisins in 2 tbsp rum for 3–4 hours. Add to mixture before freezing.

6 Pass through a fine strainer into a bowl.

7 Freeze in an ice-cream machine, gradually adding the cream.

Whisk boiling milk into the egg yolks and sugar

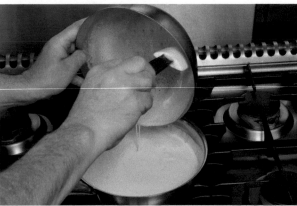
Return the mixture to the hot pan used for the milk

Test the consistency on the back of a spoon (check the temperature with a thermometer too)

Pass through a fine strainer into a cold pot

The mixture will cool down; if it was left in the hot pan it would continue to cook

Gradually add cream to the mixture in the ice cream machine

17 Rich vanilla ice cream

	Makes >	10 portions
Whipping cream		250 ml
Milk		250 ml
Pasteurised egg yolk		100 g
Caster sugar		150 g
Vanilla pod		1
Unsalted butter		65 g

1 Make a crème anglaise in the normal manner using the cream, milk, egg yolks, sugar and vanilla pod.

2 Stir in the unsalted butter and churn down in the sorbetiere.

Professional tip

The ice cream will have a better flavour if the custard is chilled down quickly and matured in a refrigerator at 3°C for 12 hours prior to freezing down.

18 Lemon curd ice cream

	Makes >	6–8 portions
Lemon curd		250 g
Crème fraiche		125 g
Greek yoghurt		250 g

1 Mix all ingredients together.

2 Churn in ice-cream machine.

Caramel, lemon curd and peach ice creams

19 Caramel ice cream

Makes >	8 portions
Crème anglaise	
Milk	500 ml
Egg yolks	5
Caster sugar	25 g
Whipping cream	100 ml
Inverted sugar (trimoline)	25 g
Caramel	
Glucose	20 g
Caster sugar	100 g
Butter	10 g
Boiling water	40ml

1 Make the crème anglaise in the normal manner, then add the inverted sugar.

2 To make the caramel, melt the glucose in a thick-bottomed pan.

3 Add half the sugar to the melted glucose and heat until a caramel colour starts to appear.

4 Gradually add the remaining sugar and continue to cook until a golden caramel is obtained.

5 Add the butter and the boiling water to arrest the cooking of the sugar and dilute the caramel down.

6 Add the caramel to the crème anglaise and freeze down in the sorbetiere.

Professional tip

Always have a frozen metal container in the deep freezer to transfer the ice cream into. This will prevent the base of the ice cream melting.

Health and safety

Remember that caramel is extremely hot; be very careful when pouring onto the crème anglaise.

20 Peach ice cream

Makes >	8 portions
Milk	250 ml
Caster sugar	175 g
Orange rind	1
Lemon rind	1
Stabiliser (Trimoline)	25 g
Single cream	250 ml
Peach purée	250 ml
Lemon juice	10 ml

1 Slowly bring the milk, sugar, rinds and stabiliser to the boil.
2 Remove from the heat and leave to cool slightly.
3 Add the cream and leave to cool.
4 When cold, add the peach purée and lemon juice. Leave overnight to mature.
5 Pass, then churn in the ice-cream machine.
6 Place into a frozen container. Store in the freezer.

21 Chocolate ice cream

Makes >	12 portions
Cream	250 ml
Milk	250 ml
Egg yolks	100 g
Glucose	20 g
Trimoline	40 g
Sugar	100 g
Plain chocolate couverture callets	125 g
Chocolate chips (optional)	

1 Bring the cream and milk to the boil together.
2 Mix the egg yolks, glucose, Trimoline and sugar together until smooth.
3 Pour the cream and milk over the egg mixture while whisking.
4 Return to the stove and cook until it coats the back of a spoon.
5 Pass. Add the chocolate pieces. Mix until combined and chill in an ice bain-marie.
6 When cooled, churn.
7 Fold in chocolate chips if required.

Professional tip

Use a stem blender to emulsify the chocolate into the hot crème anglaise.

22 Malt ice cream and caramel sauce

Makes >	12 portions
Milk	400 ml
Trimoline	25 g
Horlicks	150 g
Egg yolks, pasteurised	120 g
Caster sugar	100 g
Double cream	400 ml
Caramel sauce	
Condensed milk	200 ml

1 Bring the milk and Trimoline to the boil together. Whisk in the Horlicks until dissolved.
2 Whisk the egg yolks and sugar together until white.
3 Add half the boiling milk to the eggs, then return all to the saucepan. Cook well without allowing the mixture to boil.
4 Add the cream and cool over ice. Churn.
5 To make the sauce, place the condensed milk in a double boiler with a lid. Bring the water to the boil and cook gently for 40 to 50 minutes, stirring occasionally.
6 When the sauce is thick, with a light caramel colour, remove from the heat. Beat until smooth.

23 American orange ice cream

Makes >	8–10 portions
Milk	300 ml
Caster sugar	250 g
Stabiliser (Trimoline)	25 g
Orange rind	1
Lemon rind	1
Orange juice	250 ml
Lemon juice	10 ml
Single cream	250 ml

1 Slowly bring the milk, sugar, stabiliser and rinds to the boil.
2 Remove from the heat and leave to cool. When cold, add the juice and cream.
3 Pass, then churn in an ice-cream machine.
4 Place into a frozen container, seal and freeze.

24 Chocolate sorbet

Makes >	8 portions
Water	400 ml
Skimmed milk	100 ml
Sugar	150 g
Ice-cream stabiliser	40 g
Cocoa powder	30 g
Dark couverture	60 g

1 Combine the water, milk, sugar, stabiliser and cocoa powder. Bring to the boil slowly. Simmer for 5 minutes.
2 Add the couverture and allow to cool.
3 Pass and churn.

Fruits of the forest, apple and chocolate sorbets

25   Apple sorbet

	Makes >	8–10 portions
Granny Smith apples, washed and cored		4
Lemon, juice of		1
Water		400 ml
Sugar		200 g
Glucose		50 g

1 Cut the apples into 1 cm pieces and place into lemon juice.

2 Bring the water, sugar and glucose to the boil, then allow to cool.

3 Pour the syrup over the apples. Freeze overnight. Blitz in a food processor.

4 Pass through a conical strainer, then churn in an ice-cream machine.

Professional tip

For best results, after freezing, process in a Pacojet.

Try something different

Fruits of the forest sorbet: use a mixture of forest fruits instead of apples.

26 Lemon sorbet

Makes >	8 portions
Water	500 ml
Caster sugar	250 g
Washed whole lemons	4
Italian meringue	250 g

1 Boil the water and sugar.
2 Pour onto the juice and zest of the lemons.
3 Leave to infuse overnight.
4 Strain and freeze in the sorbetiere.
5 At the point when the appareil starts to thicken, add the Italian meringue and continue to freeze until the desired consistency is achieved.

Note

A sorbet is a water ice with the addition of an Italian meringue. Appareil is the culinary term used for a base mixture.

27 Grapefruit water ice

Makes >	8–10 portions
Water	250 ml
Sugar	100 g
Grapefruit juice	250 ml
Orange juice	100 ml
White wine	5 ml
Lemon (juice of)	1

1 Bring the water and sugar to the boil.
2 Add the rest of the ingredients, mix well and cool.
3 Churn and place into a frozen container.
4 Freeze until required.

28 Vanilla water ice

Makes >	4 portions
Water	255 ml
Caster sugar	110 g
Glucose	25 g
Vanilla pod, scraped	1

1 Mix all the ingredients together and bring to the boil. Allow to cool fully.

2 Pass and then churn.

29 Lemon water ice

Makes >	8 portions
Water	500 ml
Sugar	250 g
Washed whole lemons	3

1 Pour onto the juice and zest of the lemons.

2 Leave to infuse overnight.

3 Strain and freeze down in a sorbetiere.

30 Champagne water ice

Base syrup	
Water	500 g
Sugar	500 g
Lemon, zest of	1
Orange, zest of	1
Vanilla pod	½

1 Boil all the ingredients of the syrup and infuse for 12 hours, then strain.

2 For every 750 ml Champagne add 650ml syrup and 10g lemon juice.

3 Ensure a density of 17° Baumé.

4 Freeze down in a sorbetiere.

31 Pistachio and chocolate crackle ice cream (using a Pacojet)

Makes >	12 portions
Milk	500 ml
Cream	500 ml
Egg yolks	10
Caster sugar	200 g
Ice-cream stabiliser	50 g
Pistachio compound	40 g
Chocolate crackle crystal	50 g

1 Bring the milk and cream to boil in a saucepan.

2 Combine the egg yolks, sugar and stabiliser in a bowl. Whisk until the mixture is very pale and leaves a trail when the beaters are lifted. Gradually whisk in the milk mixture.

3 Return the mixture to the pan and cook it over a very low heat, or cook it in the top of a double boiler, stirring constantly until the custard is thick enough to coat the back of a wooden spoon.

4 Remove the custard from the heat to cool, stirring it from time to time to prevent a skin forming. Once the mixture has cooled, stir in the pistachio compound, pour into the Pacojet containers and freeze down.

5 Once frozen, place the container in the Pacojet machine and process.

6 When the ice cream is creamed-down, fold in the chocolate crackle crystal and serve.

Professional tip

Never fill the Pacojet containers to the top, as the mixture will spill out during processing.

32 Savoury ice cream using maltodextrin and a Pacojet

Makes >	8 portions
Base mixture	
Milk	565 g
Whipping cream	260 g
Maltodextrin	235 g
Salt	8 g
Monosodium glutamate	3 g

1 Boil the milk, cream and pour onto the other ingredients. This is to be used as the base mixture.

2 Whatever product you add to the savoury base is then treated in the same manner as in previous recipes, and frozen in Pacojet containers.

3 Churn in the Pacojet machine.

Flavours

Various savoury ice creams may be made by adding to the base mixture.

For smoked salmon ice cream add 500 g chopped smoked salmon, pinch cayenne and zest of 1 lemon.

For red pepper ice cream add 500 g sweet red peppers flavoured with garlic and 10 g balsamic vinegar.

Note

Maltodextrin comes in a white powder and has all the properties of sugar, without the sweetness. Sugar is an essential ingredient in ice cream; it controls the texture and smoothness.

33 Lemon and ginger sorbet (using a Pacojet)

Makes >	12 portions
Sugar	260 g
MSK sorbet stabiliser	60 g
Water	500 g
Lemon juice	400 g
Shredded ginger	50 g

1 Combine the sugar with stabiliser and add to the water. Heat this mixture until the sugar is dissolved.

2 Add the lemon juice and allow mixture to cool.

3 Freeze in Pacojet containers.

4 Churn in the Pacojet machine.

5 Fold in the shredded ginger before serving.

34 Blackberry sorbet (using a Pacojet)

Makes >	12 portions
Sugar	150 g
MSK standard sorbet stabiliser	50 g
Water	500 g
Blackberry purée	500 g
MSK malic acid	4 g

1 Combine the sugar with stabiliser and add to the water. Heat this mixture until the sugar is dissolved.

2 Add the blackberry purée and malic acid and allow to cool.

3 Freeze in Pacojet containers.

4 Churn in the Pacojet machine and serve.

Bombe glacée

Pâte à bombe (small scale)

	Makes >	10 portions
Pasteurised egg yolk		200 g
Sugar syrup at 32° Baumé (250 g sugar + 250 ml water = 32° Baumé)		250 g
Lightly whipped cream		250 g

1 Blend the egg yolks with the syrup, whisk over a heated bain-marie as for a Genoese sponge until aerated and continue to whisk until the mixture is completely cold over ice.

2 Add the whipped cream and any flavourings.

Pâte à bombe (large scale)

	Makes >	15 portions
Caster sugar		250 g
Water		62 ml
Pasteurised egg yolks		250 g
Lightly whipped cream		500 g

1 Boil the water and sugar to hard ball stage.

2 Whisk the egg yolks until aerated and gradually pour on sugar solution.

3 Whisk until cold, then fold in the whipped cream and flavouring.

Creating the bombe

1. Stand the bombe mould in a bowl packed with ice.
2. Fill the bombe mould to the top with either ice cream or water ice.
3. Scoop out the centre with a warm spoon.
4. Fill with pâte à bombe and any flavouring.
5. Skim top of bombe with ice cream or sorbet and freeze.
6. To remove the bombe mould, dip into warm water, turn upside down and release the screw in the base, and turn out onto a round of sponge.

Variations

- **Bombe Nesselrode:** lined with chestnut ice cream, interior filled with vanilla pâte à bombe.
- **Bombe Sultane:** lined with chocolate ice cream, interior filled with pâte à bombe praline.
- **Bombe Othello:** lined with praline ice cream, interior filled with vanilla pâte à bombe with diced peaches.
- **Bombe diplomate:** lined with vanilla ice cream, interior flavoured with maraschino and glacé fruits.

Fill the mould to the top, then scoop out the centre

Fill the centre with pâte à bombe and then cover with some of the ice cream removed earlier

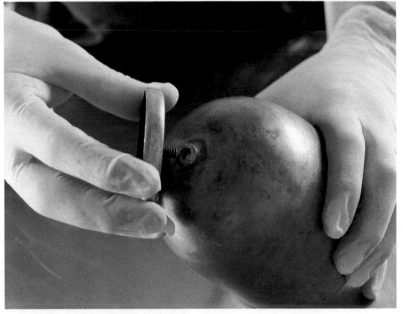

To remove from the mould after freezing, release the screw in the base

36   Orange brandy granita

Makes >	10 portions
Orange juice	500 ml
Brandy	40 ml
Stock syrup at 30° Baumé (equal quantities of sugar and water will give 30° Baumé)	100 ml
Water	175 ml

1   Mix all the ingredients together.

2   Pour into a gastronorm tray and place in the freezer.

3   Fork up to produce crystals.

4   Serve in frozen glasses.

37   Orange and Cointreau iced soufflé (cream-based)

Makes >	8 portions
Oranges (juice and zest)	3
Oranges (juice only)	8 (400 ml)
Whipping cream, half whipped	800 ml
Egg yolks	100 g
Caster sugar	75 g
Gelatine, soaked in cold water	2 leaves
Cointreau	20 ml
Italian meringue	
Caster sugar	100 g
Glucose	10 g
Water	80 ml
Egg whites	100 g

1   Use a Microplane grater or similar to zest three oranges. Bring a small pan of water to the boil, add the zest and simmer for 5 minutes. Refresh the zest in cold water and reserve.

2   Juice all the oranges (11 in total). Measure the juice; pass and reduce until you have 200 ml. Leave to cool.

3   Whip the cream to soft peaks and chill.

4   Make a sabayon by whisking the egg yolks and sugar over a bain-marie until the mixture reaches 75°C. Mix the sabayon in a food processor until it goes cold.

5   Make the Italian meringue by carefully boiling the sugar, glucose and water in a clean stainless steel pan until it reaches 110°C. Whisk the egg whites until they form soft peaks. Pour the sugar mixture (which should now be at 121°C) onto the whites gradually and keep mixing until the meringue is tepid.

6   Drain and melt the gelatine. Heat it in the Cointreau until it becomes a liquid. Pass.

7   Fold the gelatine into the sabayon.

8   Fold the Italian meringue into the sabayon mixture, then fold in the cream.

9   Add the zest and orange juice.

10   Line a soufflé mould with silicone paper to form a collar 2 cm high. Fill it to the top with the mixture. Freeze overnight. Remove the paper collar.

11   Decorate with crème Chantilly and orange confit.

Try something different

A fruit-based iced soufflé can be made from fruit purée, whipped cream and Italian meringue.

38 Raspberry parfait

Makes >	6–8 portions
Egg yolks, pasteurised	80 g
Caster sugar	60 g
Gelatine, soaked	1½ leaves
Raspberry liqueur	10 ml
Lemon juice	10 ml
Raspberry purée	120 g
Whipped cream	150 ml
Italian meringue	
Caster sugar	150 g
Glucose	20 g
Water	80 ml
Egg whites	200 g

1 Make up the Italian meringue at 121°C (see Recipe 37).

2 Combine the egg yolks and caster sugar over a pan of simmering water, to make a sabayon.

3 Drain the gelatine and dissolve it in the liqueur and lemon juice.

4 Fold the gelatine mixture into the sabayon, then fold in the raspberry purée.

5 Fold in half the Italian meringue, then fold in the whipped cream.

6 Place into prepared moulds lined with sponge and freeze.

Variation: Kirsch cream parfait

Water	70 g
Caster sugar	300 g
Pasteurised whole egg	300 g
Pasteurised egg yolk	100 g
Lightly whipped cream	1 litre
Kirsch	50 ml

1 Boil the water and sugar to soft ball stage.

2 Whip whole eggs and egg yolks to a sabayon.

3 Slowly add the boiled sugar solution, whip until cold.

4 Lightly fold in the whipped cream and kirsch and deposit into mould lined with sponge.

5 Freeze.

39 | Peach Melba

Makes >	4 portions	10 portions
Peaches	2	5
Vanilla ice cream	125 ml	300 ml
Melba sauce or raspberry coulis (see Chapter 3)	125 ml	300 ml

1. Poach the peaches. Allow to cool, then peel, halve and remove the stones.

2. Dress the fruit on a ball of the ice cream in an ice-cream coupe, or in a tuile basket.

3. Finish with the sauce. The traditional presentation is to coat the peach in Melba sauce or coulis, and decorate with whipped cream. In this picture, the peach is garnished with crushed fresh pistachios and covered with a caramel cage; the basket is then placed carefully onto a base of coulis.

Peach Melba was created by Auguste Escoffier at the Savoy Hotel, London, in honour of the visiting Australian opera singer, Dame Nellie Melba.

Professional tip

If using fresh peaches, dip them in boiling water for a few seconds, cool them by placing into cold water, then peel and halve.

Try something different

Fruit Melba can also be made using pear or banana instead of peach. Fresh pears should be peeled, halved and poached. Bananas should be peeled at the last moment.

40 | Pear belle Hélène

1. Serve a cooked pear on a ball of vanilla ice cream in a coupe.

2. Decorate with whipped cream. Serve with a sauceboat of hot chocolate sauce (Chapter 3, Recipe 32).

Alternatively, present the ingredients on a plate as shown here.

41 Crème caramel

Makes >	4–6 portions	10–12 portions
Caramel		
Sugar, granulated or cube	100 g	200 g
Water	125 ml	250 ml
Cream		
Milk, whole or skimmed	0.5 litres	1 litre
Eggs	4	8
Sugar, caster or unrefined	50 g	100 g
Vanilla essence, or a vanilla pod	3–4 drops	6–8 drops

1 Prepare the caramel by placing three-quarters of the water in a thick-based pan, adding the sugar and allowing to boil gently, without shaking or stirring the pan.

2 When the sugar has cooked to a golden-brown caramel colour, add the remaining quarter of the water, reboil until the sugar and water mix, then pour into the bottom of dariole moulds.

3 Prepare the cream by warming the milk and whisking on to the beaten eggs, sugar and essence (or vanilla pod).

4 Strain and pour into the prepared moulds.

5 Place in a roasting tin half full of water.

6 Cook in a moderate oven at 150–160°C for 30–40 minutes.

7 When thoroughly cold, loosen the edges of the crème caramel with the fingers, shake firmly to loosen and turn out on to a flat dish or plates.

8 Pour any caramel remaining in the mould around the creams.

A crème caramel decorated with poached kumquats

Professional tips

For best results, leave overnight before demoulding.

Adding a squeeze of lemon juice to the caramel will invert the sugar, thus preventing recrystallisation.

Faults

This caramel was baked at too high a temperature. The custard has curdled.

42 Chocolate mousse

Makes >	8 portions	16 portions
Stock syrup at 30° Baumé (equal quantities of sugar and water will give 30° Baume)	125 ml	250 ml
Egg yolks, pasteurised	80 ml	160 ml
Bitter couverture	250 g	500 g
Gelatine	2 leaves	4 leaves
Whipping cream, whipped	500 ml	1 litre

1 Boil the syrup.

2 Place the yolks into the bowl of a food mixer and whisk to a sabayon. Pour over the boiling syrup and whisk until thick. Remove from the mixer.

3 Add all the melted couverture at once and fold it in quickly.

4 Drain the gelatine, melt it in the microwave and fold it into the chocolate sabayon mixture.

5 Add all the whipped cream at once and fold it in carefully.

6 Place the mixture into prepared moulds. Refrigerate or freeze immediately.

> **Professional tip**
>
> Specialist plastic moulds may be used, as for the example in the first photo.

43 Raspberry or strawberry mousse (in glasses)

Makes >	7 portions
Gelatine	2 leaves
Strawberry or raspberry purée	250 g
Lemon juice	10 g
Kirsch	10 g
Whipping cream	400 g
Granulated sugar	125 g
Water, to saturate	
Egg whites, pasteurised	80 g
Cream of tartar	Pinch

1 Place 7 clean wine glasses in the refrigerator to chill.

2 Soften the leaf gelatine in cold water.

3 Heat the fruit purée and add the lemon juice and kirsch.

4 Dissolve the gelatine in the hot purée and stand the mixture over a bowl of ice water, stirring occasionally, to prevent the purée from setting on the sides of the bowl.

5 Lightly whisk the cream to a light peak.

6 Saturate the sugar with water and boil to 115°C.

7 While the sugar boils, whisk the egg whites with the cream of tartar to full peak in a mixing machine.

8 When the egg whites form a tight meringue and the boiled sugar has reached 121°C, gradually pour the sugar over the meringue.

9 Whisk on slow speed until cold and shiny. (The mixture must be cold before the next step.)

10 Using a spatula, lightly mix the whipped cream and cold Italian meringue together. Fold through the partially set purée. Do not overwork the mixture.

11 Place the mixture into a piping bag with a star tube and neatly pipe into the chilled glasses.

12 Decorate with chocolate cut-outs or run-outs.

44 Orange mousse with biscuit jaconde

Makes >	6 portions
Patterned biscuit jaconde (Chapter 8, Recipe 5)	1 sheet
Orange mousse	
Orange juice	200 g
Egg whites, pasteurised	30 g
Caster sugar	30 g
Gelatine, soaked in ice water	2 leaves
Lemon juice	few drops
Whipping cream	225 g
Orange segments, to garnish	
Glaze (jelly)	
Stock syrup (Chapter 3, Recipe 24)	50 ml
Gelatine, pre-soaked and drained	1 leaf
Orange juice	100 ml

1 Reduce the orange juice by half.

2 Whisk the egg whites and half of the sugar over a bain-marie until firm peaks form (this is Swiss meringue). Remove from the heat and whisk in the rest of the sugar.

3 Drain the gelatine and dissolve it in the reduced orange juice. Add a few drops of lemon juice.

4 Fold the meringue into the juice, then fold in the cream.

5 Line individual rings with the sponge to ¾ height. Place a disc of sponge in the base of each ring. Fill the lined moulds to the top and level off.

6 Chill to set.

7 To make the glaze, heat the stock syrup. Add the gelatine and stir until it dissolves. Add the orange juice and pass through muslin.

8 When the mousses are cold, spoon the glaze over the top. Return to the fridge to set.

9 Remove from the moulds and decorate with orange segments.

45 Blackcurrant delice

Makes	2
Vanilla or jaconde sponge discs	4
Cassis syrup	250 ml
Gelatine, soaked in ice water	10 leaves
Cassis liqueur	66 g
Blackcurrant purée	666 g
Italian meringue	340 g
Whipping cream, ¾ whipped	666 g
Glaze	
Stock syrup at 30° Baumé (equal quantities of sugar and water will give 30° Baumé)	150 ml
Gelatine, soaked in ice water	3 leaves
Blackcurrant purée	150 ml

1 Place two torte rings on a cake board. Place a sponge disc inside each to form a base.

2 Moisten the sponge with cassis syrup.

3 Drain the gelatine. Warm it with the cassis liqueur and one quarter of the blackcurrant purée, until the gelatine has dissolved.

4 Add the rest of the blackcurrant purée. Fold this mixture into the Italian meringue (see Recipe 4).

5 Fold in the whipped cream.

6 Half fill each ring with the mousse mixture. Place a smaller sponge disc on top of the filling and moisten it with syrup.

7 Fill to the top with more mousse. Level the top.

8 Allow to set in the fridge overnight.

9 To make the glaze, warm the syrup. Add the gelatine and stir until it dissolves.

10 Add the blackcurrant purée. Pass through a fine chinois.

11 Carefully mask the top of each delice with the glaze. Allow to set in the fridge.

12 Carefully remove the rings using a blowtorch. Garnish with blackcurrants *en branche*.

Faults

The uneven surface of this mousse was caused by allowing the fruit base mixture with gelatine to over-gel before adding the whipped cream and Italian meringue, thus causing an uneven texture. A second reason could be that the cream was over-whipped.

46 Empress rice (riz imperatrice)

Makes >	12 portions
Topping	
Gelatine, soaked in ice water	3 leaves
Stock syrup (Chapter 3, Recipe 24)	200 ml
Raspberry purée	200 ml
Rice bavarois	
Milk	750 ml
Vanilla pod (or use essence)	½
Carolina rice, washed	85 g
Gelatine, soaked in ice water until limp	5½ leaves
Caster sugar	165 g
Diced glacé fruits, macerated in kirsch	75 g
Whipping cream, ¾ whipped	340 ml
To decorate (suggestions)	
Raspberries	12
Chantilly cream	
Moulded chocolate	

1 To make the topping, drain the gelatine. Boil the stock syrup and add the gelatine and raspberry purée. Stir over ice.

2 Pour the topping into the bottom of individual (7 cm) rings sealed with cling film, to a depth of 3 mm. Set in the fridge.

3 Boil the milk with the vanilla. Rain in the washed rice. Cover and simmer until the rice is tender.

4 Drain the gelatine and mix with the caster sugar.

5 Remove the rice from the heat and add the sugar and gelatine. Mix well.

6 Place in a large bowl and stir over ice. When the mixture begins to gel, fold in the glacé fruits, then the whipped cream.

7 Pour into the moulds and allow to set in the fridge overnight.

8 Remove from moulds onto plates and decorate.

Variation

Maltaise rice: Follow the recipe above with the following changes:
- Use orange jelly instead of raspberry purée for the topping.
- Instead of vanilla, boil two cubes of orange sugar with the juice of 1 orange and 30 ml of Grand Marnier – use this to flavour the milk.
- Use glacé orange instead of mixed fruits.
- Decorate with orange segments.

47 Strawberry Charlotte

Makes >	2 Charlottes
Biscuit à la cuillere	2 strips
Sponge, vanilla or jaconde	2 discs
Kirsch syrup (½ kirsch, ½ syrup)	200 ml
Strawberry mousse (as for blackcurrant, Recipe 45, but using strawberry purée)	
Strawberries, diced	100 g
Whipping cream, half whipped	200 ml

1. Line the inside of two torte rings with a strip of biscuit à la cuillere to ¾ of the height of the ring.

2. Place a disc of sponge inside each ring to form a base. Moisten the sponge with kirsch syrup.

3. Fold the diced strawberries into the strawberry mousse.

4. Half fill each ring with mousse. Place a second disc of sponge on top.

5. Fill to the top of each ring with mousse and level off.

6. Allow to set in the fridge for at least 12 hours.

7. Spread the top with cream. Make a pattern using a serrated knife.

8. Remove the torte ring with the aid of a blow torch.

9. Decorate with crème Chantilly and cut-out chocolate shapes.

Makes >	2 tortes
Biscuit or sponge bases	4
Egg yolks, pasteurised	60 g
Sugar	150 g
Gelatine	3 leaves
Mascarpone cheese	600 g
Double cream	200 g
Coffee syrup	100 ml
Rum	40 ml
Cocoa powder	

1 Cut the biscuit or sponge bases into shape: cut two to the size of the torte ring, and two to the same shape, but slightly smaller.

2 Mix the egg yolks and sugar. Cook over a bain-marie to 75°C, to form a sabayon.

3 Soak the gelatine in iced water. Drain and add it to the sabayon.

4 Beat the cheese well. Add the sabayon.

5 Lightly whip the cream and fold it into the mixture.

6 Place a large biscuit or sponge base into each torte ring, on a board. Soak the base with a mixture of coffee syrup and rum.

7 Half fill each ring with the cheese mixture.

8 Place the smaller circles of biscuit or sponge on top of the filling. Again, soak with syrup and rum.

9 Fill the rest of the ring with the cheese mixture, to a level top.

10 Chill in the fridge overnight.

11 Dust with cocoa powder and decorate with inverted cooked meringue blobs.

49 Raspberry and chocolate truffle cake

Makes >	3 cakes
Chocolate genoise (Chapter 8, Recipe 1)	1 sheet
Mousse	
Dark chocolate, melted	600 g
Caster sugar	300 g
Water	100 ml
Glucose (boiled to 118°C)	50 g
Eggs, pasteurised	200 g
Egg yolks, pasteurised	120 g
Gelatine, soaked in cold water	10 leaves
Raspberry purée	750 ml
Double cream, whipped	900 ml
Glaze	
Stock syrup	250 ml
Raspberry syrup	250 ml
Gelatine	5 leaves

1 To make the mousse, melt the chocolate over a bain-marie. Boil the sugar, water and glucose until it reaches 118°C.

2 Meanwhile put the eggs and yolks in a food mixer and start mixing. When the sugar reaches the required temperature pour over the eggs and whisk until thick and cold. Melt the gelatine in 100 ml of the raspberry purée, strain; add to the egg mixture and take out of the machine.

4 Fold in the chocolate followed by the rest of the raspberry purée and the whipped cream.

5 Line the ring with the sponge, soak with some stock syrup, and fill up with the mousse, level off and freeze.

6 To serve, pour the glaze over the top and leave to set. Turn out with the blow torch.

7 Decorate with chocolate.

Floating islands (îles flottante or oeufs à la neige)

Modern presentation

Classical presentation

Makes >	12 portions
Compote	
Strawberries	500 g
Caster sugar	50 g
Champagne	125 ml
Anglaise	
Vanilla pods, split	3
Double cream	750 ml
Egg yolks	240 ml
Caster sugar	160 g
Poached meringue	
Egg whites	250 ml
Caster sugar	500 g
Lemon juice	2 drops

To make the anglaise

1 Add the split vanilla pods to the cream. Bring to the boil slowly.
2 Whisk the egg yolks and sugar together.
3 Pour half the boiling cream over the egg yolk mixture.
4 Return this to the pan of cream. Cook to 84°C.
5 Pass and chill over ice.

To make the poached meringue

1 Whisk the egg whites, lemon juice and 125 g of the sugar at a medium speed until soft peaks form.
2 Increase the speed and rain in 250 g of the sugar.
3 Fold in the remaining sugar by hand.
4 Pipe or spoon into prepared spherical moulds. Poach in a bain-marie, or steam, until firm. Alternatively, shape into quenelles and lightly poach in a water syrup, finishing with a coating of cold crème anglaise and crushed praline.

To make the compote

1 Cut any large strawberries in half.
2 Place all the fruit in a clean pan and sprinkle with caster sugar.
3 Stir over heat until hot and starting to produce liquid.
4 Douse with Champagne and chill over ice.

To serve (modern presentation)

1 Fill glasses with a layer of compote followed by a layer of anglaise.
2 Top each glass with a meringue.
3 Garnish with pink bubble sugar.

51 Crème brûlée

Makes >	4 portions	10 portions
Milk	125 ml	300 ml
Double cream	125 ml	300 ml
Natural vanilla essence or pod	3–4 drops	7–10 drops
Eggs	2	5
Egg yolk	1	2–3
Caster sugar	25 g	60 g
Demerara sugar		

1 Warm the milk, cream and vanilla essence in a pan.
2 Mix the eggs, egg yolk and caster sugar in a basin and add the warm milk. Stir well and pass through a fine strainer.
3 Pour the cream into individual dishes and place them into a tray half-filled with hot water (bain marie).
4 Place in the oven at approx. 160°C for about 30–40 minutes, until set.
5 Sprinkle the tops with demerara sugar and glaze under the salamander or by blowtorch to a golden brown.
6 Clean the dishes and serve.

Try something different

Sliced strawberries, raspberries or other fruits (such as peaches, apricots) may be placed in the bottom of the dish before adding the cream mixture, or placed on top after the creams are caramelised.

A bain marie is not always used. As an alternative, use a shallow mould and cook at 100°C.

52 Vanilla crème brulee (gastro tray method for large-scale production)

Makes >	16 portions
Whole milk	450 ml
Double cream	700 ml
Vanilla pods	2
Egg yolks	20
Caster sugar	180 g

1 Bring the milk, cream and split vanilla pods to the boil.
2 Leave to infuse for 15 minutes.
3 Re-warm, then pour onto the egg yolks and sugar. Whisk well then strain into a gastro tray.
4 Bake at 100°C until set.
5 Remove, then blend with a hand blender, ensuring not to create bubbles.
6 Using a sauce dispenser, fill into bowls, and leave to set.
7 Sprinkle surface with caster sugar and caramelise using a blowtorch.

53   Unmoulded crème brulée with spiced fruit compote

Makes >	16 portions
Crème brulée	
Whipping cream	750 g
Milk	250 g
Vanilla pods	2
Egg yolks, fresh	300 g
Sugar	150 g
Demerara sugar	
Spiced fruit compote	
Plums	500 g
Peaches	500 g
Star anise	2
Vanilla pods	2
Cinnamon sticks	2
Cloves	2
Caster sugar	100 g
Red wine	150 ml
Stock syrup (see Chapter 3, Recipe 24)	100 ml

To make the crème brulée

1 Prepare individual stainless steel rings: cover the bases with 2 layers of cling film and bake them to seal.

2 Boil the cream, milk and vanilla pods.

3 Whisk the egg yolks and sugar together.

4 Pour the boiling liquid onto the egg mixture. Pass, then stir well.

5 Using a dropper, pour into prepared individual rings. Bake in a fan oven at 100°C for 30 minutes.

6 Remove from moulds onto plates.

7 Sprinkle demerara sugar on top and caramelise with a blow torch. Place slices of dried strawberries below the caramelised sugar, to decorate.

To make the compote

1 Cut up the fruit and break up the spices.

2 Spread the fruit and spices over a roasting tray. Sprinkle with caster sugar and three-quarters of the wine.

3 Roast in a moderate oven until the fruit starts to soften.

4 Carefully remove the fruit. Deglaze the pan with more wine and a little stock syrup.

5 Pass the sauce back onto the fruit. Chill.

54 Crème beau rivage

Makes >	10 portions
Crème renversée	
Whole egg	250 g
Caster sugar	50g
Vanilla essence or pod	4 drops or 1 pod
Milk	500 ml
Praline cream	
Whipping cream	200 ml
Crushed praline	80 g

For the crème renversée

1 Whisk together the whole egg, sugar and vanilla essence.

2 Boil milk and pour onto the eggs, chill.

3 Pour into a savarin mould lightly greased with melted butter and coated with crushed praline (see Chapter 10, Recipe 22)

4 Place in a bain-marie of hot water and bake at 150°C until the surface of the custard feels firm.

5 Chill down and refrigerate for 12 hours.

6 Dip the mould into hot water and turn out onto a round serving dish.

7 Decorate with cornets made from cigarette paste (see Chapter 9, Recipe 8) filled with praline cream.

For the praline cream

1 Lightly whip the cream.

2 Fold through the sieved crushed praline.

Professional tip

Crushed praline is very hygroscopic (it attracts moisture from the atmosphere) so it needs to be stored in airtight containers to prevent it sticking together.

55 Gateau MacMahon

Makes >	10 portions
Almond sweet paste (sablé paste)	
Butter	300 g
Icing sugar	150 g
Caster sugar	62 g
Ground almonds	62 g
Vanilla essence	
Soft flour	125 g
Eggs	2
Soft flour	375 g
Strawberry bavarois	
Strawberry purée	500 ml
Gelatine	7 leaves
Lemon juice	
Lightly whipped cream	500 ml
Caster sugar	180 g

1 For the sablé paste, cream together the butter, sugars, almonds and vanilla essence using a paddle beater.

2 Add 125g of soft flour and the eggs.

3 Take off the machine and add 375g of soft flour; do not over-work.

4 Chill the pastry in the refrigerator.

5 Roll out two discs the size of a torte ring 25cm diameter and 5 cm high; using the torte ring cut out into two circles on the baking tray.

6 Mark one disc into eight sections and lightly bake both.

7 Once baked cut the marked disc into sections and place the second inside the torte hoop ring.

8 Place the torte hoop ring in the deep freeze.

9 Prepare the strawberry bavarois, fill the torte hoop to the top and level off giving a flat and smooth surface. Place in the refrigerator.

10 Dust four of the cut sablé shapes with icing sugar and coat the other four with red glazing jelly.

11 Remove the torte ring with the aid of a blow torch.

12 Neatly arrange the sablé pastry on the top of the bavarois, alternating white (dusted with icing sugar) and red (spread with glazing jelly).

For the bavarois

1 Heat the fruit purée with the sugar, do not boil.

2 Add the pre-soaked gelatine and lemon juice.

3 Partially set over ice.

4 Lightly fold in the whipped cream.

5 Deposit into the mould.

Professional tip

When pouring gelatine mousses and bavarois into moulds lined on the bottom with sponge, it is good working practice to place them in the freezer on the tray. By doing this it helps to prevent any leaks.

56 Petits pots au chocolat

Makes >	8 portions
Milk	500 ml
Plain couverture	200 g
Egg yolks	3
Whole eggs	2
Caster sugar	75 g

1 Boil the milk and emulsify onto the couverture using a stem blender.

2 Whip the eggs and sugar together, pour on the chocolate milk.

3 Pour into small pots or white egg cups.

4 Bake in a bain-marie at 150°C until set.

5 Chill down and refrigerate.

6 Serve topped with a compote of cherries, raspberries and a small quenelle of whipped vanilla cream.

Note

Recipes 56 to 59 are examples of pre-desserts. Pre-desserts are small portions or miniature versions of desserts from the main menu. They are served in high-class restaurants, normally before the main dessert, and may be hot or cold.

57 Passion fruit and white chocolate posset with Champagne jelly

Makes >	10 portions
Champagne jelly	
Stock syrup (made from equal quantities of sugar and water)	200 ml
Gelatine leaves	4
Pink Champagne	200 ml
Posset	
Whipping cream	400 g
Caster sugar	100 g
Lemons	3
White couverture	150 g
Passion fruits	2

For the jelly

1 Boil the stock syrup, then add pre-soaked gelatine.

2 Pour onto the Champagne and pour into the base of cold martini glasses. Allow to set in the refrigerator.

For the posset

1 Boil the cream with the sugar, lemon zest and leave to infuse for ten minutes.

2 Re-boil and strain onto the white chocolate, emulsify with the stem blender.

3 Add the lemon juice and pulp of the passion fruits.

4 Cool down and pour onto the set Champagne jelly.

5 Decorate with passion fruit purée and chocolate cut-out.

58 Honey chocolate panna cotta, banana and passion fruit jam with almond streusel

Makes >	10 portions
Honey chocolate panna cotta	
Milk	250 ml
Glucose	25 g
Gelatine	15 g
Honey-flavoured milk couverture	425 g
Whipping cream	500 ml
Banana and passion fruit jam	
Sugar	300 g
Passion fruit purée	100 g
Banana purée	200 g
Diced banana	300 g
Almond streusel	
Butter	100 g
Brown sugar	100 g
Soft flour	100 g
Ground almonds	100 g

For the panna cotta

1 Boil the milk with the glucose.

2 Add the gelatine.

3 Partly melt the chocolate in the microwave.

4 Pour the milk onto the chocolate and emulsify with a stem blender.

5 Add the cold cream and stem blend.

6 Store in the refrigerator for further use.

For the banana and passion fruit jam

1 Make a direct caramel with the sugar.

2 Add the hot passion fruit purée and then the banana purée.

3 Cook so it is well blended then take off the heat and add the diced banana.

For the almond streusel

1 Cut the butter into small cubes. Sieve the dry ingredients. Add the butter and mix using the paddle until homogenous.

2 Refrigerate.

3 Push the paste through a large drum sieve 4 mm wide to create regular size pieces.

4 Refrigerate or freeze for further use.

5 Bake at 150–160°C until golden.

Assembly of the dish

1 Spoon the banana and passion fruit jam into the bottom of a martini glass and freeze.

2 Pour over the panna cotta and refrigerate to set.

3 Dress the top with the almond streusel and chocolate decoration (see Chapter 10 on decorative items).

59   Snow egg with Cointreau cream

Makes >	10 portions
Snow egg	
Egg whites	8
Cream of tartar	pinch
Caster sugar	175 g
Cointreau cream	
Milk	250 ml
Whipping cream	250 ml
Egg yolks	6
Caster sugar	50 g
Cointreau concentrate	25 g

For the snow egg

1 Whisk the egg whites with the cream of tartar to firm peaks.

2 Gradually add the caster sugar in a steady stream.

3 Continue to whisk until the meringue leaves upright peaks.

4 Transfer the meringue into a piping bag and pipe into a buttered 3 cm metal ring.

5 Steam the meringue for 3 minutes until the meringue is cooked through.

6 Chill the meringue down, then trim off the top of the meringue with a thin bladed knife to form a perfect cylinder shape.

7 Remove the metal ring.

For the Cointreau cream

1 Bring the cream and milk to the boil.

2 Whisk the egg yolks and sugar together.

3 Pour half the boiling cream and milk over the egg yolk mixture, mix.

4 Return the egg mixture to the remaining cream and milk and cook to 84°C until the mixture coats the back of a wooden spoon.

5 Once cooked, pass through a chinois into a clean bowl and add the Cointreau.

6 Chill over ice and refrigerate.

To serve

1 Place the steamed cylinder of meringue into a dish.

2 Coat with the chilled Cointreau cream.

3 Sprinkle with crushed praline.

Professional tip

- The use of cream of tartar gives a more stable meringue and helps to prevent graining.
- Always scald the bowl and whisk in hot water to remove any traces of fat. If there is any fat present it will have an effect on the foaming properties of the egg white, resulting in runny meringue.

This is a pre-dessert. For the full-sized version, see Recipe 50.

Test yourself

Level 2

1 Why is important to bake a crème caramel in a bain-marie of water at a low temperature?

2 Name two different bases that can be used for a bavarois.

3 Name two cold desserts that include caramelised sugar.

Level 3

4 Explain the difference between a water ice and a sorbet.

5 Name two pieces of equipment that can be used to measure the sugar concentration in syrups.

6 Name four gelatine-based desserts, giving a brief description of each.

7 Name three iced confections that can be made without using a sorbetiere.

8 Name the legislation that governs ice cream manufacture. What are the key points of hygiene control in this legislation?

8 Biscuit, cake and sponge products

This chapter covers:
→ **NVQ Level 2** Prepare, cook and finish basic cakes, sponges and scones
→ **VRQ Level 2** Produce biscuit, cake and sponge products
→ **NVQ Level 3** Prepare, cook and finish complex cakes, sponges, biscuits and scones
→ **VRQ Level 3** Produce biscuits, cakes and sponges.

In this chapter you will:
→ Prepare and cook biscuit, cake and sponge products using correct techniques and equipment, and safe and hygienic practices (level 2)
→ Prepare and cook biscuit, cake and sponge products to the recipe specifications, in line with current professional practice (level 3).

Recipes in this chapter

Introduction

Cakes are often associated with special occasions such as a birthday, wedding, Christmas or a Sunday tea time. They are not desserts to be eaten after a main course but more often with a cup of tea mid-afternoon.

Virtually all types of cake and sponge are made using butter, eggs, sugar and soft (low gluten) flour – the difference between them is in the proportions of ingredients used. For example a pound cake is so named because it uses equal amounts of the main ingredients (1 lb butter, 1 lb eggs, etc) – this will produce a denser, firmer texture than found in a sponge which contains more eggs and far less butter and flour, giving a much lighter product. Air is an essential ingredient in both, and this is achieved by whisking eggs and sugar (sponges) or beating fat with sugar (cakes).

Some cakes may need additional help incorporating air in the form of baking powder or bicarbonate of soda (depending on what the recipe states). Both of these will produce carbon dioxide gas when brought into contact with heat and moisture (see 'Chemical aeration' below).

Successfully making a sponge is all about damage limitation. Warmed eggs whisked with sugar will create a stable foam, but if the flour is not added carefully (see 'Folding in' below) the air will be knocked out and the volume lost, resulting in a heavy close texture. Sponges are often moistened before filling and not intended to be eaten on their own.

The word 'biscuit' means twice cooked (bis-cuit) and dates back to a time when the idea was to make them last as long as possible – for example, ship's biscuits, which had to stay edible over many weeks or months. These biscuits were made by baking, slicing and baking again so as to remove all moisture and prevent the growth of bacteria – an early, but usually unsuccessful attempt at convenience food. Ship's biscuit or hardtack was notorious for being so hard and inedible that without the worms with which they were infested making lots of small holes, they would have been impossible to break. It is said most sailors preferred to eat them at night so they couldn't see what they were eating.

Techniques

Whisking

When using this technique the idea is to physically add air into eggs (for some recipes these are separated) and sugar. The first rule to remember is that cold eggs will not aerate very well, so it is necessary to either heat the sugar in the oven or warm the eggs and sugar over hot water (not boiling). Second, it is best to whisk at a medium speed – vigorous whisking will create larger but more unstable air bubbles which will burst more easily, thus providing less volume and a heavier result. When a stable foam has been achieved this is called the 'ribbon stage' (the test is to lift up the whisk and trail a figure of eight on top). After this the dry ingredients are carefully added (for example, folded in).

Beating or creaming

This is sometimes referred to as the sugar batter method.

This technique also involves physically adding air, this time into fat and sugar. Like eggs, cold fat will not aerate successfully. As the fat will more than likely have come from the fridge, cut into small even pieces, add the sugar and mix together, soften the fat/sugar over hot water, and beat the mixture. The finished mixture should increase in volume by approximately 40 per cent and will be lighter in colour. The eggs (which should be at room temperature) are then beaten into the aerated fat and sugar in two or three lots, and finally the dry ingredients are folded in.

Folding in

The final stage for both of the techniques above is to add the dry ingredients by 'folding in'. The aim is to mix in the dry ingredients while losing as little volume as possible. This is achieved by 'cutting in' with a rubber spatula (large mixes are best done using one hand), 'folding' and turning the bowl at the same time.

The all-in-one

As suggested, this method consists of adding all the ingredients together and mixing them. Sometimes, the fat used in these recipes is oil and they depend on chemical raising agents to give them lift as they are not physically aerated. Products made using this method tend to have less volume and a denser texture, but should be moist in the centre.

Flour-batter

This method is a combination of whisking and creaming. It involves whisking the eggs and sugar, creaming the fat and half the flour, and then folding everything together.

Chemical aeration

If cakes are made using equal quantities of ingredients, as illustrated by the pound cake, as long as the ratios are correct and properly aerated the ingredients will not need any extra help. However, this can be an expensive way of producing cakes, so to reduce costs they can be bulked out by adding more of the less expensive ingredients (flour and sugar). In these cases, because the flour has been increased they will then need more moisture (usually milk) and will also need extra help to rise. This is where baking powder and bicarbonate of soda come in. Heat from the oven and moisture in the cake will allow these raising agents to produce carbon dioxide gas, thus helping the cake to rise.

The above techniques are essential for producing a good result, but it should be remembered that while these cover the main principles, individual recipes will vary and may not be exactly consistent with the techniques above.

The main ingredients

Eggs

All recipes are based on a medium size three egg, which should of course be fresh. Free range eggs are thought to contain a stronger albumen and a deeper coloured yolk. Whole eggs, yolks and whites pasteurised by passing them through an infra-red beam are now available in cartons and are measured by weight or volume. For recipes made using the creaming method, the eggs should be at room temperature to avoid splitting the mixture.

Sugar

Sugar obviously sweetens, but it also plays a vital role in the aeration of cakes and sponges. Sugar enables fat and eggs to retain more air when beaten or whisked.

For most recipes, caster sugar is used, occasionally icing sugar, but not granulated sugar because it can give a gritty texture. Icing sugar is used to finish several products (dusting). Brown sugars like muscovado or demerara will add caramel flavours and colour to products and are used in some cake recipes, but they do not aerate well and should not be used in whisked egg sponges.

Fat

There are special cake fats available which are neutral in flavour. These will aerate better than traditional fats, but are more commonly used for commercial large volume production (high-ratio cakes). Oil is used in some recipes, but is not a substitute for solid fat. Where a solid fat is required the recipes in this chapter use butter because it will give a better flavour.

Flour

The types of flour used in this chapter are soft flour (low gluten), self-raising (medium strength plus the addition of a raising agent) and plain (medium strength). Weaker (low gluten) flours are always used in sponges and give the products a lighter, shorter texture. Flour should always be sieved to remove any foreign bodies and to aerate it, helping it to disperse and mix more easily with other ingredients. It is important to store flour in a cool, dry environment, protected from contamination.

Baking powders

Bicarbonate of soda or baking soda can be used as a raising agent only if the other ingredients include an acid, such as those found in citrus juices or buttermilk. Without an acid there would be no chemical reaction and consequently no lift. The acid also helps to counteract any unpleasant aftertaste.

Baking powder, of which there are several different kinds, is a complete raising agent which already contains an acid. It consists of one part alkali (bicarbonate of soda) to two parts acid (usually cream of tartar). When hydrated and heated, baking powder will release CO_2 causing the products to expand and rise.

Always add baking powder to the flour and then sieve twice onto paper so it is evenly distributed.

Do not buy in large quantities unless using a lot. Baking powder becomes less effective if it is left on the shelf for too long.

Chocolate

Products made with chocolate will only be as good as the quality of the chocolate used. Never use a compound or baker's chocolate – they will contain a low percentage of cocoa solids and too much sugar and hydrogenated vegetable fats. The chocolate (couverture) recommended is one where the majority of the fat is cocoa butter with a minimum of 60 per cent cocoa mass.

Combinations

Flavours and textures should be compatible, complimentary and provide contrast. Opera is a good example – chocolate, coffee and rum all work well together, as do the spices with red wine in a Christmas cake, or strawberries with kirsch or an orange liqueur. An effective combination of textures is illustrated by the Himmel cheese torte in Recipe 18 which contrasts a light shortcake with a soft rich filling, offset by the acidity of the raspberries.

Essential equipment

Equipment used for these products includes:
- planetary mixers (three-speed electric mixers)
- digital scales
- food processor
- fine-mesh sieves
- fine-wire balloon whisks
- range of Teflon®-coated baking tins (loaf, genoise, cake, etc.)
- plain and fluted pastry cutters
- heavy-duty baking sheets (small, medium and large)
- silicone baking mats
- disposable piping bags and range of plain and star piping tubes
- cooling racks
- range of small equipment such as stainless steel bowls, rubber and plastic spatulas, palette knives, scissors etc.

Key points

- As with all pastry work, always weigh and measure accurately.
- All equipment must be prepared and ready before the final mixing. Likewise, make sure an oven is set at the correct temperature and available. In particular, sponges need to be baked immediately and will lose volume if left standing.
- The fat used to grease the tins or moulds should be soft (never melted), so it does not run down the sides of the cake tin and collect in a puddle on the bottom.
- The shape, volume and density of the products will affect the baking time and temperature.
- Allow to cool slightly before removing from moulds, but do not leave for too long in the tins or they will 'sweat'.
- All ovens behave differently. What is the correct temperature for one may be higher or lower for another; get to know your oven and make a note of the correct temperature and top/bottom heat settings (if these are a feature).

1 Genoise nature (plain sponge)

Makes >	1 × 16 cm sponge	3 × 16 cm sponges
Eggs	4	12
Caster sugar	100 g	300 g
Soft flour	100 g	300 g
Melted butter (not hot)	25 g	75 g

1 Make sure the mixing bowl is clean, dry and has no traces of grease.

2 Prepare genoise tins by brushing with soft butter and flouring (bang out the excess). Set the oven at 185°C.

3 Place the eggs and sugar in the mixing bowl, stir over hot water until warm.

4 Whisk to the 'ribbon' stage.

5 Sieve the flour and any other dry ingredients onto paper and melt the butter.

6 Carefully fold in the flour.

7 Finally take out a small amount of mixture and add to the butter and mix, add this back to the rest of the mixture and fold in.

8 Deposit equal amounts of mixture into the tins, level the top and place in oven immediately for approximately 25 minutes.

9 When cooked the sponge should spring back when gently pressed.

10 Cool slightly then turn upside down to remove from tins.

11 When cold wrap in cling film, label, date and store in the freezer until needed.

Video: genoise,
http://bit.ly/19YsamX

Whisk the eggs and sugar together over hot water

Carry on whisking as the mixture warms up

When the mixture is ready, it will form ribbons and you can draw a figure eight with it (at this stage it is known as a sabayon)

Fold in the flour

Add part of the flour mixture to the butter

After baking, turn upside down to remove from tins

Variations

- Chocolate genoise – replace 50 per cent of the flour with 25 per cent cocoa powder and 25 per cent cornflour.
- Coffee genoise – add some strong coffee essence to the foam just before folding in the flour.

Professional tips

- Taking out some of the mixture to add to the butter will prevent it from 'dropping' through, allowing it to be mixed in more easily and minimising volume loss.
- A fatless sponge can be made exactly as above but leaving out the butter. The shelf life will be shorter.
- Genoise is best made the day before it is needed, as it can be cut much more easily.

Faults

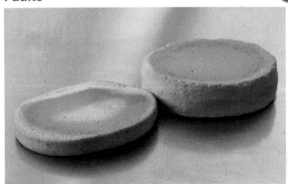

The smaller cake has a low volume and a heavy texture because it was overmixed

There are three reasons why a genoise sponge might have a low volume and a heavy texture:

1 Weak foam – not whisked enough, ingredients not warmed first, traces of fat in the bowl.
2 Air knocked out when adding dry ingredients, dry ingredients not folded in correctly.
3 Left standing before baking – lack of preparation, moulds not prepared, oven not set at the correct temperature.

Note

Genoise sponges should be light and springy, golden brown with an even texture and a fine crumb.

2 Roulade sponge

Makes >	2 sheets	4 sheets
Eggs	8 (450 ml)	16 (900 ml)
Egg yolk	2 (40 ml)	4 (80 ml)
Caster sugar	260 g	520 g
Soft flour	170 g	340 g

1 Make sure the mixing bowl is clean, dry and free from grease.

2 Line the baking sheets with silicone paper cut to fit.

3 Set the oven at 230°C.

4 In a mixing bowl place the eggs and sugar, stir over hot water until warm.

5 Whisk to the 'ribbon' stage and sieve the flour onto greaseproof paper.

6 Carefully 'fold in' the flour.

7 Divide equally between the baking sheets and spread evenly with a drop blade palette knife, place immediately in the oven for between 5 and 7 minutes.

8 Immediately the sponge is cooked, turn out onto sugared paper, place a damp clean cloth over it and lay the hot baking sheet back on top, leave to cool (this will help keep the sponge moist and flexible as it cools).

Try something different

A chocolate version can be made by substituting 30–40 per cent of the flour for cocoa powder. For a coffee sponge, add coffee extract to the eggs after whisking.

A roulade sponge may also be made using a 'split-egg' method, separating the eggs.

Professional tip

Each sheet should be left on the paper on which it is cooked, individually wrapped, labelled, kept flat and stored in the freezer to stop it from drying out and losing flexibility.

You should be able to bend a roulade sponge

Faults

There are two reasons why a roulade sponge might become hard and crisp, instead of being pliable:

1 Baked at too low a temperature for too long.

2 Mixture spread too thin.

3 | Swiss roll

Makes >	2 16 cm × 24 cm Swiss rolls	3 16 cm × 24 cm Swiss rolls
Eggs	8	12
Caster sugar	220 g	330 g
Self-raising flour	80 g	120 g
Cornflour	30 g	45 g

1 Follow steps 1–8 of the recipe for roulade sponge.

2 Allow to cool slightly before carefully removing paper and spread with warmed jam, immediately roll up as tight as possible, and leave to cool completely.

4 | Flourless chocolate sponge

Makes >	10 portions
Egg yolks	230 g
Sugar	180 g
Inverted sugar	30 g
Cocoa powder	100 g
Cornflour	50 g
Egg whites	300 g
Cream of tartar	
Sugar	150 g

1 Mix together the egg yolks, inverted sugar and 180 g of sugar.

2 Whisk until well aerated.

3 Sieve the cocoa powder and cornflour.

4 Whisk the egg whites with cream of tartar, using a little sugar at the beginning of the process and the rest at the end.

5 Transfer the egg yolk sabayon to a mixing bowl. Using a whisk, beat in half of the meringue, then fold in the cocoa powder and cornflour. Fold in the other half of the meringue.

6 Pipe out in round discs and bake at 230°C for 10–15 minutes.

Patterned biscuit jaconde

This sponge is used as a lining for desserts such as charlottes and petit gateau. The term 'biscuit' is used whenever egg whites and yolks are whisked separately or, as in this case, additional egg whites are whisked and added.

Makes >	4 medium baking sheets
Decorating paste (coloured)	
Melted butter	100 g
Icing sugar	100 g
Egg whites	50 g
Soft flour (sieved)	50 g
Colour	
Or chocolate decorating paste	
Melted butter	80 g
Icing sugar	80 g
Egg whites	80 g
Soft flour	60 g
Cocoa powder (sieved)	30 g

Biscuit jaconde	
Soft flour	100 g
Ground almonds	250 g
Melted butter	75 g
Eggs	500 ml
Icing sugar	375 g
Egg whites	375 ml
Caster sugar	50 g

For the paste

1 Place all ingredients in a bowl and mix to a paste.

2 Spread thinly onto a silicone baking mat and using a comb scraper make a pattern.

3 Set in the freezer.

For the sponge

1 Set the oven at 230°C.

2 Sieve the flour and ground almonds onto paper and melt the butter.

3 After first warming, whisk the eggs and icing sugar to the ribbon stage.

4 Separately whisk the egg whites and caster sugar to a meringue.

5 Fold the meringue into the eggs and sugar, then fold in the dry ingredients.

6 Finally add a small amount of mixture to the butter and mix, add this back to the rest and fold in.

7 Divide between the patterned trays and spread evenly.

8 Place in oven for approximately 6 minutes.

9 Immediately the sponge is cooked, turn out onto sugared paper, place a damp clean cloth over it and lay the hot baking sheet back on top.

10 When cooled, carefully peel back the mat.

11 If not using immediately, wrap individual sheets in cling film, label and freeze, making sure they are laid flat.

Professional tip

It is usual to pattern the paste with a comb scraper to create stripes or waves, but an effective alternative can be achieved by simply making circular movements with your finger or the handle of a spoon.

Note

Jaconde sponges should be light and flexible, they should not be too thick and have a clearly defined pattern.

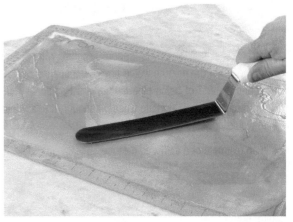

Spread the decorating paste

Make the pattern, then freeze

Spread a layer of sponge over the frozen pattern

6 | Sponge fingers (biscuits à la cuillère)

The literal translation of this is 'spoon biscuits', which comes from a time when the mixture would have been shaped between two spoons instead of being piped. They are traditionally used to line the mould for a Charlotte Russe, although the 'spooned' version would not lend itself to that.

This is an example of a split-egg method.

Makes >	Approx 60 × 8cm fingers
Egg yolks	180 g
Caster sugar	125 g
Vanilla essence	few drops
Soft flour	125 g
Cornflour	125 g
Egg whites	270 g
Caster sugar	125 g

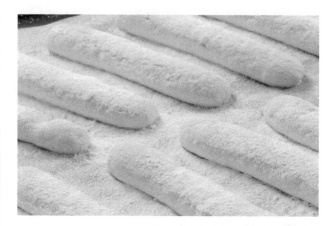

1 Prepare a baking sheet by lining with silicone paper cut to fit, and set the oven at 160°C. Have ready a piping bag fitted with a medium plain tube. Scald two mixing bowls to ensure they are clean and grease free.

2 Whisk the yolks, sugar and vanilla over a bain-marie until warm, continue whisking off the heat until a thick sabayon-like consistency is reached.

3 Sieve the flours onto paper.

4 In a second mixing bowl whisk the whites with the sugar to a soft meringue.

5 Add the whisked yolks, to the meringue and start folding in, add the flour in 2 or 3 portions, working quickly but taking care not to overwork the mixture.

6 Using a plain piping tube immediately pipe onto the prepared baking sheet in neat rows.

7 Dust evenly with icing sugar and place straight in the oven for approximately 25 minutes.

8 When cooked slide the paper (and biscuits) onto a cooling rack.

9 When cool remove from paper and store in an airtight container, at room temperature or leave on the paper and store in a dry cabinet.

Whisk the egg yolks and sugar to a sabayon

Whip the egg whites until they are firm

Add some egg whites to the sugar mixture

Mix lightly and pour back into the egg white container

Fold in the sieved flour

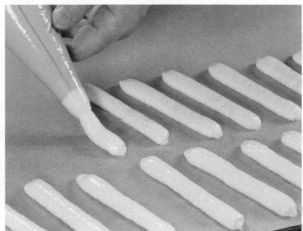

Pipe out in finger shapes

Professional tips

- It is easier to pipe the fingers all the same length if a template marked with parallel lines is placed under the silicone paper.
- This mixture will collapse more easily than most so it is crucial to work quickly once mixed.

Variation

For a chocolate version, instead of 125 g each of soft flour and cornflour use 120 g soft flour, 60 g cornflour and 70 g cocoa powder.

To make **Othellos**, use the above recipe to make small, domed sponges. Hollow them out and fill them with crème mousseline. Sandwich pairs together and coat with coloured fondant.

Note

A common problem with this recipe is over-mixing and/or not working fast enough or being disorganised, which results in biscuits which collapse.

Sponge fingers should be pale in colour, very light in texture, be dusted with icing sugar and have a rounded shape. They should also be identical in length and width.

7 Strawberry gateau

This is based on the classic French gateau 'le fraisier'.

	Makes >	1 Gateau (12 portions)
Plain genoise sponge (Recipe 1)		1 × 16 cm
Stock syrup (Chapter 3, Recipe 24) flavoured with Grand Marnier or kirsch		50 ml
Strawberry jam		50 g
Fresh strawberries		125 g
Crème pâtissière (Chapter 3, Recipe 12)		400 g
Gelatine		5 leaves
Double cream		800 ml
Grand Marnier or kirsch		80 ml
Pink colour		
Marzipan		80 g

1 Split the genoise equally, place the base on a cake board and moisten with syrup, spread with a thin layer of strawberry jam.

2 Reserve 3 or 4 strawberries for decoration, hull the rest and cut in half.

3 Place a deep stainless steel ring over the sponge and stand the halved strawberries all the way around, cut side against the ring.

4 To make the filling, beat the crème pâtissière until smooth, soak the gelatine and whip the cream.

5 Squeeze the gelatine, add to the liquor, and heat to dissolve, pass onto the crème pâtissière and beat vigorously to blend in. Finally, fold in the cream.

6 Place the filling in a piping bag and pipe into the ring almost to the top then place on the other half of the sponge and moisten with syrup. Press down evenly, set in the fridge.

7 Colour the marzipan and roll out 2–3mm thick, cut to the same size as the gateau and lay on the top (it is a good idea to spread a little jam on top of the sponge so the marzipan sticks).

8 Using melted chocolate or royal icing pipe on a decorative border.

9 Decorate with the reserved strawberries which have been dipped in crack sugar.

Professional tips

- The marzipan is traditionally piped with the words 'le fraisier' and coloured green.
- When soaking gelatine, always use iced water to prevent the sheets from breaking up.
- The finish on the marzipan in the photograph has been achieved using a patterned rolling pin, but alternatively it could be marked using the back of a knife or the edge marked with marzipan crimpers.

Try something different

If preferred this dish could be filled with Chantilly cream or crème mousseline (see Chapter 3, Recipes 11 and 13).

Note

A good-quality strawberry gateau should have a moist sponge and a good balance between sponge and filling. The filling should be smooth and lump free, the strawberries even sized and the decoration (piping) restrained and tasteful.

8 Coffee gateau

	Makes >	1 × 16 cm gateau
Plain genoise sponge (Recipe 1)		1 × 16 cm
Stock syrup (Chapter 3, Recipe 24) flavoured with rum		50 ml
Coffee buttercream (Chapter 3, Recipe 17)		750 g
Coffee marzipan		100 g
Fondant		500 g
Crystallised violets		
Chocolate squares		

1 Carefully split the sponge into three, line up the three pieces.

2 Place the sponge base on a cake card and moisten with rum syrup.

3 Pipe on an even layer of buttercream, no thicker than that of the sponge.

4 Place on the next layer of sponge, moisten with the syrup and repeat to give three layers of sponge and two of buttercream, moisten the top with syrup.

5 Put in the fridge for 1–2 hours to firm up.

6 Work some coffee essence into the marzipan, roll out to 2 mm thick and lay over the gateau, work the sides to prevent any creases.

7 Warm the fondant to blood heat, flavour with coffee essence and adjust consistency with syrup.

8 Place the gateau on a wire rack with a tray underneath to catch the fondant.

9 Starting in the centre and moving outwards, pour over the fondant to completely cover, draw a palette knife across the top to remove the excess.

10 Add some melted chocolate to some of the fondant, adjust the consistency and squeeze through muslin, decorate gateau by piping on a fine line design.

11 Finish the sides with squares of chocolate and the top with crystallised violets.

Professional tips

- Mark the sponge by cutting a 'v' on the side before splitting horizontally; when re-assembling, line up the marks so it goes back together exactly as it came apart.
- Turn the sponge upside down before splitting so the base becomes the top, this is the flattest surface and will give the best finish.
- It is best practice to use a genoise that was made the day before – fresh sponges do not cut well and are susceptible to falling apart.
- Fondant should never be heated above 30°C, as the shine will be lost.

Try something different

- Instead of enrobing with fondant, the top and sides can be covered with buttercream, the sides can be either comb-scraped or masked with toasted nibbed/flaked almonds or grated chocolate. The top can be piped with buttercream and/or decorated with coffee marzipan cut-out shapes.
- To add another texture, place a disc of meringue or dacquoise on the bottom layer. Dacquoise is an Italian meringue with the addition of toasted ground hazelnuts spread or piped onto a silicone mat and baked at 180°C for 15–20 minutes. Cut out the desired shape half way through cooking.

Note

A good-quality coffee gateau should have a moist sponge and a good balance between sponge and filling (as a guide the thickness of the sponge and the depth of the buttercream should be equal). The coffee flavour should not be in question, and the decoration should reflect and compliment the coffee theme. (It is sometimes easy to get carried away, so it is good to remember when decorating, 'less is definitely more'.)

Chocolate gateau

Makes >	1 × 16 cm gateau
Chocolate genoise sponge (Recipe 1)	1 × 16 cm
Stock syrup (Chapter 3, Recipe 24) flavoured with rum	50 ml
Chocolate buttercream (Chapter 3, Recipe 17)	750 g
Apricot jam	120 g
Chocolate glaze (Chapter 3, Recipe 40)	500 ml

1 Carefully split the sponge into three, line up the three pieces (the top is usually the bottom when it was baked as this will be the flattest surface).

2 Place the sponge base on a cake card and moisten with rum syrup.

3 Pipe on an even layer of buttercream no thicker than that of the sponge.

4 Place on the next layer of sponge and repeat to give three layers of sponge and two of buttercream, moisten the top with syrup.

5 Put in the fridge for 1–2 hours to firm up.

6 Beat the apricot jam and rub through a sieve, then carefully spread over the top and sides of the gateau, to give a protective layer.

7 Pour over the glaze as in the previous recipe, take off the excess with a palette knife.

8 Place back in the fridge to set.

9 Decorate by spreading a strip of patterned acetate with dark chocolate and placing around the sides when the chocolate is at setting point.

10 The top can be decorated with moulded chocolate.

Variations

Instead of covering with chocolate glaze, the top and sides can be masked with the buttercream, the sides comb scraped or left plain and part or completely covered with chocolate flakes, and the top with piped buttercream and/or decorated with chocolate cigarettes.

Note

The fillings used in this and the previous recipe both use buttercream made with eggs, boiled sugar and unsalted butter. This is not to be confused with butter icing, which is a much heavier product.

The two gateau recipes above would usually be served with afternoon tea and not eaten as a dessert.

10 Cupcakes

Cupcakes have become hugely popular in recent years, not least because of their versatility and variety.

Makes >	12 vanilla cupcakes
Plain flour	120 g
Baking powder	1 ½ tsps
Pinch of salt	
Unsalted butter (soft)	40 g
Caster sugar	140 g
Whole milk	120 ml
Egg	1
Vanilla extract	¼ tsp

1 Sieve the flour, baking powder and salt into a bowl, add the butter and sugar.

2 Mix to a sandy consistency and gradually add ½ the milk.

3 Whisk the eggs, vanilla and rest of the milk then pour into the flour, etc.

4 Continue beating until smooth.

5 Spoon into the paper cases until ⅔ full.

6 Bake at 170°C for 20–25 minutes, test with a skewer.

7 Cool on a wire rack.

8 Coat with vanilla butter icing and sprinkle with hundreds and thousands.

Variations

● An alternative topping to butter icing can be made by replacing 70 per cent of the butter with cream cheese.

● Cupcakes can be decorated with cut-out shapes made from sugar paste.

● **Coconut cupcakes:** Replace 50 g flour with desiccated coconut and use coconut milk instead of whole milk.

● **Lemon cupcakes:** Replace vanilla extract with the finely grated zest of a lemon and hollow out a bit of the sponge and put in a small amount of lemon curd.

● **Chocolate cupcakes:** Replace 20 g of the flour with cocoa powder, and use some chocolate spread to flavour the butter icing.

11 Rich fruit cake

In the UK fruit cakes are traditionally used as a base for celebration cakes such as Simnel, Christmas and for weddings.

Makes >	1 × 16 cm	1 × 21 cm	1 × 26 cm
Butter	150 g	200 g	300 g
Soft brown sugar	150 g	200 g	300 g
Eggs	4	6	8
Black treacle	2 tsp	3 tsp	1 tbsp
Soft flour	125 g	175 g	275 g
Salt	6 g	8 g	10 g
Nutmeg	3 g	4 g	5 g
Mixed spice	3 g	4 g	5 g
Ground cinnamon	3 g	4 g	5 g
Ground almonds	75 g	100 g	125 g
Currants	150 g	200 g	300 g
Sultanas	150 g	200 g	300 g
Raisins	125 g	150 g	225 g
Mixed peel	75 g	100 g	125 g
Glacé cherries	75 g	100 g	125 g
Grated zest of lemon	½	¾	1
Oven temperatures			
	150°C	140°C	130°C
Approximate cooking times			
	2 hours	3 hours	4 ½ hours

1 Cream the butter and sugar until soft and light.

2 Break up the eggs and beat in gradually.

3 Add the black treacle.

4 Sieve all the dry ingredients together and fold in.

5 Finally fold in the dried fruit and lemon zest.

6 Deposit into buttered cake tins lined with silicone paper.

7 Level the mix and make a well in the centre.

8 Check the cakes during baking, turn to make sure they are being cooked evenly.

9 Test to see if cooked with a metal skewer or needle.

10 Allow to cool completely before wrapping and storing in an airtight container.

Professional tips

- This is a dense mixture. To prevent the outside becoming overcooked insulate the cake tins by standing on newspaper, and tying several layers of newspaper around the sides.
- The dried fruit can be pre-soaked in brandy or rum the day before, or, as the cake matures it can be given a 'drink' every so often.
- Take a spoon and make a well in the centre of the mixture, this will help stop the cake doming as it bakes.
- Unless filling the oven with cakes, it is advisable to place a tray of water in the oven when baking. This will create steam and allow to cake to expand before the crust sets.
- To test, insert a needle in the centre, when the cake is cooked the needle should come out clean and hot.
- After cooling, wrap in paper or foil and place in an airtight container and leave to mature for 3–4 weeks before covering with marzipan and decorating with icing.

Note

- If using square cake tins increase the quantities by ¼.
- This cake can be made less rich by cutting down on the fruit and spices – these are all 'carried' ingredients and will not affect the cake as long as the basic ingredients (butter, sugar, eggs, flour) are not tampered with.

Faults

A   B   C   D

Cake A is domed ('cauliflower top'). This occurs if:
- the flour used is too strong
- oven is too hot and/or too dry,
- there is not enough fat
- the ingredients are over-mixed after adding flour.

Cake B has a sunken top ('M' fault). This is caused by adding too much baking powder and/or too much sugar.

Cake C has a sunken top and sides ('X' fault). This occurs if the mixture is too wet or too much baking powder is added.

Cake D has a low volume. This occurs if:
- too much fat is added
- the mixture is too dry
- there is not enough aeration.

Note

A good-quality fruit cake will:
- have straight sides, a flat top, and good height
- have an even distribution of fruit
- be moist and dark.

12 Plain scones

Makes >	16 scones
Plain flour	450 g
Baking powder	25 g
Pinch of salt	
Butter	225 g
Caster sugar	170 g
Sour cream	300 ml

1 Sieve the flour, baking powder and salt.

2 Cut the butter into small pieces and rub into the flour to achieve a sandy texture.

3 Dissolve the sugar in the cream.

4 Add the liquid to the dry ingredients and cut in with a plastic scraper, mix lightly and do not overwork. Wrap in cling film and chill for 1 hour.

5 Set the oven at 180°C and line a baking sheet with silicone paper.

6 Roll out 2 cm thick on a floured surface, cut out with a plain or fluted cutter.

7 Brush with milk or eggwash and bake at 180°C for approximately 15–20 minutes.

8 After 15 minutes, pull one apart to test if they are cooked.

9 Allow to cool and dust with icing sugar before serving.

Variation

Fruit scones: Add 50 g sultanas to the basic mix, or try adding dried cranberries or apricots as alternatives.

Professional tip

After cutting out the scones, turn upside on the baking sheet – this will help them rise with straight sides.

Note

Scones are traditionally served at afternoon tea with jam and butter or clotted cream, and are best served on the day they are made.

13 Banana cake

This is a good way to use up overripe bananas.

	Makes >	3 cakes
Ripe bananas (peeled)		450 g
Caster sugar		450 g
Plain flour		450 g
Baking powder		20 g
Eggs		4
Salt		1 tsp
Sunflower oil		350 ml
Melted butter		120 g

1 Line loaf tins with silicone paper and set the oven to 160°C.
2 Mash the bananas and sugar.
3 Sieve the flour and baking powder twice.
4 Whisk the eggs, salt, oil and butter.
5 Combine all the ingredients and mix well.
6 Pour into loaf tins until ¾ full.
7 Bake for 35–40 minutes, test with a skewer.
8 Remove from tins and allow to cool.
9 Store in an airtight container for at least 24 hours.

Professional tip

When lining a tapered loaf tin with silicone paper, just line the base and the two long sides, the ends can be left and easily loosened by running a knife along. This will save a lot of overlapped, creased paper and frustration.

14 Victoria sandwich

	Makes >	2 × 18cm cakes
Butter		250 g
Caster sugar		250 g
Soft flour		250 g
Baking powder		10 g
Eggs		5
Vanilla extract		½ tsp
Jam to fill		

1 Cream the butter and sugar until soft and light.

2 Sieve the flour and baking powder twice.

3 Mix together the eggs and vanilla extract.

4 Beat the eggs gradually into the butter and sugar mixture.

5 Fold in the flour.

6 Deposit into buttered and floured cake tins and level.

7 Bake at 180°C for approximately 15–20 minutes.

8 Turn out onto a wire rack to cool.

9 Spread the bottom half with softened jam.

10 Place on the top sponge and dust with icing sugar.

Beat the sugar and butter together

Place the mixture into buttered cake tins

Flatten the top before baking

Variations

- In addition to jam, the sponge can be filled with either butter icing or Chantilly cream.
- The official Women's Institute version specifies the cake is filled with jam only and dusted with caster not icing sugar.

Note

This is a classic afternoon tea cake named after Queen Victoria. Although traditionally made in two halves, a slimmer version can be made by using a single sponge and splitting it.

15 | Dobos torte

Traditional Dobos torte

This is a Hungarian speciality which consists of seven layers of biscuit. Six are sandwiched with chocolate buttercream. The seventh is coated with caramel, cut into triangles and placed on top.

Makes >	1 × 20cm torte
Biscuit discs	
Butter	120 g
Icing sugar	120 g
Soft flour	120 g
Eggs	2
Vanilla extract	½ tsp
Chocolate buttercream (Chapter 3, Recipe 17)	500 g
Caramel	
Water	100 ml
Granulated sugar	250 g
Glucose	25 g

1 Cream the butter and sugar until soft and light.

2 Sieve the flour.

3 Mix together the eggs and vanilla and beat into the butter and sugar mixture.

4 Mix in the flour.

5 Spread the paste thinly onto a silicone mat and bake at 180°C until evenly coloured.

6 Turn the paste over onto the hot baking sheet, remove the mat and quickly cut out the seven discs.

7 Reserve a disc for the top.

8 Build the torte by piping on thin layers of chocolate buttercream.

9 Coat the top and sides with buttercream, comb scrape the sides and level the top.

10 Make the caramel.

11 When the caramel reaches a pale amber, place the biscuit on a lightly oiled marble slab, carefully pour over the caramel, trim around the edge and quickly cut into segments using a large oiled knife.

12 Finish by piping small bulbs of buttercream on top and set the caramel biscuit at an angle.

13 The bottom edge can be finished with grated chocolate or roast chopped almonds.

- The biscuits are very fragile so it is advisable to make extra.
- When cutting out the biscuits and the caramel top it is essential to work quickly.
- Do not cut out the biscuits on the baking mats.

Note

A good-quality Dobos torte will have:

- six even layers
- a good ratio of biscuit to buttercream
- a neat light caramel top, even-sized segments without any breaks or chips
- light chocolate-flavoured buttercream.

Modern Dobos torte

Try something different

For a more modern interpretation, the layers of biscuit and buttercream can be replaced with a layer each of chocolate and orange mousse on a base of almond (jaconde) sponge. The outside and top are coated with Chantilly cream.

16 Sachertorte

This is a classic Austrian chocolate cake from Vienna. The long fought legal battle over who had the original recipe between the Hotel Sacher and the patisserie Demel has helped to make this cake famous. The hotel finally won in 1950, but both establishments make their own versions.

Makes >	1 × 20 cm torte
Eggs	6
Soft unsalted butter	125 g
Caster sugar	125 g
Dark chocolate (melted)	125 g
Plain flour	125 g
Apricot jam (approx.)	200 g
Chocolate glaze (Chapter 3, Recipe 40)	200 ml

1 Set the oven at 170°C and butter a 20 cm genoise tin, line the base with a disc of paper, re-butter and dust with flour.

2 Separate the eggs.

3 Cream the butter with 100g of the sugar until soft and light.

4 Add the egg yolks gradually.

5 Add the chocolate gradually.

6 Sieve the flour and fold in.

7 In a separate bowl whisk the egg whites with the remaining 25 g sugar to soft peaks.

8 Fold the meringue into the cake mixture, do not over-mix.

9 Deposit into the prepared cake tin and bake for approximately 45 minutes.

10 Test with a metal skewer.

11 Leave to cool slightly before turning out onto a cooling rack.

12 Rest overnight in an airtight container.

13 Cut in half and fill with apricot jam, assemble.

14 Boil the remaining jam, strain and brush over the top and sides.

15 Place on a wire rack with a tray underneath, pour over warmed chocolate glaze, remove excess with a palette knife, allow to set.

16 Finish by piping the word 'Sacher' on top using some of the remaining glaze.

Note

A good-quality Sachertorte will have:
- a smooth dark shiny glaze evenly coating the cake
- the word 'Sacher' written on the top
- rich moist chocolate cake with one layer of apricot jam.

17 Opera

Makes >	a 12 × 15 cm cake
Plain biscuit jaconde (Recipe 5)	3 sheets
Rum and coffee stock syrup	150 ml
Coffee buttercream (Chapter 3, Recipe 17)	800 g
Ganache (Chapter 3, Recipe 19)	800 g
Chocolate glaze (Chapter 3, Recipe 40)	500 ml

1 Build the opera on a flat baking sheet covered with silicone paper.

2 Place on first sheet of sponge, brush with syrup.

3 Pipe on an even layer of coffee buttercream and level.

4 Place on second layer of sponge and repeat, this time using ganache.

5 Place on final layer of sponge, cover with paper, place on another baking sheet and gently press.

6 Chill in the fridge to set.

7 Pour over the chocolate glaze and quickly smooth with a palette knife, leave to set.

8 If presenting as individual slices, using a large knife dipped in hot water, or a guitar, carefully cut into 8 × 3 cm rectangles.

9 Finish by piping on the word 'Opera'.

Try something different

● Decorate with a few flecks of gold leaf.

● If you do not feel confident enough to write the word 'Opera' you can decorate with piped dots decreasing in size, spun with chocolate, or tastefully decorate with moulded chocolate decorations.

18   Himmel cheese torte

Makes >	2 × 20cm tortes
Himmel bases	
Butter	675 g
Caster sugar	340 g
Soft flour	675 g
Vanilla extract	½ tsp
Egg yolks	9
Filling	
Cream cheese	900 g
Caster sugar	225 g
Vanilla extract	½ tsp
Double cream	570 ml
Raspberry jam	100 g
Raspberries	400 g
Icing sugar to dust	

1 Cream the butter and sugar until soft and light.

2 Sieve the flour.

3 Mix together the egg yolks and vanilla, beat into the butter and sugar mixture.

4 Mix in the flour.

5 Divide the mixture between six buttered 20 cm flan rings set on silicone mats, level and bake at 180°C for approximately 20 minutes.

6 Transfer to cooling racks.

7 For the filling, mix together the cream cheese, sugar and vanilla (do not overwork), set aside.

8 Whip the cream to firm peaks, add to the cream cheese and fold in until evenly mixed.

9 Set a biscuit base on a cake card and spread with jam, place on second biscuit.

10 Pipe the filling in concentric circles, fill the gaps with the fruit.

11 Invert the final biscuit layer on top and gently press down.

12 Smooth the sides and dust the top with icing sugar.

Professional tips

- Add a couple of gelatine leaves to the filling for a slightly firmer set.
- To achieve an even spread, the biscuit base can be piped inside the flan rings.
- Transfer the biscuit bases onto a cooling rack by sliding a cake card underneath – take care, they are very fragile!
- The top can be pre-cut before placing on, this will make it much easier to serve.

Variation

Any red berries can be used in this recipe.

19 Red wine cake

Makes >	2 × 16 cm cakes
Butter	250 g
Caster sugar	250 g
Soft flour	250g
Cocoa powder	10 g
Ground cinnamon	10 g
Clove powder	½ tsp
Vanilla extract	½ tsp
Eggs	4
Red wine	150 ml
Dark chocolate chips	125 g
Syrup	
Red wine	125 ml
Caster sugar	40 g
Water	50 ml

1 Cream the butter and sugar until soft and light.

2 Sieve the flour, cocoa and spices onto paper.

3 Whisk together the eggs and vanilla, beat into the butter and sugar mixture.

4 Fold in the dry ingredients.

5 Stir in the red wine and chocolate chips.

6 Divide between buttered and floured savarin rings, or other suitable cake tins.

7 Place on a baking sheet and bake at 170°C for approx. 35 minutes.

8 When cooked allow to cool slightly before removing from tins.

9 Boil the syrup ingredients, pour into the cleaned savarin ring and invert the cake back in to soak.

10 Remove again from tin, glaze with boiled redcurrant jelly and spin with white chocolate to finish.

Try something different

- This cake can alternatively be made in a round or square tin, but it will take slightly longer to bake.
- Instead of spinning with white chocolate, it can be decorated using half circles of moulded chocolate placed around the sides.
- Red wine cake can be an interesting alternative to a fruit cake at Christmas.

Professional tips

- To avoid ending up with a pile of crumbs, use a cake card to slide the cake back in the tin to soak.
- Make sure the redcurrant glaze is the correct consistency – too thick and it will tear the sponge, too thin and it will soak in. For the best results coat twice.
- A little red colour can be added to the glaze for a better finish.

Note

Be careful when measuring the clove powder – it is a very strong flavour and too much will overpower.

20 | Lemon drizzle cake

Makes >	2 × 16cm cakes
Butter	250 g
Caster sugar	400 g
Grated zest of lemons	3
Soft flour	380 g
Baking powder	10 g
Eggs	4
Vanilla extract	½ tsp
Milk	25 ml
Syrup	
Lemons, juice of	3
Caster sugar	100 g

1. Cream the butter, sugar and zest until soft and light.
2. Sieve the flour and baking powder twice.
3. Mix together the eggs and vanilla extract.
4. Beat the eggs into the butter and sugar mixture.
5. Fold in the flour.
6. Add milk to achieve a dropping consistency.
7. Deposit into buttered and floured cake tins.
8. Bake at 165°C for approximately 45 minutes.
9. Boil the lemon juice and sugar.
10. When the cake is cooked, stab with a skewer and pour over the syrup.
11. Leave to cool in the tin.

Note

This cake can be finished with lemon icing and decorated with strips of crystallised lemon peel.

21 | Chocolate pecan brownie

Makes >	24 slices
Butter	330 g
Dark chocolate	330 g
Soft flour	150 g
Cocoa powder	50 g
Eggs	6
Caster sugar	450 g
Vanilla extract	½ tsp
Roasted and chopped pecans	100 g
Topping	
Whipping cream	250 ml
Glucose	25 g
Dark chocolate	250 g

1. Cut the butter into small pieces and melt with the chocolate over hot water.
2. Sieve the flour and cocoa powder onto paper.
3. Warm the eggs, vanilla and sugar and whisk to the ribbon stage.
4. Fold the chocolate and butter into the eggs followed by the flour, cocoa and pecans.
5. Pour into a silicone-lined deep tray, smooth top and bake at 165°C for approximately 30 minutes. The mixture should still be soft in the middle.
6. For the topping, boil the cream and glucose, whisk in the chocolate.
7. Allow to cool before pouring over the brownie, spread evenly and at setting point use a comb scraper to produce wavy lines.
8. When set cut into even sized rectangles.

Variation
The mixture can be baked in individual dariole moulds and served as a dessert with cream or ice cream.

22 Spiced hazelnut shortbread

Makes >	Approximately 100 pieces
Butter	500 g
Icing sugar	375 g
Eggs	2
Vanilla seeds	1 pod
Soft flour	750 g
Clove powder	5 g
Whole hazelnuts	250 g
Chopped hazelnuts	250 g
For dusting	
Caster sugar	25 g
Clove powder	pinch
Ground cinnamon	5 g

1 Cream together the butter and icing sugar until light and soft.

2 Whisk the eggs and vanilla and beat into the butter and sugar.

3 Sieve the flour and clove powder and fold in with the hazelnuts.

4 Line a plastic or stainless steel tray with silicone paper and press the mixture in, wrap in cling film and place over a second tray and chill under weights.

5 Remove from tray, peel off the paper and cut into batons.

6 Lay flat on a baking sheet lined with silicone paper.

7 Bake at 170°C for approximately 15 minutes.

8 Transfer to a cooling rack.

9 While still hot, dust with the spiced sugar.

Variation

These shortbreads can be cut smaller and used as petits fours secs.

Professional tips

- Be careful when adding the clove powder.
- To cut make sure the paste is chilled and use a sharp large chopping knife to avoid dislodging the whole nuts.

Test yourself

Level 2

1 What are the differences between a cake and a sponge in terms of texture and use?

2 State the storage procedures for each of the following:
   a genoise sponge
   b fruit cake
   c fresh cream gateau
   d langue de chat biscuits.

3 What are the possible reasons for the following faults associated with cakes:
   a cauliflower top
   b 'M' fault?

Level 3

1 List the ingredients used to make each of the following:
   a dacquoise
   b biscuit jaconde.

2 What are the differences between mechanical and chemical aeration? Give examples of products which use each technique.

3 Describe the split-egg method of making sponges. Name two products made in this way.

9 Petits fours

This chapter covers:

→ **VRQ Level 3 Produce petits fours**

In this chapter you will produce and finish petits fours (level 3).

Recipes in this chapter

Introduction

Petits fours refer to small-sized confections that are usually eaten in one or two bites and are served at the end of a large dinner, or as an accompaniment to tea or coffee with afternoon tea. Originally a French term, the name literally translates as 'small oven' due to the fact that traditionally cakes in France were baked in coal-fuelled ovens which were hotter than normal ovens and quite difficult to control to the correct temperature. After the initial baking of large cakes, the oven was then cooled and it was at this stage the small petits fours confections were baked. All petits fours contain a very high percentage of sugar and are said to be hygroscopic. This term means that they will absorb the moisture from the atmosphere quickly and should always be kept in airtight containers to prolong their shelf life. Fruits au caramel and fruits déguisés can be prepared in advance and then dipped into the boiled sugar solution, which is boiled to hard-crack temperature 156°C, just before service. This is essential otherwise the surface will become sticky and start to dissolve.

There are many alternative names that can be used on menus for the collective term of petits fours. These may include:

● friandes
● gourmandises
● sweetmeats
● frivolities
● mignardises.

Types of petits fours

Generally petits fours can be divided into three categories: glacé, sec and confiserie variée.

Glacé

These are petits fours that have a glazed finish on the outer surface which can be achieved by using fondant, boiled sugar, chocolate, etc. A particular type of fruit base, petit four glacé, is referred to as a 'déguisé' which translates as a 'disguise' fruit. This is achieved by neatly filling small fruits with coloured marzipan flavoured with kirsch, pistachio compound, praline, coffee extract and other suitable products, then dipping them in boiled sugar cooked to hard-crack temperature. The finished confection will now have a crunchy coating of sugar on the outside with a soft textured marzipan fruit inside.

Sec

Petits fours secs are 'dry' confections (unlike petits fours glacés) which require further embellishment once baked to give them a decorative finish. Examples of petits fours secs are cats' tongues, macaroons and Dutch biscuits.

Confiserie variée

This term covers all the other petits fours confections that do not come under the glacé or sec categories. Petits fours in this category include truffles, fudges, caramel mou and pralines.

Food allergies related to petits fours

Some customers can have a very serious adverse reaction to certain foods. Such a reaction can occur within a few minutes of exposure to the allergen; typically the lips and tongue tingle and swell. They may also experience abdominal cramps and diarrhoea, and sometimes the person vomits.

There may also be wheezing and shortness of breath followed, in rare cases, by cardiovascular failure and collapse, leading to death if very prompt action is not taken.

Petits fours are normally served on a tray with several different varieties. There are many different ingredients in their make-up and some could cause an allergic reaction. The pastry chef should train staff in product knowledge – they should know exactly what ingredients are used in the dishes and the ones that may cause an allergic reaction and communicate these products to the service staff so that they are aware of the make-up of each petits fours dish.

In any catering establishment, should a customer enquire if a substance or food to which they are allergic is included in the recipe's ingredients, it is vital that staff give an accurate and clear answer (for example, a person who is allergic to nuts would need to know if there were walnuts in the chocolate fudge, marzipan filling in the fruits déguisés, baton almonds in the chocolate rochers, flaked almonds in the Florentine squares or ground almonds in the macaroons). Failure to provide the correct information could have dire consequences.

Commodities that may trigger an allergic reaction in petits fours

- Nuts: cashew, pecan, walnut, ground almonds, nib almonds, brazil nuts, pistachio nuts, ground hazelnuts, praline, gianduia, caramelized nuts
- Sesame seeds, caraway seeds, poppy seeds
- Milk, eggs, cream, yoghurt, crème fraiche
- Chocolate, coffee, oranges, red fruits
- Yeast, wheat, soya, sugar.

Presentation of petits fours

When petits fours are received by a customer at the end of the meal, they need to appeal to their senses of sight and smell, even before taste. For this reason, you need to consider presentation early in the design, include different types of petits fours and ensure that there is a wide variety of colours, textures, shapes and flavours in the presentation. Remember, petits fours are the last memory of a dining experience.

Video: freestyle piping for petits fours, http://bit.ly/142tgsF

Storage of petits fours

Because petits fours are generally made from large concentrations of sugar (such as tuiles, dipped fruits in caramel, chocolates, cats' tongues) they tend to absorb moisture from the atmosphere because they are hygroscopic, so therefore need to be stored carefully. To keep biscuits crisp and prevent a sticky surface from forming on boiled sugar products, store them in airtight containers. Chocolate petits fours, however, need to be stored in cool surroundings approximately 15°C, away from any humidity, as this could result in sugar bloom and the absorption of strong smells from foods such as onions and garlic. Once removed from moulds, place the chocolates in silicone-lined gastronorm trays with lids. If the chocolates are exposed to heat or direct sunlight this could result in fat bloom.

Shelf life of handmade chocolates

The shelf life of chocolates generally depends on the water activity in the ganache being used as a filling. The less water (cream is classified as water) in the recipe, the longer the shelf life. Always use UHT cream in the making of ganache as this is the most sterile cream to use and should give a longer shelf life.

What is fondant and how is it processed?

Fondant is made by boiling sugar, water and glucose to the soft ball stage (115°C). Once it reaches this temperature it is then poured onto a lightly oiled marble slab and worked with a palette knife entrapping air into the mixture until it becomes a white opaque and solid mass. Commercial fondant is purchased ready made in a white opaque block. To process, it is broken down into chucks and placed over a bain-marie of simmering water. Stock syrup is added to thin down the fondant to the desired consistency and the temperature is taken to 37°C. If the fondant is taken above this temperature, then the sugar crystals within the fondant break down, resulting in them not reflecting the light from the surface, and thus creating a dull appearance. Fondant can be flavoured with chocolate, raspberry, vanilla or mint, or piped into starch moulds (these are made by compacting cornflour in a container and then making indentations) which are allowed to set for a few days, then enrobed in chocolate and served as petits fours (for example, peppermint creams).

Marzipan (almond paste) in petits fours

Marzipan is widely used in the production of petits fours (for example, fruits déguisés, fondant glacé and cut-out pieces of marzipan lightly grilled to give a caramelised surface). There are two methods of producing marzipan: boiling or cooking, and raw (uncooked). For cooked marzipan, sugar and water is boiled to 116°C and mixed with ground almonds and egg yolk. The mixture is then worked on a table top lightly dusted with icing sugar until it is smooth. Raw marzipan is made by mixing together ground almonds and icing sugar and using egg to bind the ingredients together. It is worked on a table top lightly dusted with icing sugar to form a smooth paste. There are commercial marzipans especially made for marzipan modelling (see marzipan figures in Chapter 10). Marzipan is also used for covering fruit cakes (for example, to finish the top of simnel cake, which is traditionally eaten at Easter) and can also be placed inside (for example, inside a stollen before baking).

Professional tip

Always keep marzipan covered during processing to prevent the surface from drying out and always keep covered when being stored.

Petit Genoese fondants (fondant dips)

Either pain de Gêne or heavy genoise can be used for this recipe.

Makes >	50 pieces
Heavy sponge (pain de Gêne or heavy genoise – see below)	500 g
Marzipan	150 g
Apricot glaze	50 g
Fondant	500 g
Stock syrup	
Pain de Gêne (almond sponge)	
Makes	**2 × 46 cm × 30 cm trays**
Butter	750 g
Sugar	750 g
Ground almonds	750 g
Soft flour	150 g
Cornflour	150 g
Whole eggs	750 g

OR heavy genoise	
Soft flour	455 g
Caster sugar	340 g
Milk powder	42 g
Salt	6 g
Margarine	340 g
Egg	425 g
Glycerine	35 g
Glucose	115 g
Water	175 g
Soft flour	70 g
Baking powder	15 g

To make pain de Gêne

1 Cream the butter and sugar until aerated and white in colour.

2 Sieve the ground almonds, soft flour and cornflour together onto a sheet of greaseproof paper.

3 Gradually add the whole eggs to the creamed butter and sugar mixture.

4 Carefully fold in the sieved dry ingredients.

5 Deposit into trays lined with silicone paper and bake at 180°C for 30 minutes.

To make heavy genoise sponge

1 Sieve the caster sugar, milk powder, salt and 455 g of flour together.

2 Place into a mixing bowl with the margarine and mix on speed 1 to a crumbly consistency. (It is very important that the mixture does not form into a paste.)

3 Add the egg and glycerine gradually, cream for 5–7 minutes on speed 2. Scrape down as required.

4 Dissolve the glucose and water, add half to the above ingredients on speed 1.

5 Blend with a few turns of the beater. Do not over-mix as this will toughen the gluten in the flour.

6 Add the remaining 70 g of sieved flour and baking powder and blend in.

7 Add the remaining liquid and mix to a smooth batter.

8 Deposit into a tray lined with silicone paper and bake at 195°C for approximately 30 minutes.

To prepare the heavy sponge for petit Genoese fondants

1 Remove the top crust from the sponge.

2 Clear surplus crumb.

3 Cut the sponge, sandwich with buttercream and brush with kirsch-flavoured stock syrup.

To prepare and apply the marzipan

1 Soften the marzipan in readiness for rolling.

2 Roll the marzipan out thinly using icing sugar to prevent sticking.

3 Spread apricot glaze on the sponge, or on the correct size of rolled out marzipan.

4 Apply the marzipan to the sponge.

To cut the heavy sponge into the correct size

1 Using metal bars or rulers, cut the sponge to the required shape and size with the marzipan face down on the work bench. (Choose shapes that limit the amount of waste.)

2 Turn the shapes over so that the marzipan is facing upwards and lightly glaze the surface with apricot glaze – this forms a key for the fondant to stick to.

To prepare the fondant

Prepare the fondant by adding stock syrup and warming gently over a bain marie of hot water to bring the temperature of the fondant to 37°C with the correct fluidity.

The dipping procedure

1 Prepare the work surface with the fondant pan in the centre, the sponge pieces to one side and the dipping wire on the other side with a drip tray underneath.

2 Dip each piece, marzipan surface down, into the fondant using a dipping fork, remove from the fondant, allow surplus to run off, and then place on wire to drain. (Place each piece side by side on the wire and do not allow any fondant drips to pass over previously dipped pieces.)

To decorate

Any fondant piped lines used in the decoration should be very thin, especially when stronger colours are used. A good piping mix to use is equal quantities of melted chocolate and piping jelly, mixed to the required consistency with stock syrup.

To serve

1 Allow fondants ample time to set.

2 Place in suitable paper cases and display neatly.

3 Any surplus fondant dripped through the wires and free of crumb may be returned to the pan.

Professional tips

When preparing fondant it is important not to take it above 37°C otherwise the finished product will have a dull appearance.

You can make 20 petit Genoese fondants in four colours, starting from one container of white fondant, by following this sequence:

● White: Dip five pieces in the white fondant.

● Pink: Colour the fondant with pink colouring, then dip five more pieces.

● Orange: Add yellow colouring to the pink fondant, then dip five more pieces.

● Brown: Add chocolate colouring to the orange fondant and dip the last five pieces.

2   Carolines glacés

This is long piped choux paste, filled with crème diplomat and dipped in coloured fondant. Choux paste and crème diplomat can be found in Chapter 3 (Recipes 5 and 14).

1. Pipe choux pastry into straight lines 2.5 cm long.
2. Bake at 180°C until golden brown.
3. Once cold, fill with crème diplomat.
4. Dip the surface of the Carolines in coloured fondant.

3   Fruits déguisés

Makes >	30 pieces
Fruits and nuts for filling (dates, prunes, apricots, glacé cherries, walnuts)	30
Marzipan	100 g
Granulated sugar	400 g
Glucose	60 g
Water to saturate the sugar	

1. Cut the fruits in half and fill or sandwich together with marzipan that has been coloured and flavoured with liqueur.
2. Allow to dry overnight.
3. Press a cocktail stick into a firm part of the fruit and dip into the boiled sugar solution (follow the same sugar boiling method as for fruits au caramel, Recipe 4).
4. Stand onto a silicone mat and remove cocktail stick.

4 Fruits au caramel

Satsuma segments need to be peeled and the skins allowed to dry prior to dipping.

Professional tip

Because sugar is hygroscopic (it attracts moisture from the atmosphere) the dipping into the boiled sugar process should be carried out just before service time. This will give the dipped fruits a crunchy outer sugar surface.

Safety tip

Forward planning should always be carried out prior to the commencement of sugar boiling, and always have a clean working area ready to process the product.

Faults

The cause of a thick base, commonly known as 'feet on dipped fruits' is caused by dipping the fruit in boiled sugar which has thickened as it has cooled down. This results in a thick outer coating on the fruit which runs to the base once the fruit is placed onto the non-stick surface.

Another cause is not wiping the dipped fruits against the side of the pan to remove excess sugar syrup prior to placing on a non-stick surface.

Makes >	50
Granulated sugar	400 g
Glucose	60 g
Water to saturate the sugar	
Fruits suitable for dipping (grapes, cape gooseberries, cherries, strawberries, orange segments)	50 pieces

1 Place the sugar in a small saucepan and just saturate with water, add the glucose and gradually bring to the boil following the principles of sugar boiling.

2 When the sugar reaches 155°C hard crack, dip the pan into a container of cold water to stop the sugar from cooking any further.

3 Remove the pan of sugar from the cold water and, with the aid of tweezers, dip the fruits in the sugar solution ensuring that the fruits have an even coating.

4 Stand the fruits upright on a silicone mat.

5 | Madeleines

Makes >	45 pieces
Caster sugar	125 g
Eggs	3
Vanilla pod, seeds from	1
Flour	150 g
Baking powder	1 tsp
Beurre noisette	125 g

1 Whisk the sugar, eggs and vanilla seeds to a hot sabayon.
2 Fold in the flour and the baking powder.
3 Fold in the beurre noisette and chill for up to 2 hours.
4 Pipe into well-buttered madeleine moulds and bake in a moderate oven.
5 Turn out and allow to cool.

6 | Cats' tongues

Makes >	Approximately 40
Icing sugar	125 g
Butter	100 g
Vanilla essence	3–4 drops
Egg whites	3–4
Soft flour	100 g

1 Lightly cream the sugar and butter, add the vanilla essence.
2 Add the egg whites one by one, continually mixing and being careful not to allow the mixture to curdle.
3 Gently fold in the sifted flour and mix lightly.
4 Pipe on to a lightly greased baking sheet using a 3 mm plain tube, 2½ cm apart.
5 Bake at 230–250°C for a few minutes.
6 The outside edges should be light brown and the centres yellow.
7 When cooked, remove on to a cooling rack using a palette knife.

7 Brandy snaps

Makes >	Approximately 20
Strong flour	225 g
Ground ginger	10 g
Golden syrup	225 g
Butter	250 g
Caster sugar	450 g

1. Combine the flour and ginger in a bowl on the scales. Make a well.
2. Pour in golden syrup until the correct weight is reached.
3. Cut the butter into small pieces. Add the butter and sugar.
4. Mix together at a slow speed.
5. Divide into 4 even pieces. Roll into sausage shapes, wrap each in cling film and chill, preferably overnight.
6. Slice each roll into rounds. Place on a baking tray, spaced well apart.
7. Flatten each round using a fork dipped in cold water, keeping a round shape.
9. Bake in a pre-heated oven at 200°C until evenly coloured and bubbly.
10. Remove from oven. Allow to cool slightly, then lift off and shape over a dariole mould.
11. Stack the snaps, no more than 4 together, on a stainless steel tray and store.

8 Cigarette paste cornets

Makes >	Approx. 30
Icing sugar	125 g
Butter, melted	100 g
Vanilla essence	3–4 drops
Egg whites	3–4
Soft flour	100 g

1 Proceed as for steps 1–3 of Recipe 6 (cats' tongues).

2 Using a plain tube, pipe out the mixture onto a lightly greased baking sheet into bulbs, spaced well apart. Place a template over each bulb and spread it with a palette knife.

3 Bake at 150°C, until evenly coloured.

4 Remove the tray from the oven. Turn the cornets over but keep them on the hot tray.

5 Work quickly while the cornets are hot and twist them into a cornet shape using the point of a cream horn mould. (For a tight cornet shape it is best to set the pieces tightly inside the cream horn moulds and leave them until set.) If the cornets set hard before you have shaped them all, warm them in the oven until they become flexible.

The same paste may also be used for cigarettes russes, coupeaux and other shapes.

9 Almond biscuits

Makes >	30
Butter	200 g
Icing sugar	80 g
Soft flour	200 g
Salt	pinch
Whole egg	25 g
Ground almonds	100 g
Granulated sugar	100 g

1 Rub together the butter, icing sugar, flour and salt until the mixture resembles fine breadcrumbs.

2 Add the egg and work to a smooth paste.

3 Roll into a cylinder 2 cm in diameter.

4 Brush with egg white and roll in a mixture of equal quantities of granulated sugar and ground almonds.

5 Refrigerate to firm up the paste.

6 Cut into discs and bake at 180°C until lightly golden.

Variation

Almond biscuits can also be flavoured with lemon zest or chopped glacé cherries.

10 Piped biscuits (sablés à la poche)

	Makes >	20–30
Caster or unrefined sugar		75 g
Butter or margarine		150 g
Egg		1
Vanilla essence		3–4 drops
or		
Grated lemon zest		
Soft flour, white or wholemeal		200 g
Ground almonds		35 g

1 Cream the sugar and butter until light in colour and texture.

2 Add the egg gradually, beating continuously, add the vanilla essence or lemon zest.

3 Gently fold in the sifted flour and almonds, mix well until suitable for piping. If too stiff, add a little beaten egg.

4 Pipe on to a lightly greased and floured baking sheet using a medium-sized star tube (a variety of shapes can be used).

5 Some biscuits can be left plain, some decorated with half almonds or neatly cut pieces of angelica and glacé cherries.

6 Bake in a moderate oven at 190°C for about 10 minutes.

7 When cooked, remove on to a cooling rack using a palette knife.

11   Poppy seed tuiles

Makes >	80 tuiles
Glucose	50 g
Granulated sugar	500 g
Water	200 g
Soft flour	150 g
Butter, melted	250 g
Almonds, nibbed	250 g
Poppy seeds	250 g

1. Add the glucose and sugar to the water. Bring to the boil.
2. Remove from the heat. Stir in the flour and allow to stand for a few minutes.
3. Add the rest of the ingredients and mix well.
4. Place ½ teaspoon portions of the mixture onto a silicone mat, with space between them. Batten out with a fork.
5. Bake at 180°C for 3–4 minutes. Remove from the oven and invert immediately into tuile moulds.

Variation
Sesame seeds can be used instead of poppy seeds.

Tuiles (left to right: banana, poppy seed, sesame seed, chocolate, coconut)

12 Fruit tuiles

Makes >	25 tuiles
Fruit coulis	
Berries	225 g
Icing sugar	225 g
Tuiles	
Fruit coulis	100 g
Icing sugar	200 g
Plain flour	50 g
Butter, melted	80 g

1 To make the fruit coulis, macerate the berries and sugar together until the fruit starts to leak. Liquidise, then pass.

2 To make the tuiles, beat all the ingredients together.

3 Chill the mixture for 1 hour, then spread on a silicone mat.

4 Bake at 180°C for 6 to 8 minutes, until translucent.

13 Coconut tuiles

Makes >	50 tuiles
Icing sugar	300 g
Flour	100 g
Desiccated coconut	100 g
Egg whites	250 g
Butter, melted, not hot	200 g
Vanilla essence	

1 Mix the dry ingredients together well in a food processor.

2 Mixing on a slow speed, gradually add the egg whites, then the melted butter and vanilla essence.

3 Pipe onto silicone mats, in bulbs about 2½ cm in size.

4 Tap the tray to flatten the bulbs.

5 Bake at 205–220°C.

6 Mould inside a tuile plaque.

14 Chocolate tuiles

Makes >	60 tuiles
Cocoa powder	30 g
Sugar	450 g
Water	250 g
Pectin	10 g
Chocolate paste, made with 100% cocoa chocolate	150 g
Unsalted butter	150 g

1 Place the cocoa powder and 360 g of the sugar in the water. Bring to the boil.

2 Whisk in the pectin and the remaining sugar.

3 Simmer on a medium heat for 10 minutes.

4 Remove from the heat and add the chocolate paste and butter. Mix together.

5 Place ¼ teaspoon portions of the mixture onto silicone mats, well spaced apart.

6 Bake at 200°C for 9 to 10 minutes, until they start to bubble. To check that they are done, place one onto marble and test that it snaps when broken.

7 Mould in a tuile plaque.

15 Macaroons using Italian meringue

Makes >	40
Ground almonds	195 g
Icing sugar	195 g
Egg white	75 g
Italian meringue	
Caster sugar	190 g
Water	50 ml
Egg white	75 g
Cream of tartar	pinch

1 Sieve the icing sugar and ground almonds together.

2 Beat in the egg white until a smooth paste is formed with no lumps.

3 Make an Italian meringue cooked to 121°C.

4 Beat the Italian meringue into the almond mixture until a shiny surface is achieved.

5 Place into a piping bag and pipe 1cm small rounds onto a non-stick mat.

6 Tap the tray to even out the mixture.

7 Leave to stand in a warm place for 30 minutes to form a crust on the surface.

8 Bake at 150°C for 10 minutes

Fillings

Possible fillings include:

- Ganache
- Lemon cream
- Flavoured crème mousseline
- Buttercream.

16   Rothschilds

Makes >	25
Caster sugar	125 g
Ground almonds	125 g
Strong flour	25 g
Egg whites	5
Caster sugar	25 g
Cream of tartar	pinch

1. Sieve the 125 g sugar, almonds and strong flour together onto a sheet of greaseproof paper.
2. Whisk the egg whites, the 25 g of sugar and cream of tartar to a firm meringue.
3. Gradually mix the meringue into the dry ingredients.
4. Deposit into a piping bag and pipe out into 3 cm batons and sprinkle with nib almonds.
5. Bake at 180°C for 20 minutes.

17   Chocolate Viennese biscuits

Makes >	20
Butter	125 g
Icing sugar	50 g
Egg white	20 g
Soft flour	130 g
Bitter cocoa powder	25 g
Salt	pinch
Vanilla essence	drop

1. Cream the butter and sugar together until white.
2. Gradually beat in the egg white.
3. Fold in the flour and cocoa powder. Add the salt and vanilla essence.
4. Pipe into 3 cm lengths onto a greased and floured tray.
5. Bake at 180°C.
6. Dust with icing sugar or spin with chocolate.

18 | Dutch biscuits

Makes >	50
White paste	
Soft flour	300 g
Icing sugar	100 g
Vanilla	4 drops
Butter	200g
Egg yolk	1
Chocolate paste	
Soft flour	285 g
Icing sugar	100 g
Bitter cocoa power	20 g
Butter	200 g
Egg yolk	1
Chocolate colour to darken if required	

For the white paste

1 Sieve the flour and icing sugar. Add the vanilla.

2 Rub in the butter until a breadcrumb texture is achieved.

3 Bind with the egg yolk and refrigerate until the mixture is firm

For the chocolate paste

1 Sieve the flour, icing sugar and cocoa powder together.

2 Rub in the butter until a breadcrumb texture is achieved.

3 Bind with egg yolk and refrigerate until the mixture is firm.

Using the white and dark Dutch biscuit mixture, shape into various designs. Cut into petits fours size and bake at 180°C for 10 minutes

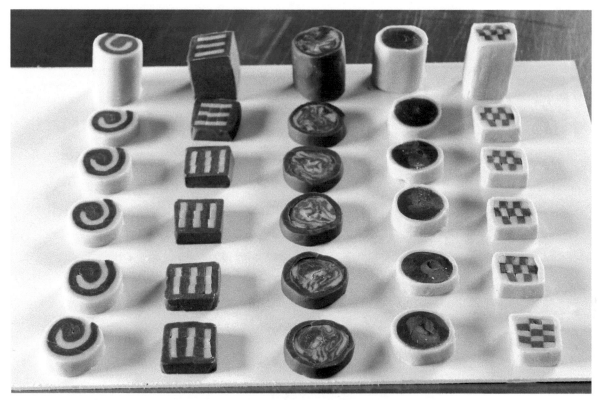

The patterns are formed (top) and then the biscuits are sliced up before baking

Design 1

1 Roll the white paste into a cylindrical shape 1.5 cm diameter.

2 Brush the surface of the white paste with water.

3 Place onto a square sheet of thinly rolled out chocolate paste the same width as the white cylinder.

4 Roll the white cylinder in the chocolate paste forming a thin coating of chocolate paste around the cylinder.

5 Refrigerate until firm and then cut into thin discs ready for baking.

Design 2

1 Mix equal amounts of white and chocolate paste together to form a marbled effect.

2 Roll into a cylindrical shape 1.5 cm diameter.

3 Brush the surface with water.

4 Place onto a square sheet of thinly rolled out white or chocolate paste the same width as the cylinder.

5 Roll the marbled cylinder in the white or chocolate paste forming a thin coating around the cylinder.

6 Refrigerate until firm and then cut into thin discs ready for baking.

Design 3 (Battenburg design)

1 Roll out two sheets of paste one white and one chocolate 5 mm thick.

2 Brush the surface of one paste with water and then attach the second.

3 Cut the paste in half, placing one on top of the other to give four alternating coloured pastes.

4 Refrigerate to set the pastes.

5 Cut one slice 5 mm thick and place on the working surface, brush the surface with water.

6 Cut a second slice, rotate this before placing on the first slice to create two layers of alternating lines of white and dark paste.

7 Repeat this process until you have four layers (all alternating).

8 Brush the surface.

9 Place onto a square sheet of thinly rolled out white or chocolate paste.

10 Roll in the desired coloured paste and refrigerate.

11 Cut into thin squares revealing a battenburg design ready for baking.

Base mixture

	Makes >	20
Ground almonds		90 g
Soft flour		15 g
Egg white		90 g
Cream of tartar		pinch
Caster sugar		120 g

1. Sieve the ground almonds and flour together onto a sheet of greaseproof paper.

2. Whisk the egg white with the cream of tartar to full foam, adding the sugar gradually to form a firm meringue.

3. Fold the meringue through the dry ingredients.

4. Spoon into a piping bag with a plain nozzle.

Batons au chocolat

Pipe into 3 cm lengths onto a greased and floured baking tray, sprinkle with nib almonds and bake at 180°C. When cold, spin with chocolate or dip on an angle into melted chocolate and sandwich together with a ganache.

Boules de neige

Using a plain nozzle, pipe out small domed bulbs of the japonaise mixture on to a greased and floured baking tray, sprinkle with nibbed almonds and bake at 180°C. When cold, sandwich together with ganache and dust the surface with icing sugar to represent snow balls.

20 Chocolate truffles (general purpose ganache)

Makes >	60 truffles
Whipping cream	175 g
Milk	75 g
Butter	75 g
Inverted sugar (Trimoline)	75 g
Powder sorbitol (optional, gives a better shelf life to the ganache)	12 g
Milk couverture	165 g
Plain couverture	333 g
Alcohol concentrate	10 g

1. Boil the cream, milk, butter, inverted sugar and sorbitol to 80°C.
2. Partially melt both chocolates and gradually add the boiled liquid, working from the centre of the chocolate, forming a good elastic emulsion.
3. Finish with a stem blender to achieve a smooth and shiny ganache. Add the alcohol concentrate.
4. Pour into a bowl and leave to cool and crystallise. Periodically stir the outside mixture to the centre.
5. When the ganache is firm, pipe out into the desired shapes.

Video: ganache, http://bit.ly/15fKB31

21 Caramel truffles

Makes >	120 truffles
Caster sugar	325 g
Trimoline	100 g
Whipping or double cream	500 g
Plain chocolate	575 g
Milk chocolate	75 g
Butter	10 g
Chocolate spheres	120

1. Place the sugar and Trimoline together in a pan and take to a caramel, being mindful that it will turn from caramel to burnt quickly.
2. Remove from the heat and slowly add the boiled cream. Return to the heat to dissolve the set caramel.
3. Once dissolved, add the chocolate and emulsify using a stem blender.
4. Add the butter.
5. Remove and allow to chill naturally.
6. Once the mixture has reached room temperature, place it in a disposable piping bag.
7. Snip off the end of the bag and carefully pipe the mix into the desired chocolate spheres.
8. Allow to set and carefully close the top of each sphere with melted chocolate (see Recipe 31).
9. Once set, roll in desired chocolate, allow to set and then serve.

Professional tip

Trimoline is an inverted sugar syrup.

22 | Cut pralines

	Makes >	80–100
General purpose ganache (Recipe 20)	2 kg	
Couverture to enrobe		

1 Place a sheet of acetate in the bottom of a praline frame (see Appendix).

2 Prepare general purpose ganache and pour into the praline frame, tap to level.

3 Leave the ganache for 12 hours to crystallise and firm up.

4 Remove the ganache from the praline frame and coat one side in tempered couverture.

5 Place the couverture side down onto the guitar base (see Appendix) and cut through the ganache.

6 Lift the slab of ganache off the guitar base using a metal take-off sheet, raise the wire and wipe clean.

7 Place the slab of ganache back on the guitar in the opposite direction and cut through once again, giving uniform cut squares.

8 Using a dipping fork, place the cut praline, chocolate-coated side, into tempered couverture.

9 Lift out of the couverture using a dipping fork and tap on the side of the container to remove excess chocolate.

10 Carefully deposit the praline onto a sheet of silicone paper.

11 The top of the pralines can be decorated by placing transfer sheets, textured sheets on top or by giving a ripple effect by using the dipping fork on the surface.

Allow the ganache to firm up in the frame, then coat the surface with tempered couverture using a paint roller

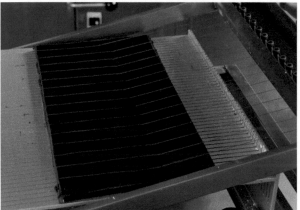

Place the couverture side down onto the guitar base and cut through the ganache

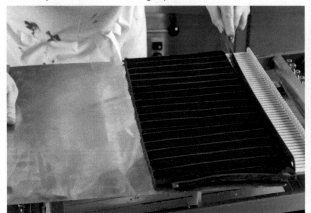

Lift the slab of ganache off the guitar base using a take-off metal sheet

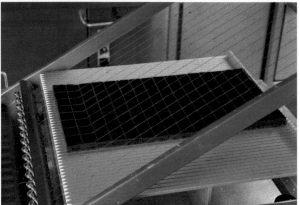

Wipe the guitar clean, then carefully place the slab of ganache back on the guitar in the opposite direction and cut through once again, creating uniform squares

Video: cut pralines,
http://bit.ly/14HxDVU

Professional tip

Layered pralines can be made by pouring general purpose ganache into a frame, allowing to set then topping with a layer of pâte des fruits. Once completely set, process in the normal manner giving a two-layered praline

Video: layered pralines,
http://bit.ly/1edLkSa

23 | Nougat Montelimar

Makes >	50–60 pieces
Granulated sugar	350 g
Water	100 g
Honey	100 g
Glucose	100 g
Egg white	35 g
Glacé cherries	50 g
Pistachio nuts	50 g
Nibbed almonds	25 g
Flaked almonds or flaked hazelnuts	25 g

1 Place the sugar and water into a suitable pan, bring to the boil and cook to 107°C.

2 When the temperature has been reached, add the honey and glucose and cook to 137°C.

3 Meanwhile, whisk the egg whites to full peak in a machine, then add the syrup at 137°C slowly, while whisking on full speed.

4 Reduce speed, add the glacé cherries (cut into quarters), chopped pistachio nuts, and the nibbed and flaked almonds.

5 Turn out onto a lightly oiled tray or rice paper and mark into pieces while still warm.

6 When cold, cut into pieces and place into paper cases to serve.

Professional tips

Instead of using rice paper, the tray may be dusted with neige-décor.

24 | Marshmallows

Makes >	Approximately 50 pieces
Granulated or cube sugar	600 g
Egg whites	3
Leaf gelatine, soaked in cold water	35 g

1 Place sugar into a suitable saucepan with 125 ml water and boil to soft ball stage, 140°C.

2 When sugar is nearly ready, whisk the egg whites to a firm peak.

3 Pour in boiling water and continue to whisk.

4 Squeeze the water from the gelatine and add.

5 Add colour and flavour if desired.

6 Turn out onto a tray dusted with cornflour and dust with more cornflour.

7 Cut into sections and roll in a mixture of 5 tablespoons of icing sugar and 2 tablespoons of cornflour.

25 Turkish delight

Makes >	50 pieces
Water	500 ml
Caster sugar	450 g
Icing sugar	100 g
Lemon juice	1 tbsp
Cornflour	150 g
Cream of tartar	1 tsp
Rose water	1 tbsp
Grenadine	1 tbsp
To finish	
Cornflour	2 tbsp
Icing sugar	5 tbsp

1 Dust a baking tray measuring 20 × 25 cm with cornflour. Pour half of the water into a heavy based saucepan and add the sugars and lemon juice. Heat until the sugar has dissolved and then bring to the boil.

2 Reduce the heat and simmer until the mixture reaches 115°C on a sugar thermometer (soft ball stage). Remove from the heat.

3 In a separate saucepan, mix the cornflour and cream of tartar together with the remaining water until the mixture is smooth. Cook over a medium heat until the mixture thickens.

4 Gradually pour the hot sugar syrup into the cornflour paste, stirring continuously. Return the mixture to the heat and simmer for about one hour, until the mixture is pale and feels stringy when a little of the cold mixture is pulled between the fingers. Stir in the rose water and Grenadine.

5 Pour the mixture into the prepared baking tin and leave to set overnight.

6 Cut into squares. Mix the cornflour and icing sugar together and toss the Turkish delight in it. Store in an airtight container, between layers of greaseproof paper.

26 Pâte des fruits

Makes >	60 pieces
Blackcurrant pulp	400 g
Caster sugar	360 g
Granulated sugar	100 g
Pectin mixture	
Pectin	30 g
Caster sugar	40 g

1 Heat the blackcurrant pulp to 50°C.

2 Add the caster sugar and then bring to the boil and skim.

3 Add the pectin mixture and cook to 105°C.

4 Remove from the heat and leave to settle for 10 seconds.

5 Pour the mixture into a prepared tray. Wrap in cling film.

6 Leave to set for 4 hours.

7 Just before serving, cut into 2 cm cubes. Roll the pieces in the granulated sugar and place into petit four cases.

Professional tip

These jellies should not be stored in the fridge.

Variations

Use other varieties of fruit purée (pulp) for different flavours of jelly.

27 Chocolate rochers

First you need to prepare the caramelised almonds (used as a base for the rochers) using one of the following methods.

Caramelised almonds (method one)

1. Pass baton almonds through egg yolk, then caster sugar and spread out onto silicone paper.
2. Bake in the oven at 160°C until golden and caramelised.

Caramelised almonds (method two)

1. Boil 400 g sugar and 160 g water to 115°C.
2. Add 1200 g baton almonds or nib almonds and mix.
3. Reheat over an open flame in a round bottomed bowl until the mixture starts to caramelise, keeping the mixture moving around the bowl at all times.
4. Add 50 g butter and mix in to separate the almonds.
5. Store in an airtight container until required.

Orange rochers

	Makes >	30
Tempered orange couverture		500 g
Cocoa butter		50 g
Caramelised almonds		600 g
Orange confit		200 g

Milk chocolate rochers

	Makes >	30
Tempered milk couverture		500 g
Cocoa butter		50 g
Caramelised almonds		600 g
Rice Krispies		100 g

White rochers

	Makes >	30
Tempered white couverture		500 g
Cocoa butter		50 g
Pine kernels		400 g
Pistachio nuts chopped		100 g
Orange confit		200 g
Rice Krispies		50 g

1. For all three rochers, mix all the ingredients together and spoon onto silicone paper.
2. Remove from the paper once set.

Professional tip

Instead of using rice paper, the tray may be dusted with neige-décor.

28 White and dark chocolate fudge

Makes >	50
Glucose	275 g
Caster sugar	800 g
Whipping cream	250 g
Couverture (white or plain)	400 g
Butter	60 g
Chopped walnuts for the dark fudge *or* pistachios for the white	150 g
Dried cranberries, for the white fudge	30 g

1 Line a shallow tray with silicone paper.

2 Boil the cream, glucose and sugar to 120°C.

3 Pour onto the melted chocolate and add the appropriate nuts and fruit for the type of chocolate used. Add the butter.

4 Pour into the prepared tray lined with silicone paper and allow to set.

5 Once set, remove from the tray and spread the surface of the white fudge with over-tempered plain couverture and the dark fudge with over-tempered white couverture. (Over-tempered is when the couverture is taken above the recommended tempering temperature so that the chocolate does not become too brittle and splinter when being cut.)

6 Just before both couvertures set, comb the surface using a toothed scraper.

7 Once the chocolate décor has set, cut into neat squares.

Professional tip

It is important that the sugar boils to the correct temperature, or the fudge will be too hard.

29 Caramel mou (soft toffee)

Makes >	60
Cream	1 litre
Inverted sugar	100 g
Sorbitol	40 g
Vanilla pods	2
Glucose	600 g
Water	50 g
Caster sugar	760 g
Butter	100 g

1 Boil the cream, inverted sugar, sorbitol, and split vanilla pods. Leave to infuse.

2 Boil the glucose, water and caster sugar to 145°C.

3 Add the strained hot cream mixture and continue to boil to 120°C.

4 Add the butter once the mixture reaches the required temperature

5 Pour into a frame and leave in a cool area until fully set.

6 Cut into squares.

Preparing the praline moulds

1 Polish the mould with cotton wool. Never use an abrasive as any marks made on the surface of the mould will be picked up on the outer surface of the chocolate.

2 Embellish the mould with tempered coloured cocoa butter, metallic powders, different coloured tempered chocolate or just leave plain.

3 Fill the mould to the top with tempered couverture and tap the mould on the table top to raise any air bubbles to the surface of the chocolate.

4 Invert the mould and tap the sides with a rubber spatula, forcing all the chocolate out of the mould, leaving a thin-coated cavity.

5 Using a chocolate scraper, clean the surface above the chocolate-coated cavities.

6 Place the mould, with the open cavities coated in chocolate, inverted onto a flat surface covered with silicone paper.

7 Leave the chocolate to crystallise (set).

8 Once the chocolate has set, invert the mould so that the open chocolate cavities are facing upwards.

9 Using a piping bag, fill the cavities with a ganache or any other suitable filling to just below the top of the mould.

10 Allow the ganache to harden.

11 Using a palette knife, spread tempered chocolate over the ganache forming a flat smooth surface enclosing the ganache.

12 Leave the chocolate to set.

13 Once set, turn the moulds upside down and tap the chocolates out of the mould.

Professional tips

Always wear cotton gloves when handling chocolates to maintain the shine on the tempered couverture surface.

Callets are the purchasing unit for couverture. They are small pieces that melt easily.

There are many ready-made products that can be used as fillings for this recipe.

The following are suitable ganache recipes for filling praline moulds.

Mandarin ganache

	Makes >	80–100
Whipping cream		700 g
Glucose		50 g
Inverted sugar (Trimoline)		60 g
Mandarin compound		50 g
Milk couverture		1200 g
Butter		100 g

1 Boil the cream with the glucose, the inverted sugar and the mandarin compound.

2 Pour onto the chocolate callets and mix well.

3 Cool down to room temperature.

4 Add the softened butter.

5 Pipe immediately into the praline moulds or chocolate spheres.

Caramel and orange ganache

	Makes >	80
Caster sugar		80 g
Glucose		30 g
Orange purée		325 g
Cream		200 g
Sorbitol		20 g
Butter		130 g
Plain couverture		650 g
Milk couverture		325 g

1 Prepare a dry caramel from the caster sugar and glucose.

2 Add the boiled orange purée gradually to the caramelised sugar solution.

3 Add the cream, sorbitol and butter, reboil.

4 Allow the cream to cool to 80°C.

5 Pour onto the partially melted chocolates and emulsify using a stem blender.

6 Cool down before filling the praline shells.

Green tea ganache

	Makes >	60
Whipping cream		400 g
Glucose		80 g
Green tea powder		40 g
Dark couverture		300 g
Milk couverture		400 g

1 Boil the cream, glucose and tea together and infuse for a few minutes.

2 Strain onto the melted chocolate and emulsify using a stem blender.

3 Cool down and pipe into the praline shells.

Embellish the mould with coloured, tempered cocoa butter and plain tempered couverture

Fill to the top with tempered couverture, then tap the mould to release any air bubbles

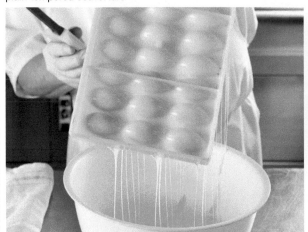

Invert the mould and tap it; most of the chocolate will drain out, leaving a thin coating in the cavity

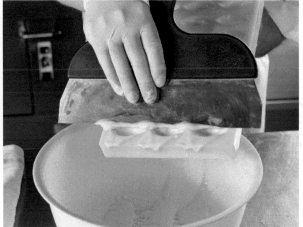

Scrape away and reserve any excess chocolate from the top of the mould

Once the chocolate has set, fill with ganache

Once the ganache has set, spread tempered chocolate over the top

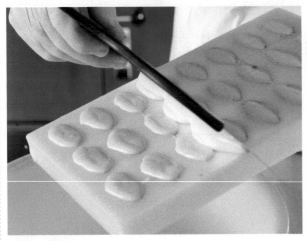

Scrape off any excess, leaving a flat, smooth surface

Once completely set, tap the chocolates out of the mould

Video: working with moulds,
http://bit.ly/14l9Ban

31 The use of convenience shells for pralines

These shells come in various forms – spheres, squares, rectangles, rounds.

1 Fill the convenience shell with a suitable ganache and seal the top with tempered couverture.

2 Can be decorated with transfer sheets, chocolate shavings, etc.

Video: dipping and decorating, http://bit.ly/1edLqcn

Fill the sphere or shell with ganache, using a closing template

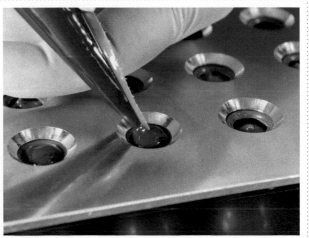

Seal the top with tempered couverture

Once set, coat in tempered couverture

To finish, dip in cocoa powder using a dipping fork

1 Line the base of a magnetic mould with transfer sheet.

2 Place lollipop frame on top of the transfer sheet.

3 Fill with tempered couverture and place lollipop stick in the indentation.

4 Decorate top of chocolate with crisp pearls, dried raspberries, Bres.

5 When the couverture has set, remove lollipops from the mould.

Variation: using transfer sheets

1 Cut strips of transfer sheets 5 cm wide.

2 Pipe round discs of tempered couverture onto the transfer sheet.

3 Place a lollipop stick in the centre of the disc of chocolate.

4 Decorate the top of the chocolate, as in the main recipe.

5 When the couverture has set, remove from the transfer sheet.

33 Frozen petits fours lollipops

1 Pipe fruit water ice into praline spheres.

2 Place lollipop sticks into the water ice.

3 Return to the freezer and freeze until solid.

4 Once frozen, seal the lollipop stick at the opening of the sphere with melted couverture.

5 Once set, dip the sphere in over-tempered couverture.

6 Serve frozen as a petit four.

Professional tip

For presentation purposes, serve in an iced socle on a bed of dry ice. (A socle is a moulded ice base for the presentation of food.) Just before serving, add warm water to the dry ice to form a mist.

Lemon feuilletée

Makes >	80
Pasteurised egg yolks	600 g
Pasteurised egg whites	720 g
Lemon juice	1 litre
Lemon zest	36 g
Sugar	1.5 kg
Melted butter	600g
Gelatine	20 leaves
Chocolate glaze	
Double cream	320 ml
Plain couverture	300 g
Butter, soft	100 g

1　Whisk the egg yolks, egg white, lemon juice, lemon zest and sugar over a bain-marie of simmering water until the mixture thickens and reaches a temperature of 80°C.

2　Add the butter and pre-soaked gelatine.

3　Pour into a sided mould lined with silicone paper and biscuit jaconde.

4　Once set, remove from the mould and coat the surface with chocolate glaze.

5　Cut into petit-four size pieces with a warm knife.

For the chocolate glaze

1　Boil the cream, add in the chocolate and mix well.

2　Blend in the butter, ensuring that there are no bubbles.

35 Florentine squares

Makes >	80
Sweet paste base	
Preparation A	
Cream	500 ml
Honey	500 g
Butter	500 g
Granulated sugar	1 kg
Preparation B	
Flaked almonds	750 g
Ground almonds	250 g
Strong flour	250 g
Chopped glacé fruit (cherries/angelica)	1 kg

1 Mix together and heat the ingredients for preparation A until it is at soft ball stage (116°C). Add all the preparation B ingredients and mix in well.

2 Turn out the mixture onto a sheet of cooked sweet paste in a sided container.

3 Place in an oven until the mixture bubbles at 180°C.

4 When cold, cut into squares and dip in chocolate if desired.

Makes>	80
Lemon juice	380 g
Egg yolks	12
Eggs	6
Sugar	300 g
Butter	250 g

1 Bring the lemon juice to the boil.

2 Cream together the egg yolks, eggs and sugar, then pour in 100 g of hot lemon juice.

3 Whisk until well blended and add the rest of the lemon juice.

4 Bring back to the boil, whisking until smooth.

5 Remove from the heat and add the butter progressively, whisking all the time. Chill until firm in texture.

6 Pipe into baked petit-four pastry cases and level the surface with a palette knife, dust with icing sugar and caramelise with a blow torch.

Test yourself

Level two

1 Briefly describe the preparation of cats' tongues biscuits.

2 Name four types of baked biscuits which may be served as petits fours.

Level three

3 Name the **three** classifications of petits fours, and give two examples of each.

4 Briefly describe why it is important to form a good emulsion when making a ganache.

5 Name four convenience products that may be used in the production of petits fours.

6 List the steps you would carry out to produce a tray of moulded pralines.

10 Decorative items

This chapter covers:

→ **VRQ level 3 Produce display pieces and decorative items**
→ **NVQ level 3 Prepare, process and finish complex chocolate products**
→ **NVQ level 3 Prepare, process and finish marzipan, pastillage and sugar products.**

In this chapter you will learn how to produce and finish display pieces and decorative items.

Recipes in this chapter ·

Quick techniques used for centrepieces			
32 Iced sugar	372		✓
33 Blowtorch sugar flower	373		✓
34 Boiled moulded sugar	374		✓
35 Pressed sugar	375		✓
Chocolate work			
36 Granite chocolate base	376		✓
37 Poured chocolate base colour with metallic powder	377		✓
38 Bubble-wrap chocolate bases	377		✓
39 Chocolate poles	378		✓
40 Cut-out chocolate free style	378		✓
41 Modelling chocolate	379		✓
42 Moulded chocolate figures	380		✓
43 Chocolate cones	382		✓
44 Chocolate spraying	383		✓
45 Modelling paste using a food processor	384		✓
Pastillage (gum paste)			
46 Pastillage	385		✓
47 Marzipan	386		✓

Chocolate

Translated, the French word 'couverture' means 'covering' or 'coating' in English. Coverture has a very high percentage of cocoa butter (at least 30 per cent) and is used to flavour patisserie products such as ice creams, mousses, ganaches, soufflés, etc. For flavouring purposes, couverture only requires melting and adding to the desired product. For moulding and setting purposes, the couverture needs to be 'tempered'. This is a process whereby the chocolate is taken through different temperatures to stabilise one particular chocolate crystal known as the 'beta crystal'. This crystal has all the characteristics of good tempered chocolate (snap, shine and retraction) enabling the prepared couverture to literally fall out of the mould it has been set in, giving a solid, shiny piece of chocolate with a perfect snap.

Types of couverture

Couverture comes in dark chocolate, milk chocolate and white chocolate varieties. You can also purchase coloured and flavoured couverture. The purchasing unit is either in callets or a solid block. For tempering purposes, callets are preferred as they melt uniformly, making tempering of the chocolate more effective.

White couverture is chocolate which does not contain the dark-coloured cocoa solids derived from cocoa beans. It only contains 30 per cent cocoa butter (the fatty substance derived from cocoa beans), milk and sugar. White chocolate is sweet, with a slight vanilla taste, and has a light flavour which is not too heavy or intense.

Milk couverture has added dried milk powder, along with cocoa butter, 40 per cent sweeteners and flavourings; and it contains a minimum of 10 per cent chocolate liquor and 12 per cent milk solids.

Plain couverture has a higher content of cocoa butter (60–70 per cent) which gives the chocolate more viscosity, and cocoa solids which give the chocolate its colour. It is more fluid than the white and milk varieties and used for decorations, moulding, enrobing and flavouring.

All three chocolates are factory-tempered before being packaged. (For small quantities it is possible to melt the chocolate whilst keeping the tempering qualities in the chocolate. This will be discussed in more detail under tempering of chocolate.)

Compound chocolate is used in food manufacturing; it gives a crisp, hard coat. It may also contain vegetable oil, hydrogenated fats, coconut and/or palm oil, and sometimes artificial chocolate flavouring. This type of chocolate does not need to be tempered in order to set. It is inferior to couverture in taste and quality, but less expensive.

What you need to know about cocoa

Cocoa bean

This was once called 'cocoa almond' or 'cocoa grain'. It is the seed that is found in the pods of cacao trees. After being treated, it is packed and sent to be sold on the international market. It is from this bean that cocoa butter, chocolate liquor, cocoa powder and cocoa nibs are extracted.

Cocoa nibs

These are roasted, shelled cocoa beans broken into small pieces. This is a very interesting product with an intense flavour – 100 per cent cocoa. It gives aroma, flavour and texture to many preparations, like sponge cakes, chocolate pralines, muffins, ice creams, cookies and cake decorations. Care should be taken not to use excess quantities so that the balance with the other ingredients is not upset.

Chocolate liquor

This is a smooth, liquid paste. In addition to being the base for other cocoa derivatives, such as cocoa butter or cocoa powder, it can be used in all types of desserts and cakes (toffee, for example). One of its main characteristics is that it contains no sugar, which gives it a slightly bitter flavour in its pure state.

Cocoa butter

Once obtained, chocolate liquor is pressed to extract the fat (cocoa butter) and separate it from the dry extract. Cocoa butter is the 'spine' of chocolate, since its proper crystallisation determines whether chocolates (couvertures) have adequate densities and melting points. We would recommend melting cocoa butter at 55°C (it begins melting at 35°C) to achieve proper de-crystallisation. Cocoa butter is used to coat with a spray gun (mixed with chocolate in greater or lesser quantity), for chocolate praline moulds, desserts, cakes and artistic pieces, or in pure form for moulds and marzipan figurines. Cocoa butter is available in various colours for decoration purposes.

Cocoa powder

Two products are extracted from pressed chocolate liquor – cocoa butter in liquid form, and dry matter, which is ground and refined to make cocoa powder. The quality of cocoa powder is a function of its finesse, its fat content, the quantity of impurities it contains, its colour and its flavour. It is very important to store it in a dry place and in an airtight container.

The main characteristics of the cacao tree and its fruit

- The majority of the world's cacao trees are concentrated around the equator.
- The cacao tree needs a hot, humid and rainy climate – the tropics are ideal for it.
- High levels of wind and sun can be damaging to the cacao tree and it must be protected from both.
- A productive tree can measure between 5 and 10 metres in height, depending on its age.
- The fruit, or 'cocoa pod', measures between 15 and 30 cm.
- Each cocoa pod holds approximately 30–40 seeds (cocoa beans).

Roasting
After being cleaned, the cocoa beans are roasted which develops the distinctive flavour of the cocoa bean.

Winnowing
After roasting, the beans are put through a winnowing machine which removes the outer husks or shells, leaving behind the roasted beans, now called nibs.

Milling – making cocoa liquor
The nibs are then ground into a thick liquid called chocolate liquor (this is cocoa solids suspended in cocoa butter). Despite its name, chocolate liquor contains no alcohol and has a strong unsweetened taste.

Pressing for the production of cocoa powder and cocoa butter
The next stage is to press the cocoa liquor and extract the cocoa butter. This leaves behind a solid mass which is then processed into cocoa powder.

Making the chocolate
The following ingredients are mixed together to make the three different types of chocolate:

White chocolate: made from the same ingredients as milk chocolate (cocoa butter, milk, sugar) but without the chocolate liquor. White chocolate must contain at least 20% cocoa butter and 14% total milk ingredients.

Milk chocolate: a combination of chocolate liquor, cocoa butter, sugar and milk or cream.

Plain chocolate: a combination of chocolate liquor, cocoa butter and sugar. Must contain at least 35% chocolate liquor.

Refining
The next step is to pass the chocolate through heavy rollers to form a fine flake. Additional cocoa butter is added at this stage and an emulsifying agent called lecithin. The mixture is now mixed to a paste.

Conching
The process of conching kneads the chocolate through heavy rollers which develops the flavour.

Tempering
The chocolate is now tempered ready for use which gives it shine, snap and retraction.

Moulding
The liquid tempered chocolate is now deposited into solid block moulds or shaped into callets ready for use.

Use in the patisserie kitchen
For small quantities, melt the couverture in the microwave following the technique for microwave tempering, or fully melt to the stated temperature for the type of chocolate and then follow the tempering procedure (see page 378).

The process of manufacturing chocolate

Tempering chocolate

As already mentioned, cocoa butter is a vital component of chocolate, since the final result depends on its crystallisation. It determines good hardness, balance, texture and shine, and it prevents excessive hardening, whitening and the formation of beads of oil on the surface.

When we melt chocolate, the cocoa butter melts and its particles separate. To achieve a perfect result we must re-bond them by cooling the chocolate (re-crystallising the cocoa butter).

Tempering allows us to manipulate chocolate and combine it with other ingredients or make artistic pieces that, when re-crystallised, regain the texture and consistency of the chocolate before it was melted.

Tempering is necessary because of the high proportion of cocoa butter and other fats in the chocolate. This stabilises the fats in the chocolate to give a crisp, glossy finish when dry.

A recipe for tempered chocolate is provided later in this chapter (Recipe 6).

Note

It is essential that a thermometer is used for the tempering process and that the chocolate moves about to develop the beta crystals.

Sugar work

Boiled sugar

Sugar is boiled for a number of purposes – pastry work, bakery and sweet-making.

Soaked sugar (approximately 125 ml water per 250 g sugar) is boiled steadily without being stirred. Any impurities on the surface should be carefully removed (otherwise the sugar is liable to crystallise). Once the water has evaporated, the sugar begins to cook – you will be notice that the bubbling in the pan will get slower. It is then necessary to keep the sides of the pan free from crystallised sugar – this can be done with a wet pastry brush. The brush should be dipped in ice water or cold water, rubbed round the inside of the pan and then quickly dipped back into the water.

The cooking of the sugar then passes through several stages, during which the temperature may be tested with a sugar thermometer or by the hand-testing method (dip the fingers into ice water, then into the sugar and quickly back into the ice water).

Degrees of cooking sugar

- **Thread (104°C)** – when a drop of sugar held between thumb and forefinger forms small threads when the finger and thumb are drawn apart.
- **Pearl (110°C)** – when proceeding as for thread, the threads are more numerous and stronger. Used for crystallising fruits, fruit liqueur making and some icings.
- **Soufflé/blown (113°C)** – if a metal loop is inserted into the sugar and blown, a small, thin, aerated ball can be formed.
- **Feather (115°C)** – if the skimmer (used for cleaning the sugar) is dipped into the sugar solution and then given a sudden jerk as if to throw the sugar away from you, long, thin, fine strings are formed. Used for fruit confit.
- **Soft ball (118°C)** – proceeding as for thread, the sugar rolls into a soft ball. Used in the production of fondant, fudge, pralines, pâte à bombe and peppermint creams.
- **Hard ball (121°C)** – as for soft ball, but the sugar rolls into a firmer ball, used for products such as Italian meringue, boiled buttercream, nougat and marshmallows.
- **Soft crack (140°C)** – the sugar lying on the finger peels off in the form of a thin pliable film, which sticks to the teeth when chewed.
- **Hard crack (155–160°C)** – the sugar taken from the end of the fingers when chewed breaks clean in between the teeth, like glass. Used for dipping fruits and the production of poured, pulled and spun sugar.
- **Caramel (176°C)** – cooking is continued until the sugar is a golden-brown colour. Used for crème caramel, caramel sauce, nougatine, croquant and praline.

Inversion or 'cutting the grain' when sugar boiling

Monosaccharides

These are single sugars, such as fructose and glucose. They are more stable and there is less chance of premature crystallisation during the boiling process.

Disaccharides

These are double sugars, such as granulated sugar (sucrose), which are made from two single sugars (fructose and glucose) which crystallise easily during sugar boiling. As granulated sugar is a key ingredient when sugar boiling and is likely to crystallise, acid is sometimes added in the form of lemon juice or tartaric acid. This breaks down double sugars into single sugars, thus making the solution more stable and less likely to crystallise. This process is known as 'inversion' or 'cutting the grain'.

Professional tips

- The purpose of glucose in sugar boiling is to add single sugars, thus making the solution more stable.
- The addition of acid will make the sugar elastic for pulling.

Sugar preparations

Poured sugar	sucre coulé
Spun sugar	sucre filé
Pulled sugar	sucre tiré
Blown sugar	sucre soufflé
Straw sugar	sucre paille
Rock sugar	sucre roche

Faults

The sugar in the jar has crystallised

The sugar poured to make these bases crystallised prematurely

Causes of premature crystallisation:

- working with dirty equipment
- working with dirty sugar
- intermittent boiling of the sugar solution
- stirring the sugar solution once boiling starts
- not removing sugar crystals from the sides of the pan which will crystallise and cause the main sugar solution to do the same
- incorrect dipping procedure for items such as fruits déguisés.

Professional tips

- Never attempt to cook sugar in a damp atmosphere, when the humidity is high. The sugar will absorb water from the air and this will render it impossible to handle.
- Work in clean conditions as any dirt or grease can adversely affect the sugar.
- The choice of equipment is important – copper sugar boilers are ideal as these conduct heat rapidly. Induction hobs are used as these cook the sugar very fast with no naked flames (which might crystallise the sugar on the sides of the pan).
- Never use wooden implements for working with or stirring the sugar. Wood absorbs grease, which can in turn ruin the sugar.
- Make sure the sugar is cooked to temperature according to the specific purpose of the product being made.
- If you are colouring the sugar, it is advisable to use powdered food colourings as these tend to be brighter. Before using, dilute with a few drops of 90 per cent-proof alcohol or water. Add the colourings to the boiling sugar when the sugar reaches 140°C and then continue to cook the sugar to the desired temperature. For poured sugar, if you want a transparent effect, add the colour while the sugar is cooking.
- Once the sugar is poured on to a silicone mat and it becomes pliable, it should be transferred to a special, very thick and heat-resistant plastic sheet.
- To keep the sugar pliable, it should be kept under infra-red or radiant heat lamps.

- For a good result with poured sugar, use a small gas jet to eliminate any air bubbles while you pour it.
- Ten per cent calcium carbonate (chalk) may be added to sugar before pouring to give an opaque effect and to improve its shelf life. This should be added at a pre-mixed ratio of 2 parts water to 1 part calcium carbonate and added to the boiled sugar solution at 140°C. The sugar is then further boiled to 160°C.
- To keep completed sugar work, place in airtight containers, the bottom of which should be lined with a dehydrating compound, such as silica gel, carbide or quicklime. Pulled sugar pieces can be vacuum-sealed for storage.
- If you are using a weak acid, such as lemon juice or cream of tartar, to prevent crystal formation, it is advisable to add a small amount of acid towards the end of the cooking. Too much acid will over-invert the sugar, producing a sticky, unworkable product.
- The best sugar to use is granulated sugar, straight from a 1 kg bag.

There is a range of commercial products on the market that greatly assist the pastry chef in the production of specialised sugar work – isomalt, for example. This product is not as hygroscopic as normal sugar solutions, thus enabling finished goods to be stored relatively easily. It can be used several times over and has a long shelf life. Such commercial products are simple to use, quick and labour saving.

Equipment used in sugar boiling

Refer to the Appendix for lists and detailed explanations of specialist equipment.

Marzipan (almond paste)

There are two methods of producing marzipan – raw and cooked (see recipes in this chapter).

Marzipan can be purchased ready-made, as it is much more of a consistent product. There are various types on the market, such as that suitable for diabetics, organic and specialist modelling marzipan. Almonds are the key ingredient in marzipan and are grown principally in California and Spain with small quantities coming from other countries around the Mediterranean. Mediterranean almonds are generally considered to be superior to Californian because they are cultivated in more natural and wilder surroundings with more favourable climatic conditions.

High-grade marzipan generally has a lower sugar content (around 35 per cent), whereas marzipans used for figures or decorations will have a higher sugar content (up to 70 per cent) to make them more suitable for use as a modelling mass. Due to the high sugar content in marzipan, it dries very quickly when exposed to the air and should be kept covered at all times.

Marzipan is used extensively in pastry work. It can be rolled out the same as pastry (but using icing sugar instead of flour) to cover cakes. It can be left smooth or textured using a marzipan roller (as in the photo of strawberry gateau, Chapter 8, Recipe 7). Other uses include the filling of fruits déguisés (Chapter 9, Recipe 3), or the covering of pain de Gêne for fondant dips (Chapter 9, Recipe 1). Marzipan with a high almond percentage can be used as a base for almond tuiles, lightly toasted shapes used as petit fours and for modelling.

Pastillage (gum paste)

Pastillage is a white paste that is made from icing sugar, cornflour, lemon juice and gelatine or gum tragacanth, which makes the pastillage pliable. The paste is rolled out very thinly, dusting the working surface using cornflour in a muslin bag, and then cut into various shapes using templates of set designs. It is then left to dry until solid on flat boards, turning over periodically. Once fully dry, the pastillage is stuck together using royal icing into the shape of the template – square boxes, heart shaped caskets, etc. Centrepieces can be made by combining pastillage cut-out shapes with pulled sugar. As the pastillage is set firm, it can only be used for decorative purposes. (See Recipe 46 for a pastillage centrepiece.)

1 Nougatine sheets

Caster sugar	500 g
Fondant	500 g
Glucose	500 g
Flaked almonds	150 g

1 Cook the fondant, sugar and glucose to a blond caramel colour.

2 Roast the flaked almonds on a baking tray in the oven and mix into the sugar when cooked.

3 Pour onto a non-stick mat and cool down until cold.

4 Crush in a food processor until it becomes a sandy texture, pass through a sieve and keep in a sealed container until required.

5 When required, place a plastic cut-out template onto a non-stick mat.

6 Dust over the template with the nut powder, remove the template.

7 Bake at 180°C for approximately 10 minutes.

8 When cold and firm, remove from the mat.

9 Can be used to decorate plated desserts and tortens.

2 Bubble sugar

Makes >	750 g
Water	100 ml
Fondant	450 g
Glucose	300 g
Unsalted butter	20 g
Colouring	

1 Heat the water, fondant and glucose to 150°C.

2 Remove the thermometer. Add the butter and colouring. Swirl to mix the ingredients.

4 Pour onto a silicone mat and leave to set.

5 Break into pieces. The pieces can then be stored in an airtight container until needed.

6 When needed, blitz the pieces of sugar in a mixer.

7 Sieve them over a silicone mat in a slightly uneven layer.

8 Melt in the oven at 165°C.

9 Allow to set. The sugar sets very thin, with bubbles throughout.

10 Store in an airtight container and keep dry. Use to decorate sweet dishes.

3 Dried fruits

Water	500 ml
Granulated sugar	300 ml
Lemon juice	25 ml
Glucose	50 g
Fruit (lemons, limes, oranges, apples, pears)	

1 Boil the water, sugar, lemon juice and glucose together to make a syrup.

2 Prepare the fruit. Leave the skin on. Remove the cores from apples and pears. Slice the fruit very thinly using a meat slicer. Brush slices of apple or pear with lemon juice to prevent browning.

3 Pass each slice through the hot syrup. Lay them on silicone mats.

4 Leave to dry in the oven overnight, or use a commercial dehydrating cabinet (see Appendix).

4 | Piped sugar spirals

	Makes >	approximately 50
Water		100 ml
Fondant		500 g

1 Place the water in a copper sugar boiler. Add the fondant.
2 Cook until a pale caramel forms.
3 Allow to stand for 5 minutes.
4 Pipe onto a silicone mat in spirals.
5 Allow to set.
6 Warm the spirals and pull them up to the desired height.

5 | Spun sugar

Granulated sugar	500 g
Water to saturate	
Liquid glucose	60 g

1 Place the sugar into a sugar boiler.
2 Add water until the sugar is just saturated. Stir with a metal spoon to distribute the water, ensuring that all the sugar is moistened.
3 Gently dissolve the sugar, removing any scum as it rises to the surface.
4 Add the liquid glucose and boil to 160°C.
5 Once the temperature is reached, arrest the cooking by placing the pan into a bowl of cold water.
6 Remove the sugar pan from the cold water and leave until the boiled sugar syrup slightly thickens.
7 Using a sawn-off whisk, spin the sugar over a lightly oiled steel forming thin strands of brittle sugar.

Professional tip

The fondant used in Recipe 4 can also be used to make spun sugar.

Tempering chocolate techniques

Chocolate that is to be moulded or used in decoration needs to be tempered to achieve snap, shine and retraction in the finished product.

Temperatures for the three types of couverture:

Initial melted temperature	Finished working temperature
Plain couverture 45°C	31°C–32°C
Milk couverture 40°C	30°C–31°C
White couverture 40°C	28°C–30°C

Note

The melting and working temperatures are given as a guideline. Some brands of chocolate may vary. Always check the tempering instructions on the packaging.

Table-top method

1 Melt carefully to the specific temperature for the type of couverture, avoiding steam, moisture and over-heating. (The use of a chocolate melting tank is ideal for this – see Appendix.)

2 Once the couverture has reached the melting temperature, remove from the heat source.

3 Pour 70 per cent of it on to a very clean and dry marble surface/slab. Work continuously by spreading outwards and pulling back to the centre with a step palette knife until the couverture starts to thicken and the formation of the good beta-crystal is developed.

4 Quickly add the couverture back to the remaining 30 per cent, stirring continuously dispersing and seeding the beta-crystal into the liquid chocolate until it reaches its finished working temperature (31°C–32°C for plain, 30°C–31°C for milk and 28°C–30°C for white).

5 Check the finished temperature with a digital probe. If the chocolate is still too warm, pour a small amount once more onto the marble and repeat the steps above until the chocolate reaches the desired temperature.

Injection (or seeding) method

1 Take 30 per cent of couverture callets of the total weight being tempered.

2 Melt the remaining 70 per cent of the couverture following stages 1 and 2 of the table-top method above.

3 Remove the container of melted chocolate and stand it on the table top with a folded cloth underneath (this is to prevent the chocolate from setting on the base).

4 Gradually add the remaining 30 per of the chocolate callets to the melted couverture, stirring continuously until the finished working temperature is achieved (31°C–32°C for plain, 30°C–31°C for milk and 28°C–30°C for white).

5 For large-scale production, a wheel tempering tank is used (see Appendix).

Microwave method (for small quantities)

As previously discussed, couverture is packaged already tempered.

1 Take 500 g couverture and place into a plastic bowl.

2 Warm the chocolate in short intervals in a microwave oven set at 50 per cent until the chocolate partially melts and there are still signs of solid chocolate.

3 Remove from the microwave and continue to stir the chocolate until the solid chocolate pieces melt down. Any solid pieces are still tempered and the gentle mixing and melting will seed the newly melted chocolate.

Note

Mycryo cocoa butter is a commercial product for use in tempering chocolate. It can save time. To use, follow the manufacturer's instructions.

Professional tips

- If the working temperature is exceeded by more than 3°C, the process will have to be repeated as the couverture will not be correctly tempered and faults will occur.
- If the temperature of the chocolate drops during processing, it can be gently reheated with a heat gun to the working temperature without causing any detrimental effects to the characteristics of the chocolate.
- Always keep water away from melted couverture and never store in humid conditions.
- The ideal room temperature for working with chocolate is 18°C with 20 per cent humidity.
- Chocolate products should be stored in dry places at 15°C–16°C and at 20 per cent humidity.
- Chocolate absorbs all odours and should therefore be stored well covered.
- The higher its fat content, the faster chocolate melts in your mouth.
- In tempering, it is essential to check the temperature with a thermometer and to perform the 'paper test'. This is done by dipping a piece of paper in the tempered chocolate. The tempering is optimal if, in about 2 minutes, it has crystallised with a flawless, uniform shine and without stains or fat drops on the surface.
- A glossy surface is a sign of good tempering.
- Two important points are to be followed to achieve good tempered couverture – correct temperature and continuous movement of the chocolate. Movement develops beta crystals in the chocolate.

This couverture was tempered and spread over textured sheets (see Appendix) to give the desired finish

Faults

The two most common faults in prepared couverture are fat bloom and sugar bloom.

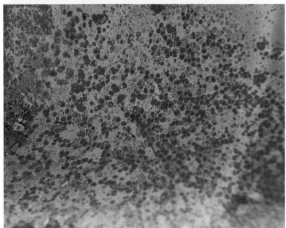

The picture at the top shows fat bloom. This is caused by:

- poor tempering of the chocolate
- incorrect cooling methods
- covering a confectionery product that is too cold
- warm storage conditions.

The picture underneath shows sugar bloom. This is caused by:

- storage of chocolates in damp conditions
- working in humid conditions
- using hygroscopic ingredients
- products packaged which have high liquid content and stored in a warm area. (Vapour is given off which is entrapped in the packaging, creating a layer of moisture on the surface of the chocolate.)

7 | Chocolate spirals

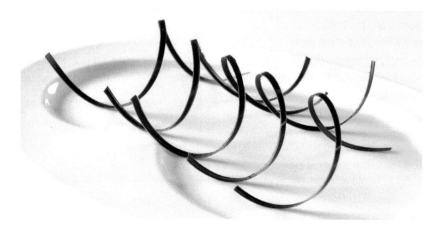

1. Apply tempered couverture onto an acetate strip.
2. Comb the chocolate with a grooved scraper to form parallel lines.
3. Allow the chocolate to partially set.
4. Twist the acetate and lay the strip inside a hollow drainpipe.
5. Once fully set, remove from the acetate and separate into curls.

Spread tempered couverture onto an acetate strip

Comb into parallel lines

Once partially set, twist the acetate and lay it inside a hollow pipe

Once fully set, remove from the pipe and peel away from the acetate

8 Chocolate run-outs

1 Draw two parallel lines 5 cm apart on silicone paper.
2 Take a small quantity of tempered plain couverture or melted compound chocolate.
3 Add drops of stock syrup until the chocolate slightly thickens.
4 Pipe fine line designs between the two parallel lines.
5 Allow to set, then remove using a palette knife.

9 Transfer-sheet swirls

1 Thinly spread tempered couverture onto the reverse side of a transfer sheet.
2 Allow to partially set, then mark with a thin-bladed knife, avoiding cutting through the plastic sheet. (A ruler or pastry cutter may be used to create various shapes.)
3 Place a sheet of silicone paper on top of the chocolate.
4 Roll up and allow to set.
5 Unfold the plastic sheet to release the chocolate swirls.

Spread tempered couverture over the acetate transfer sheet

When partially set, mark with a knife, but do not cut through the acetate

Cover with silicone paper, roll up and leave to set

Brush the acetate sheet with coloured cocoa butter; once set, mark with a cocktail stick and brush over with metallic powder

Method one

1 Brush coloured melted cocoa butter onto a plastic guitar sheet, allow to set.
2 Mark with a cocktail stick cutting through the coloured cocoa butter.
3 Brush over with metallic powder.
4 Thinly cover with tempered coverture.
5 When the couverture has almost set, cut into squares using pastry wheels.

Method two

1 Using a wood grainer, spread plain tempered couverture onto a plastic guitar sheet.
2 Once set, spread with tempered white couverture.
3 When white couverture is set cut into desired shapes.

Apply tempered couverture to the acetate sheet with a wood grainer

Method three

1 Using crinkled cling film, dip into tempered plain couverture and dab onto the surface of a sheet acetate.
2 Once set, spread with tempered white couverture.
3 When white couverture is set, cut into desired shapes.

Dab the acetate sheet with cocoa butter in different colours

11 Plastic sheet cut-out shapes

1 Take one plastic guitar sheet and brush with coloured melted cocoa butter. Allow to set.

2 Thinly spread with tempered couverture.

3 Place second guitar sheet on top and spread the chocolate thinly by smoothing out using a ruler on the surface of the second plastic sheet.

4 When the chocolate has almost set, make indentations into the surface of the top plastic sheet using pastry cutters.

5 Once the chocolate has set, carefully remove the top plastic sheet.

6 Remove the shapes from the first plastic sheet using a palette knife.

7 This method gives a shiny surface on both sides of the chocolate.

Spread tempered couverture over a plastic guitar sheet (shown here without decoration)

Place another sheet over the top, and spread the couverture thinly between the sheets by pressing down with a ruler

Make indentations in the couverture, pressing through the top sheet but not breaking it

12 Chocolate swirls using frozen alcohol

1. Place tempered plain coverture into a piping bag.
2. Fill a measuring jug with frozen alcohol such as kirsch or Cointreau.
3. Pipe spirals of tempered couverture into the frozen alcohol.
4. Once the couverture has set, lift out of the alcohol and use as a decoration.

Pipe spirals of tempered couverture into frozen alcohol

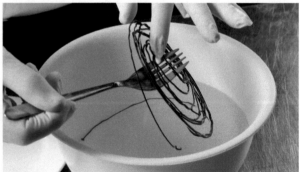

Once set, lift the couverture out of the liquid

13 Spaghetti chocolate

1 Colour white tempered couverture with yellow cocoa butter.

2 Place into a piping bag and quickly pipe very thin lines over a marble slab that has been in the freezer.

3 Using a chocolate scraper remove the solid chocolate spaghetti strands.

Note

This decoration is particularly effective on Easter display pieces.

Pipe very thin lines of coloured, tempered couverture over a frozen marble slab

Use a scraper to lift the strands off the slab

14 Chocolate fans

1 Heat either white, milk or plain couverture to 40°C (for white chocolate fans, use 600 g white couverture with 60 g vegetable oil).

2 Heat a flat tray in the oven until it is just warm (40°C to 45°C).

3 Using a step palette knife, spread the melted couverture thinly and evenly.

4 Place in the refrigerator to set solid. If you intend to keep for long periods in the refrigerator, cling film the tray to keep away the humidity.

5 When required, remove from the refrigerator and allow to come back to room temperature.

6 To create curls, either use a filleting knife pulled towards you under the chocolate or use a chocolate scraper and work away from you with your finger touching the bottom of the scraper and cutting under the chocolate in one movement.

For marbled chocolate fans

Spin the warm tray with dark couverture, then spread thinly with white couverture.

15 Chocolate discs using mats

1 Place a commercial cut-out mat template on a transfer sheet.

2 Spread with tempered couverture filling the cut-out shapes.

3 Smooth the surface with a chocolate scraper.

4 Once the chocolate has set, remove the mat leaving the shaped chocolate discs on the transfer sheet.

5 Remove the discs, leaving the transfer sheet print on the chocolate.

Place a cut-out mat template on a transfer sheet, and spread tempered couverture over it

Smooth and allow to set, then remove the mat template, leaving the shaped chocolate pieces behind

Try something different

Customised transfer sheets can be specifically made to your own requirements.

16   Spun chocolate over frozen pipes

1   Freeze a stainless steel pipe.

2   Place tempered plain coverture into a piping bag and spin over the pipes.

3   Once the chocolate has set, gently remove the chocolate pieces.

Spin tempered couverture over frozen pipes

Once set, gently remove the chocolate

It should be possible to remove whole pieces without many breaks

17 Chocolate cigarettes

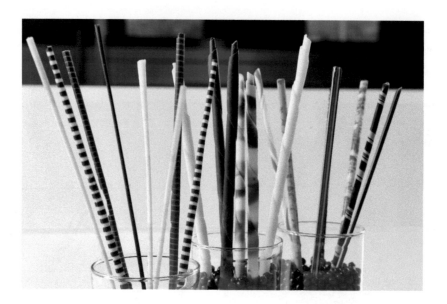

Two-tone coloured cigarettes

1 Spread tempered white couverture onto a marble slab and comb using a grooved scraper, or the plastic toothed edge from a cling film box.

2 When the chocolate has almost set, cover with a thin coat of tempered plain couverture.

3 Once the plain couverture has set, scrape up into thin two-tone cigarettes.

Multi-coloured cigarettes

1 Using crinkled cling film, dab different coloured tempered cocoa butter onto a marble slab.

2 Thinly spread with tempered white couverture.

3 Once set, scrape up into white cigarettes with a multi-coloured outer surface.

White or plain cigarettes

1 Spread either white or plain tempered couverture thinly onto a marble slab.

2 When almost set, scrape up into cigarettes.

Spread tempered white couverture on marble and comb through it

When the white couverture has almost set, pour tempered plain couverture over it

Spread the plain couverture thinly over the white

Clear away excess couverture from the edges

Working at the edge of the chocolate, scrape up thin cigarettes

Professional tip

It is advisable to use a special scraper for this task.

Never attempt to make these cigarettes in a warm kitchen.

18 Chocolate tubes

1 Take a piece of transfer sheet 10 cm wide.

2 Place tempered couverture into a piping bag and zig-zag across the sheet.

3 Pipe a straight line of couverture along one edge of the transfer sheet.

4 Allow to almost set, then fold into a tube. Secure with sellotape.

5 Once set, carefully remove the transfer sheet, releasing the chocolate tube.

Note

Chocolate tubes can be used for decoration or cut into smaller tubes and filled with sorbets or mousses.

Pipe tempered couverture in zig-zags across an acetate transfer sheet, leaving part of the sheet clear

When the couverture has almost set, roll it over in the sheet so that it forms a tube

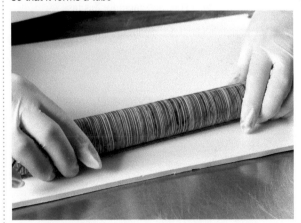

Wrap up the roll in the remaining acetate and leave to set before unrolling

19 Chocolate leaves using dipping irons

1 Place the dipping irons in the freezer overnight.
2 Once frozen, dip the surface of the irons into tempered couverture.
3 Remove the iron from the couverture and allow the couverture to set on the surface of the iron.
4 Release the chocolate off the iron using a knife, revealing the veined shape of a leaf.

20 Chocolate teardrops and tubes

1 Place acetate strips onto a flat surface.
2 Embellish using coloured tempered cocoa butter, tempered white, milk or plain couverture. Allow to set.
3 Thinly spread the acetate strip with tempered couverture.
4 When the couverture is almost set, join both ends of the acetate together and secure with a paper clip.
5 Stand inside a metal ring and shape into a teardrop.
6 Once set, remove acetate and fill the teardrop shell with mousses.

Chocolate tubes

1 For tubes, follow the same process as the teardrop but this time join the acetate to make a tube and stand inside a metal ring until set.

General purpose recipes for sugar boiling

Recipe 1

Water	350 ml
Granulated sugar	1 kg
Cream of tartar	1–2 g
Glucose	50 g

1 Place water, sugar and cream of tartar in a pan and stir on low heat. Do not boil mixture. Before it reaches boiling point, skim using a tea strainer dipped into water, and wash the sides of the pan. (This process will take approximately 20 minutes.)

2 Once the sugar is clean and still simmering, warm the glucose in the microwave and add to the sugar solution. Bring to the boil.

3 Boil for one minute, then pour into a kiln jar almost to the top. Store until required (this can be stored for up to three months).

4 Cook one kg at a time to 165°C. Skim and colour at any stage. (Mix colour with water and warm in microwave to bring together.)

5 Pour onto silicone paper, cool and break up. Store with de-humidifying product.

6 When required, soften either in a microwave or under a lamp and then pull to obtain a glossy shine.

7 Place under heated lamp and process as required.

Note

This recipe can be used for pulled sugar, blown sugar, ribbons, etc. using conventional ingredients.

Fault

This poured sugar crystallised prematurely

Causes of premature crystallisation:
- working with dirty equipment
- working with dirty sugar
- intermittent boiling of the sugar solution
- stirring the sugar solution once boiling starts
- not removing sugar crystals from the sides of the pan which will crystallise and cause the main sugar solution to do the same

Recipe 2

Granulated sugar	1 kg
Water	500 g
Glucose	200 g
Tartaric acid (see below)	

1 Bring the water and sugar to simmer. Clean as the previous recipe.

2 Add glucose.

3 Boil to 165°C, then add 12–24 drops of tartaric acid using a pipette, depending on the elasticity you need.

4 Pour onto silicone mat and pull until a glossy sheen is achieved.

5 Place under a heated lamp and process.

To prepare tartaric acid for this recipe, simply mix together equal quantities of tartaric acid and water. Heat in the microwave until the mixture becomes clear. Keep in a pipette bottle.

Recipe 3

Glucose	400 g
Fondant	600 g

1 Warm glucose in a sugar boiler.

2 Add fondant cut into chunks.

3 Heat until liquid clears.

4 Boil to 165°C.

5 Finish as in previous recipes.

Sugar syrups used in patisserie work for soaking sponges

Base syrup	
Water	1 kg
Sugar	1.4 kg
Rum syrup	
Base syrup	1 kg
Rum	400 g
Kirsch syrup	
Base syrup	1 kg
Kirsch	400 g
Grand Marnier syrup	
Base syrup	1 kg
Grand Marnier	400 g
Curaçao syrup	
Base syrup	1 kg
Curaçao	400 g
Vanilla syrup	
Water	1.25 kg
Sugar	800 g
Vanilla pods, seeds from	5

Faults in sugar preparations

Poured sugar not setting:
- sugar insufficiently boiled
- sugar thermometer giving inaccurate reading.

Uneven texture of spun sugar:
- not allowing the sugar solution to thicken slightly before spinning.

Milky-coloured pulled sugar:
- too much acid used, over-inverted
- pulling the sugar too soon.

Sugar cracking when assembling poured sugar pieces using a blowtorch:
- sugar bases are too cold.

22 Nougatine (croquant)

	Makes >	1 kg
Granulated sugar		500 g
Water		200 ml
Glucose		100 g
Flaked almonds		375 g

1 Boil the sugar, water and glucose to a light caramel.

2 Remove from the heat and immediately stir in the almonds.

3 Roll out between silicone mats on a baking sheet.

4 Cut out into shapes as required.

Variation: praline

Praline is made using the same method as nougatine, but using 150 g hazelnuts and 150 g whole almonds, in place of the flaked almonds. Pour the mixture on a non-stick mat, cool until solid, then break down to a fine powder using a food processor.

23 Meringue sticks

1 Make a cold meringue with 100 g egg white, 200 g caster sugar and a pinch of cream of tartar. (For raspberry meringue, add 50 g raspberry purée and red colour at the full meringue stage.)

2 Using a plain nozzle, pipe out straight lines 4 cm diameter onto a non-stick mat.

3 Sprinkle crushed dried raspberries onto raspberry meringue or cocoa nibs on white meringue.

4 Place in a dehydrator or an oven at 80°C for 2 hours until fully dry and crisp.

Professional tip

The meringue can also be spread out thinly onto a non-stick mat and dried as above. For decoration, break into irregular shapes.

24 Poured clear sugar centrepiece

Water	500 ml
Granulated cane sugar	1 kg
Glucose	200 g
Soluble strong powdered colours, pre-mixed (if required)	few drops

1 Clean out a copper sugar boiler with salt and lemon. Rinse but do not wipe dry.

2 Place the water in the pan, then the granulated sugar, then the glucose.

3 Bring to the boil slowly. Stir carefully – do not scrape the bottom or sides.

4 Skim off impurities.

5 Cook on a fast boil. Place a thermometer in the pan.

6 If coloured sugar is required, when the temperature of the sugar reaches 150°C, add a few drops of colour solution.

7 Cook until the temperature reaches between 155°C and 160°C. Remove from the heat and arrest cooking by placing the pan in a bowl of cold water. Allow to stand.

8 Pour the sugar out into moulds. Always pour into the centre of the mould in a steady stream.

9 Allow to set completely before trying to move or touch the sugar.

10 To assemble a shape from moulded pieces, heat the edge of each piece with a spirit burner to melt the sugar slightly. Hold it in place and it will harden, welding the two pieces together.

11 To finish, blast with cold air from a hairdryer to quickly set.

Pour the sugar into the moulds

Once set, dip the edge of each piece in hot sugar solution (or heat it with a burner) and then stick it into place

25 Pulled sugar using isomalt

Isomalt	1 kg
Water	100 ml

1 Bring the isomalt and water to the boil. Whisk to make sure that the isomalt dissolves completely.

2 Cook until the temperature reaches 165°C (not more than 20 minutes, or it will get too hot).

3 Plunge the pan into cold water to arrest the cooking.

4 Pour out and allow to set.

5 Place the sugar on a board under a lamp and allow it to reheat slowly. When it starts to run, turn it. When the sugar becomes pliable, it is ready to be pulled.

6 Pull and fold the sugar evenly, 20 to 30 times, until it is shiny. Use the lamp to keep it at the right consistency.

7 After this point, if the sugar gets cold and hard, it can be brought back by microwaving on a low heat setting in 8- to 10-second bursts.

8 Form into shape as required.

Note

Isomalt is sugar rearranged with hydrogen. It is less susceptible to moisture than sugar.

It was developed as a sweetener for diabetics, but unfortunately it acts as a laxative in large quantities.

Reheat the poured sugar slowly, until pliable, and then begin to pull it

Pull and fold repeatedly until the sugar is shiny, lighter in colour and less opaque

26 Blown sugar swan

1 Warm pulled sugar using either isomalt or traditional sugar solution.

2 Shape into a smooth ball and make an indentation using your finger.

3 Insert a warm copper tube into the indentation and press the outer sugar to seal the pipe inside of the ball.

4 Apply pressure with a sugar pump, shaping the sugar into a swan.

5 Cool using a hair drier on a cold setting until the sugar has set.

6 To remove the copper tube once the sugar has set into the shape of a swan, gently warm the tube 6 cm away from the sugar figure and gently twist to release.

7 Attach the beak made from pulled yellow and black sugar.

8 Make the wings by pulling white sugar and shaping in a leaf mould. Warm on the blow torch and attach to the body of the swan.

9 Stand on a poured sugar tin foil base.

Use a sugar pump to shape a ball of warm pulled sugar into the swan. This picture shows the shapes that form during this process. Make the wings over a leaf mould.

Release the swan from the pump and trim to shape using a hot knife

Warm each part with the blow torch so that it can be attached. Start by attaching the swan's body to a base.

Attach the wings

Attach the tail

27 Sugar spheres using silicone mould, embellished with titanium dioxide

1 Prepare the mould by joining together and holding with sellotape.

2 Boil isomalt to 165°C using 10 per cent water and fill the mould half way, squirt titanium dioxide into the centre, swirl using a knife.

3 Leave to stand, then keep topping up until full.

4 Once completely set (12 hours), remove from the mould and warm with a hair drier. This will prevent the sugar from cracking when applying heat.

5 Apply a blow torch and melt outer surface until sugar becomes clear. Cool down with hair drier.

Fill the prepared mould with boiled sugar

Remove from the mould

Run a blow torch over the surface to finish

28 Pulled sugar ribbon

1 Align two to three colour strips of malleable pulled sugar under a heated lamp.

2 Gently pull and stretch until double in length.

3 Cut the strip in half and join side by side (thus giving six strips if you started with three).

4 Once again stretch, keeping the sugar the same thickness along the strip.

5 Continue to stretch until you have a thin, shiny, multi-coloured ribbon.

6 Cut into lengths with a hot knife, and heat to bend into shape.

Poured sugar base (Recipe 30) with rock sugar (Recipe 29), pressed sugar (Recipe 35), swan (Recipe 26), sphere (Recipe 27) and ribbon (Recipe 28)

29 Rock sugar

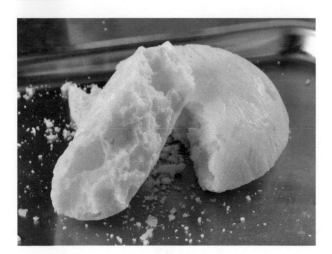

1 Use 1 litre of the sugar syrup from Recipe 21.
2 Boil to 140°C and whisk in 60 g of royal icing.
3 This causes the sugar syrup to become opaque and bubble up in the pan.
4 Pour into a bowl lined with oiled tin foil.
5 Once cold, remove from the mould and break up into small rocks used for decoration.

30 Poured sugar bases

Plain opaque sugar bases

1 Lightly grease the sides of a flan ring and stand on a non-stick mat.
2 Boil the sugar opaque using calcium carbonate (as for Recipe 31), adding colour.
3 Boil to 160°C, arrest in cold water and shake out any bubbles.
4 Pour into greased flan rings.
5 Once set, remove the flan ring.

Marbled sugar bases

1 Lightly grease the sides of a flan ring and stand on a non-stick mat.
2 Boil the sugar opaque using calcium carbonate (as for Recipe 31), adding colour.
3 Boil to 160°C, arrest in cold water and shake out any bubbles.
4 Add drops of different liquid colours to the surface of the sugar.
5 Swirl together to form a marbled effect and cast into the flan ring.
6 Once set, remove the flan ring.

Tin foil sugar bases

1 Lightly grease the sides of a flan ring and stand on crinkled tin foil, shiny side facing upwards.
2 Using Recipe 24, boil to 160°C using calcium carbonate, and keeping the sugar clear.
3 Add liquid colour and cast into the flan ring.
4 Once set, remove the flan ring and trim the tin foil along the sides of the sugar using a pair of scissors.

31   Poured sugar centrepiece using calcium carbonate

Sugar solution for pouring	
Granulated sugar	1 kg
Water	350 g
Liquid glucose	100 g
Chalk solution (to make sugar opaque)	
Powdered calcium carbonate	115 g
Water	250 g

1 Boil one litre of the sugar solution to 140°C, add one tablespoon of chalk solution (this is the calcium carbonate and water mixed together to form a paste) and liquid colour.

3 Once the sugar has reached 160°C, remove from the heat and arrest the cooking of the sugar by placing the pan into a bowl of cold water.

4 Remove the pan from the water and place onto a folded cloth on the table top.

5 Shake the pan to encourage any bubbles to rise to the surface.

6 Pour the cooked sugar solution into ready-prepared lightly greased rubber cut-out moulds, standing on non-stick mats. (See Appendix.)

7 Once the sugar has set, carefully remove the sugar pieces, ensuring that there are no finger prints on the surface of the sugar and place onto silicone paper.

8 To assemble the sugar shapes, use a blow torch to melt the sugar and stick the shapes together.

9 Parts of the sugar centrepiece can be highlighted using colours sprayed from an airbrush.

Fault !

This poured sugar base has started to crystallise. This may have happened because:

- The calcium carbonate paste was too thick, and did not dissolve into the boiled sugar solution
- The boiled sugar solution was stirred during cooking
- The sugar solution boiled intermittently
- Equipment was dirty or sugar contaminated with other foods
- The sides of the sugar boiler were not washed with cold water during the boiling process.

32 | Iced sugar

1 Take a cylindrical container with an open top and bottom and stand on a cooling wire over a tray.

2 Fill the cylinder with chunks of ice.

3 Using the general purpose sugar boiling solution, boil to 160°C, adding any colour.

4 Pour over the ice and leave to set.

5 Once the ice has melted, remove the sugar piece.

Professional tip

Because the sugar comes into contact with water, this piece needs to be used on the day it is produced.

Try something different

The same technique can be used with tempered couverture.

Pour boiled sugar over chunks of ice to make this decoration

33 Blowtorch sugar flower

1 Fill a small stainless steel bowl with granulated sugar.

2 Use a ladle to make an indentation in the centre of the sugar.

3 Using a blowtorch, melt and caramelise the sugar within the indentation.

4 Once the caramelised sugar has set, remove and brush off any excess sugar.

Make an indentation in granulated sugar and use a blow torch to melt the sugar at that point

Once set, lift out the flower of caramelised sugar

Combined sugar centerpiece: poured sugar base (Recipe 30), rock sugar (Recipe 29), iced sugar (Recipe 32), moulded sugar (Recipe 34) and blowtorch sugar flowers (Recipe 33)

34 Boiled moulded sugar

1 Fill a deep flat tray with granulated sugar.

2 Use shaped tin foil to make an indentation in the sugar.

3 Using the general purpose sugar boiling solution, boil to 160°C, adding any colour.

4 Pour the sugar solution into the indentation made by the tin foil.

5 Sprinkle sugar on top of the surface of the boiled sugar.

6 When the sugar is almost set, remove from the granulated sugar and reshape.

7 This sugar piece is good to use as a support to place other sugar items on.

Health and safety !

Ensure the sugar has fully set, and is not still liquid, before handling it. Wear rubber gloves for additional protection.

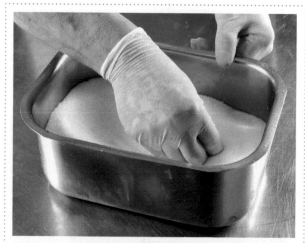

Make an indentation in granulated sugar

Pour sugar solution into the indentation

Sprinkle sugar over the top

35 Pressed sugar

1 Mix egg white with coloured granulated sugar to form a firm, moist, textured sugar solution.

2 Grease and line a small bowl with cling film.

3 Fill and press the sugar solution into the bowl.

4 Place in dehydrator set at 60°C for 4 hours until solid.

5 Turn out of the bowl.

6 Pressed sugar is good to use as a base for centrepieces.

Mix granulated sugar with egg white and food colouring

Fill a small, lined bowl with the sugar mixture and press down

Leave in a warm place for 24 hours to set, or dry in a dehydrator, then remove the pressed sugar

Break down white, milk and plain callets of couverture in a food processor. Rub over a sieve to remove fine particles of chocolate, then bind with liquid tempered couverture.

1 Place white, milk and dark chocolate couverture callets in a food processor.

2 Process until the callets are two-thirds their original size.

3 Rub the processed callets on a sieve to remove any fine particles, leaving the larger pieces of couverture.

4 Mix the processed callets with tempered liquid white couverture to bind all three coloured callets together.

5 Press into a square or round mould and leave to set.

6 Once set, remove from the mould and melt the surface and sides on a warm marble slab.

7 Use a scraper to smooth the surface and sides revealing the different coloured chocolate creating a granite effect.

8 Once the full granite effect is achieved by melting and scraping, a shine can be obtained by dipping kitchen paper in iced water and rubbing the chocolate surface several times.

Press the mixture into a mould and leave to set

Once set, remove from the mould and melt the surface and sides on a warm marble slab. Use a scraper to smooth the surface and sides.

37   Poured chocolate base with metallic powder

1   Stand a flan ring on crinkled tin foil and place a round tube ring off centre.
2   Fill with tempered liquid plain couverture.
3   Once set, remove the ring and the flan ring.
4   Invert the chocolate and remove the tin foil from the chocolate base leaving a crinkled surface on the chocolate.
5   Brush the surface with metallic powder.

Place a flan ring over crinkled tin foil (or bubble wrap) and fill with tempered couverture

Once set, remove the mould and the foil, and brush the chocolate with metallic powder

38   Bubble-wrap chocolate bases

1   Pour liquid tempered chocolate on to bubble wrap.
2   Once the chocolate has set, remove the bubble wrap, leaving the bubble wrap texture on the chocolate.

39 Chocolate poles

1 Cut a textured sheet and shape into a tube.
2 Tape the tube to secure or use a metal ring.
3 Stand the tube upright on a frozen metal tray.
4 Fill a piping bag with tempered liquid plain couverture.
5 Pipe 1 cm of tempered liquid couverture into the base of the tube and leave to set.
6 Once the chocolate has set at the base, fill the tube to the top with the remaining tempered liquid couverture.
7 Once the chocolate has set, remove the textured sheet, revealing the chocolate pole.
8 Attach the chocolate pole to the chocolate base by piping melted tempered liquid couverture into the round left from the tube ring on the chocolate base.
9 Insert the pole on top of the liquid tempered couverture and use freeze spray to set the chocolate to secure the pole.
10 The pole can now be used as a base to attach other chocolate decorations onto.

40 Cut-out chocolate free style

1 Spread tempered liquid couverture onto a sheet of silicone paper.
2 Place a second sheet of silicone paper on top of the chocolate and level the surface by using the edge of a ruler.
3 When the chocolate has almost set, cut through both sheets of silicone paper into various shapes.
4 A template can be placed on the surface of the silicone paper to give more definition.
5 Once the chocolate has fully set, remove both pieces of silicone paper from the top and bottom of the chocolate.

41 Modelling chocolate

White

White couverture	890 g
Stock syrup made from equal quantities of sugar and water	100 g
Liquid glucose	270 g

1 Melt the couverture and cool down to 26°C.
2 Warm the stock syrup and glucose until dissolved and cool down to 32°C.
3 Mix the stock syrup and glucose with the chocolate.
4 Spread on a non-stick mat and leave covered for 12 hours to firm up.
5 Use as required.
6 The modelling chocolate can be coloured or made whiter by adding titanium dioxide.

Dark

Water	100 g
Glucose	400 g
Plain couverture	1.2 kg

1 Boil water and glucose until dissolved.
2 Add to the melted chocolate.
3 Spread onto a non-stick mat and leave covered for 12 hours to firm up.
4 Use as required.

Note

Both types of modelling chocolate may be used for ribbons, roses, baskets, etc.

Method 1: single-sided mould producing half-figure shapes

1 Polish the inside of the half-figure mould with cotton wool.

2 Using a fine paint brush, highlight any areas within the mould with coloured liquid cocoa butter, white tempered couverture or plain tempered couverture.

3 Once the embellishment of the mould is set, fill with tempered liquid couverture, tap the side of the mould to release any air bubbles to the surface of the chocolate.

4 Invert the mould, once again tapping the side of the mould to remove excess chocolate, leaving a thin coating of chocolate inside the mould.

5 Using a chocolate scraper remove excess chocolate from the top of the mould and invert the mould flat side down onto a level surface lined with silicone paper.

6 Once the chocolate has set, remove the half figures from the mould.

7 Warm a flat metal tray and quickly warm the edge of one of the half figures to melt the couverture.

8 Place the other matching half figure on top and leave until fully set.

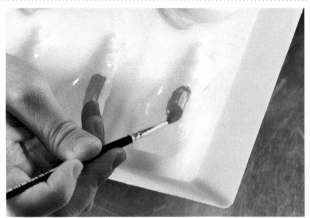

Highlight parts of the mould using coloured cocoa butter

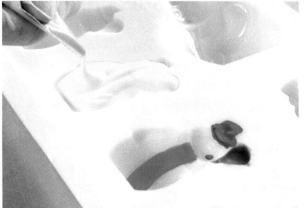

Fill the mould with tempered couverture and tap to release any air bubbles

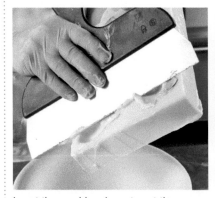

Invert the mould and empty out the excess chocolate, leaving a thin coating, then scrape any excess from the top

Invert the mould flat side down onto a level surface lined with silicone paper

Remove the half figures, quickly warm the edges on a flat warm tray and join them together

Method two: fully attached figure mould producing fully shaped figures

1 Polish the inside of the half-figure mould with cotton wool.

2 Embellish as above.

3 Join both half moulds together and clip to secure in place.

4 Fill the mould to the top with liquid tempered couverture, tap the side of the mould to release air bubbles.

5 Turn the mould upside down and tap out all the chocolate, leaving a thin coating of chocolate inside the mould.

6 Using a chocolate scraper, clean the base of the mould and stand on a flat surface lined with silicone paper until set. (Sometimes a second coat of chocolate may be required, depending on the size of the mould.)

7 Once the couverture has fully set, remove the clips and gently remove both halves of the mould, revealing a complete, hollow figure shape.

Removing a chocolate figure from a fully attached mould – the whole figure comes out, rather than a half figure

43 Chocolate cones

1. Make shaped cones from acetate sheets.
2. Fill with liquid tempered couverture.
3. Empty the chocolate out and stand the base of the cone on silicon paper.
4. When the chocolate has set, repeat the process once more.
5. When the chocolate has fully set, remove the acetate, leaving a chocolate cone.

44 Chocolate spraying

Spraying chocolate to decorate a frozen dessert giving a velvet finish

Plain couverture	200 g
Cocoa butter (melted red cocoa butter may be added to give a richer chocolate effect)	200 g

Melt both together, allow to cool to 35°C and, using a spray gun, spray onto the chocolate centrepiece.

Variations

There are two finishes that can be achieved on a centrepiece using chocolate spraying. One is a smooth matt finish and the other is a velvet finish.

To achieve a matt finish, simply spray onto the centrepiece at room temperature. To achieve a velvet finish, the centrepiece must be cold and refrigerated for half an hour prior to being sprayed.

Professional tips

- Always ensure that the spray gun is warm before putting the spraying chocolate mixture in, otherwise the cocoa butter will set inside the chamber of the gun as it is being sprayed.
- Never clean the spray gun with water, simply spray vegetable oil through the gun until the oil becomes clear.

45 Modelling paste using a food processor

1 Place either white, milk or plain couverture callets into a food processor.

2 Process at intervals, stopping the machine and scraping the couverture down into the base of the machine.

3 Continue this process until the couverture breaks down and becomes a paste.

4 At this stage, the couverture paste can be shaped and moulded into various shapes or figures.

5 Allow the finished shape to harden.

Process callets of couverture until they break down and eventually form a paste

The couverture will form a smooth paste that can be rolled and shaped

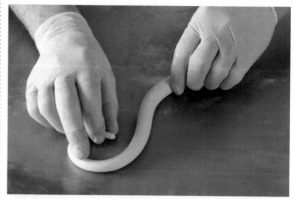

Mould the paste into the exact shape required, then leave it to harden

46 Pastillage

Makes >	500 g
Gelatine	2½ leaves
Icing sugar	450 g
Lemon juice	25 g

1 Separate the gelatine leaves and soak them in iced water.

2 Sieve the icing sugar onto paper twice.

3 Drain the gelatine and squeeze out the water. Add the gelatine to the lemon juice and warm to dissolve.

4 Place approximately half the icing sugar in a clean bowl, make a well in the centre and add the dissolved gelatine and lemon juice. Mix to a smooth paste.

5 Gradually add the rest of the icing sugar and work/knead to obtain a smooth firm paste.

6 Place the paste in a clean plastic bag and cover with a damp cloth.

7 Allow to rest for 20 minutes before using.

8 Roll out as thinly as possible on a smooth surface.

9 Lay on a dusted board. Cut out around a template, using a sharp, thin-bladed knife.

10 Leave the pieces to dry overnight, turning them over after the first 2–3 hours.

11 Assemble the pieces into a centrepiece using royal icing as the adhesive.

Professional tips

Keep the paste covered as much as possible during processing, to prevent it drying out.

Use cornflour tied in a muslin bag to dust the work surface.

All templates must be exactly the right size.

The centrepiece may be embellished by spraying it with liquid colours.

Add the dissolved gelatine to the sugar

Mixing

Kneading

Roll out thinly

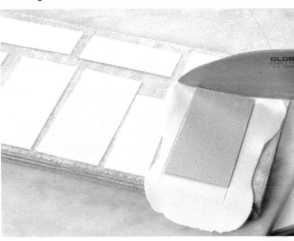
Cut out each piece around a template

47 Marzipan

	Makes >	400 g
Water		250 ml
Caster sugar		1 kg
Ground almonds		400 g
Egg yolks		3
Almond essence		2–3 drops

1 Place the water and sugar in a pan and boil. Skim as necessary.

2 When the sugar reaches 116°C, draw aside and mix in the ground almonds, then add the egg yolks and almond essence and mix in quickly to avoid scrambling.

3 Knead well until smooth.

Marzipan may be shaped into fruits, figures, etc.

Modelling with marzipan

Test yourself

1 Give four reasons why a sugar solution may crystallise.

2 Briefly describe three methods of tempering chocolate.

3 What is meant by the phrase 'sugar is hygroscopic'?

4 From the principles that have been covered in this chapter, design a chocolate centrepiece, drawing it to scale and incorporating five different chocolate techniques.

5 Briefly explain the term 'inversion', as applied to sugar solutions.

6 State four points to consider before preparing a pastillage centrepiece.

Appendix: Specialist ingredients and equipment

In this appendix, you will learn:
→ how to use specialist ingredients used in patisserie work and what effect they have within certain recipes
→ how to use different pieces of specialist equipment and how they are used to produce patisserie products.

Gels and gums

- **Agar agar** is a gelling agent derived from seaweed. It produces a firm and brittle gel that will retain its structure up to a temperature of about 75°C. It is insoluble in cold water and slowly soluble in hot water.
- **Carrageenan iota** produces soft, elastic, vegetarian gels. It is an ideal alternative to gelatine.
- **Carrageenan kappa** produces strong, rigid gels and tends to be clear when used with a sugar or syrup base.
- **Gelatine** is extracted from animals' bones and, more recently, fish skin. Sold in 2 g sheets, it is easy to control the amount precisely. The most common grades of leaf gelatine are silver and gold, each of which has a different Bloom strength. The Bloom strength is tested with a Bloom gelometer (named after Oscar T Bloom, who designed it in 1925). Gelatine is also available in powdered form.
- **Gellan gum F** is a vegetarian gelling agent produced by bacterial fermentation. Gellan gum produces highly heat-resistant gels, fluid gels and heat-stable sorbets that can be flamed at the table.
- **Guar gum** is derived from guar beans. It can best be described as a natural food thickener, and is often used as a stabiliser. Guar gum keeps ice cream smooth by preventing ice crystals from forming.
- **Gum Arabic** is a naturally occurring gum taken from the sap of the acacia tree. It is used primarily as a food emulsifier, stabiliser and thickener in many processed foods such as ice cream. It was traditionally used to glaze rout biscuits.
- **Locust bean gum** is derived from the locust or carob bean. It is used primarily as a thickener or stabiliser. Locust bean gum only forms a gel when used in conjunction with other gelling agents, such as agar agar or carrageenan. It is excellent for freeze/thaw formulations.
- **Pectin** occurs naturally in fruits such as crab apples, cranberries, gooseberries, redcurrants and in the skin of citrus fruits. Pectin is used primarily for setting jams and pâte de fruits.
- **Xanthan gum** is made by bacterial fermentation. It is widely used as a thickener and stabiliser.

Other specialist products

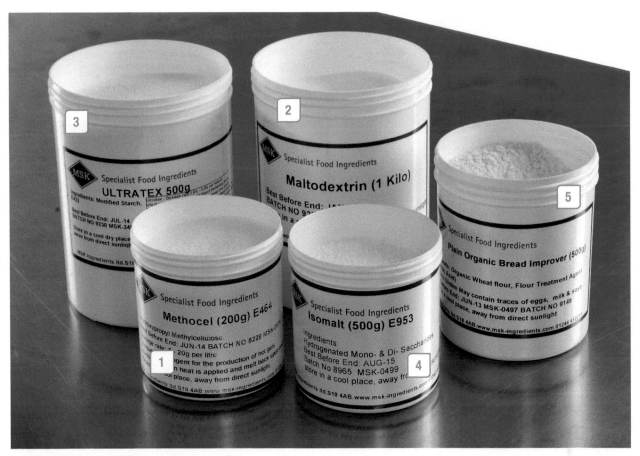

Isomalt and other specialist ingredients (see list)

1. **Methocel** is derived from cellulose fibre. Unlike other gelling agents, methocel produces gels which form when heated and will turn back to liquid when cooled.
2. **Maltodextrin** is derived from wheat starch. It is used for a wide variety of food preparations. Maltodextrin can be used as a dispersing aid for gums and gelling agents.
3. **Ultratex** is a corn starch that does not require heat to thicken liquids and sauces to the desired consistency. It adds a smooth, creamy and glossy texture to liquids, which can then be dried or frozen into a thin sheet for garnish. Ultratex has no after-taste and will preserve the taste and colour of the original liquid. Designed for easy dispersion, it will not lump like other starches.
4. **Isomalt** is a natural sweetener. It has sugar-like physical properties, and provides the taste and texture of sugar with only half the calories. Isomalt is perfect for sugar pulling and casting sweet decorations (flowers, leaves, ribbons, etc.). It has a higher resistance to humidity and stays flexible longer than regular sugar.
5. **Bread improver** is used to increase the speed of dough fermentation and reduces BFT (bulk fermentation time). Bread making with the addition of bread improver is known as activated dough development (ADD).

Spherification

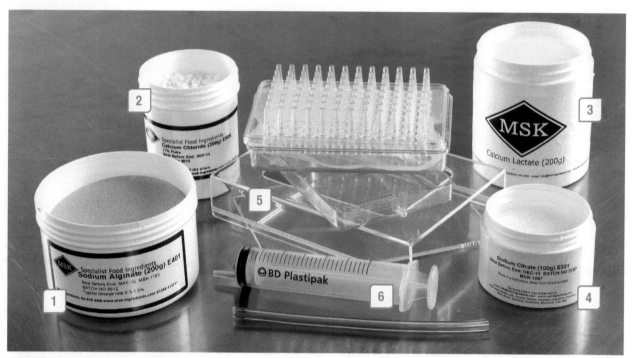

Ingredients used in spherification (see list)

1. **Sodium alginate** is extracted from brown seaweed. It is often used as a stabiliser, thickener or emulsifier. In the presence of calcium, it forms a gel. It is most commonly used with calcium lactate or calcium chloride in the spherification process.
2. **Calcium chloride** is a salt of calcium and chlorine. It is one of the primary ingredients in direct spherification. It is used in the setting bath to activate the sodium alginate.
3. **Calcium lactate** is a calcium-rich product and is perfect for reverse spherification (dipped in a sodium alginate bath) without adding any flavour at all to the end product.
4. **Sodium citrate** is the sodium salt of citric acid. It is mainly used as a food additive, usually for flavour or as a preservative. It controls acidity in the making of some spheres.
5. A **caviar box** is a specially designed tool for the creation of false 'caviar'. This handy little box allows you produce more than 96 drops per second. The box can be taken apart easily for cleaning.
6. A **syringe** is used to drop a base solution into a calcium bath to form 'caviar' pearls (see Recipe 1).

1 Rhubarb 'caviar' pearls (spherification)

Makes >	10 portions
Base solution	
Rhubarb poaching liquor	300 g
Sodium alginate	3 g
Caster sugar	20 g
Rhubarb flavour drops	2 drops
Calcium bath	
Water	500 ml
Calcium chloride	5 g

1 Whisk the poaching liquor on a machine at the lowest speed.

2 Mix the sodium alginate and the sugar together and add this mixture slowly to the liquor. Take care not to add the mixture to the centre of the bowl as it will adhere to the whisk.

3 Add the flavour drops to the mixture. Continue to whisk slowly for 10 minutes.

4 Allow the mixture to stand for a further 10 minutes. The mixture should be thickened to a syrupy texture and be completely free of lumps.

5 Place the water into a narrow container such as a beaker or measuring jug. Add the calcium chloride and stir until dissolved.

6 Draw the rhubarb solution into the syringe. Hold the syringe approximately 15 cm above the calcium chloride solution and then push the syringe plunger to create a steady stream of drops.

7 Lightly stir the water bath to help shape the cylindrical balls. Leave for 3 minutes to allow a skin to form on the balls.

8 Pour the balls into a shallow sieve and rinse under cold water. The caviar pearls are now ready to serve.

Professional tip

To maintain the beautiful shape and structure of the caviar pearls, store them in a little of the poaching liquor.

Try something different

Try serving lemon water ice topped with gin and tonic caviar pearls as a pre-dessert.

Inclusions

Inclusions (see list)

Inclusions are used as fillings or exterior decorations for deserts.

1. **Popping candy** or **crackle crystal** consists of carbonated sugar crystals which crackle and pop when placed in the mouth. Used as a topping or garnish to give a surprising but pleasant effect.
2. **Freeze-dried fruits** are fresh fruits which have been put through a freeze-drying process, removing all the excess moisture while retaining the colour and structure. These ingredients are used to add texture and flavour to desserts, cakes and chocolates.
3. **Cocoa nibs** (grue de cacao) are dry-roasted pieces of cocoa bean. They are used to make tuiles, added to ice creams and mousses, and sprinkled over desserts.
4. **Bres** (Brésilienne) is made from roasted and caramelised hazelnuts, crushed. It is used to flavour ice cream, in mousses or as a base to stand scooped ice cream on plates.
5. **Pailleté feuilletine** is crunchy biscuit that is used as a decoration. It is made into an inclusion for mousses or flans by mixing it with equal quantities of praline paste and chocolate and then rolling out to a thin biscuit.
6. **Ginger mini cubes** are small cubes of sugar-coated crystallised ginger. This is an example of an inclusion that could be added to ganache-based desserts, chocolate mousses or ice creams.
7. **Caramel fudge pieces** are small cubes of caramel fudge, used as inclusions in iced confections.
8. **Chocolate fudge brownie pieces** are used to flavour ice creams, mousses and the inclusion in gateaux.

Ice cream products

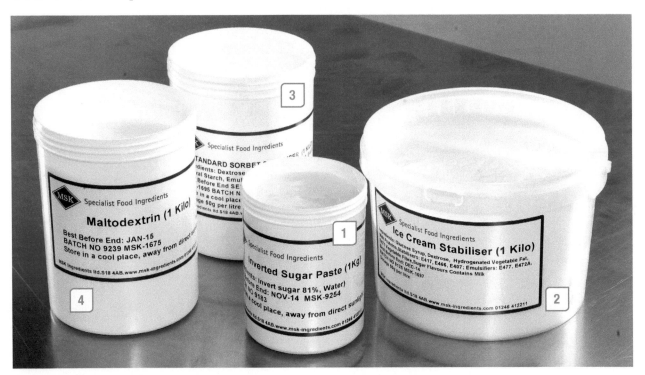

Ice cream products (see list)

1. **Inverted sugar paste** is a sweetener made of a mixture of glucose, fructose and acid. It is added to ice creams and ganaches to prevent the crystallisation of sugars. It will also enhance the texture and smoothness of fillings and ganaches.
2. **Ice cream stabiliser** is designed for the production of ice cream. The stabiliser is added to the ice cream mixture prior to churning, to produce a rich, creamy and stable result.
3. **Sorbet stabiliser** is designed for the production of fruit-based water ices. The stabiliser produces a soft scooping texture and an increased overrun which can be maintained over long periods.
4. **Maltodextrin** is derived from wheat starch. It enables savoury ice creams to be produced without the sweetness.

Chocolate products

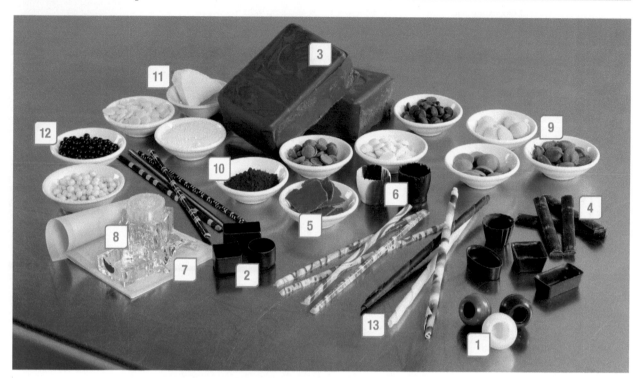

Chocolate products (see list)

1. **Praline shells** are a convenience product. They may be filled with ganaches, water ices, gianduja or caramel, and used as petits fours.
2. **Praline squares and rounds** are a convenience product used in the same way as praline shells.
3. **Chocolate couverture block** is a purchasing unit for white, milk or plain couverture.
4. **Croissant sticks** are heat-stable chocolate sticks manufactured for the production of pain au chocolat.
5. **Gianduja** or **gianduia** is a sweet chocolate containing about 30 per cent hazelnut paste. It was invented in Turin by Caffarel in 1852. It can be used to fill pralines or to flavour ice creams, creams and chocolate mousses.
6. **Chocolate cups** are used as petits fours or added to plated desserts. Possible fillings include griottines topped with ganache, mousses, liquid caramel, etc.
7. Pure **gold leaf** (24 carat) is used as a decoration, especially with chocolate desserts.
8. Pure **gold dust** (24 carat) can be sprinkled over dishes as a garnish.
9. **Chocolate couverture callets** are the purchasing units for couverture. They come in white, milk, plain, coloured and flavoured units.
10. **Bitter cocoa powder** is very dark in colour. It is used to coat truffles and dusted on desserts.
11. **Cocoa butter** is available in blocks, callets and dehydrated powder. It is used in ganache, for spraying, and to enrich sauces, fillings and mousses.
12. **Crisp pearls** are round biscuits coated in white, milk or plain chocolate, used as decorations and inserts.
13. **Chocolate cigarettes** are a commercial product used to decorate cold desserts and gâteaux.
14. **Metallic/shimmer powders** are brushed onto chocolate centrepieces and pralines to enhance their appearance (see Chapter 10, Recipe 37).
15. **Ready-mixed coloured cocoa butters** are intended for use with spray guns. They are used to decorate and colour chocolate moulds, or sprayed directly onto chocolate centrepieces. Coloured cocoa butter can also be used to create your own transfer sheets and painted into figure moulds.
16. **Freeze spray** is used to set chocolate quickly when assembling chocolate centrepieces.
17. **Commercial glazes** are chocolate based and give a very high gloss shine. Let the glaze down with 10 per cent stock syrup and warm in the microwave to 30°C. Apply over frozen mousses and they can be kept frozen without losing their shine.
18. **Metallic spray** is used to embellish chocolate centrepieces and sprayed into praline moulds prior to coating with chocolate couverture.
19. **Praline paste** is a combination of caramelised sugar and nuts, usually hazelnuts or almonds. It is used to flavour the centres of chocolates or in desserts and ice creams.
20. **Sorbitol** comes in liquid or powdered form and is used in ganaches to keep them smooth. It helps to prevent drying out due to its properties as a moisture stabiliser.
21. **Invert sugar** is sometimes added to ganaches to keep them moist and smooth.

Sugar products

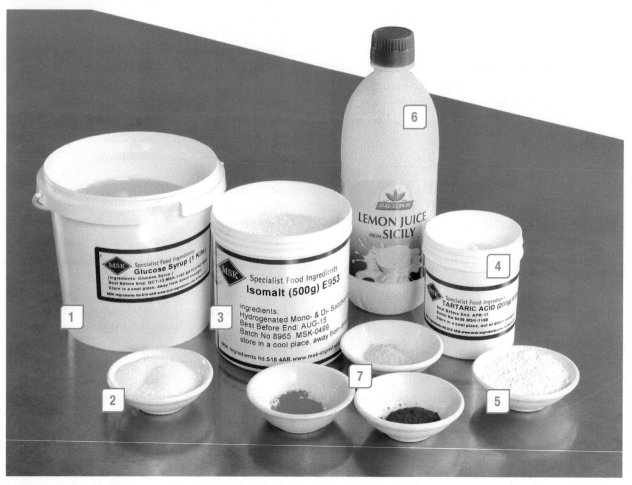

Sugar products (see list)

1. **Glucose** comes in liquid or powdered form. It is a mono-saccharide that helps to prevent premature crystallisation in boiled sugar solutions. It can also be used in ganaches to give smoothness.
2. **Granulated sugar** is the best sugar for boiling purposes. Always use sugar from a 1 kg bag to ensure purity.
3. **Isomalt** is a natural sweetener. It has sugar-like physical properties, and provides the taste and texture of sugar with only half the calories. Isomalt is perfect for sugar pulling and casting sweet decorations (flowers, leaves, ribbons, etc.).

It has a higher resistance to humidity and stays flexible longer than regular sugar.
4. **Tartaric acid** is used in sugar boiling to invert the sugar solution and make it more stable.
5. **Calcium carbonate** is powdered chalk. It is used in poured sugar work to make the sugar opaque.
6. **Lemon juice**, like tartaric acid, creates inversion in a boiled sugar solution.
7. **Powdered colours** specifically for sugar boiling are available. Mix with water and add to the boiled sugar solution.

Flavourings

Flavourings (see list)

1. **Essential oils** are pressed from plants or fruit and then purified so that they impart a very clean and sharp flavour to food. They are ideal for flavouring ganaches, chocolate and desserts.

2. **Flavour drops** are used to add intense flavour to ganaches, ice creams, desserts and fondant.

3. **Alcohol concentrates** are natural concentrates made in exactly the same way as the original spirit or liqueur. They have an enhanced flavour and higher alcohol strength, making them more economical to use.

4. **Fruit pastes** are highly concentrated to give a strong flavour. The normal dosage rate is 20–30 per cent of the product. Fruit pastes are used to flavour cream-based mousses, gâteaux, chocolate centres and ice creams.

5. **Vanilla pods** are dark brown, plump and moist pods with a sweet flavour and aroma. Vanilla powder can be used to add a visual effect to products. Vanilla is used to flavour crème anglaise, crème renversée, crème patissière, bavarois, etc. Vanilla may be stored in the deep freeze and reconstituted in the microwave.

6. **Griottines** are made using a type of morello cherry, the Oblachinska cherry, found only in the Balkans. The cherries are macerated twice in kirsch or Cointreau; their shape and flavour are preserved. Framboisines are a similar product made of raspberries, also macerated in liqueur.

Equipment for sugar work

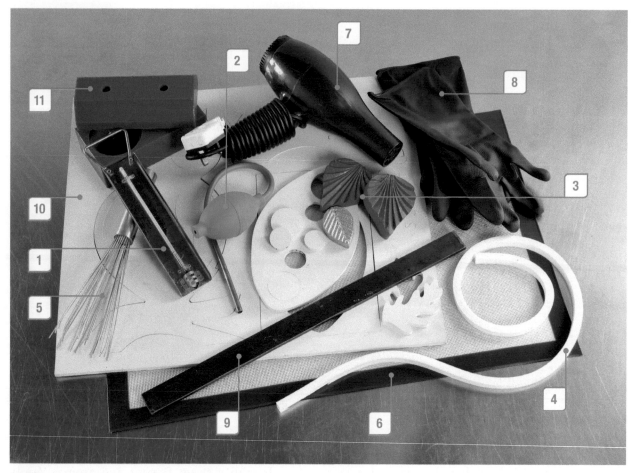

Small items of equipment for sugar work (see list)

The first photo shows some of the smaller items of equipment used in sugar work:

1. **Sugar thermometer:** used to measure the temperature of a boiled sugar solution. Calibrated in the centigrade scale.
2. **Sugar pump:** used for blowing pulled sugar into various shapes.
3. **Leaf mould:** made from silicone or metal and used to make pulled sugar leaves.
4. **Silicon band:** shaped into various designs and filled with boiled sugar solution to make parts for centrepieces.
5. **Sugar-spinning whisk:** formed by cutting off the end of a whisk. This is used for dipping into a boiled sugar solution and then spinning the sugar to form thin brittle strands.
6. **Non-stick mat:** poured sugar is cast onto the mat. Also used when pulling sugar.
7. **Hair drier:** used to cool down blown sugar.
8. **Rubber gloves:** used to protect the hands from high temperatures when pulling sugar. Also prevent moisture from the hands coming into contact with the sugar.
9. Thin **metal bars:** boiled sugar can be cast into these, to make supports for poured sugar centrepieces.
10. **Poured sugar mats:** cut-out rubber mats into which sugar is poured to form shapes and centrepieces.
11. **Sphere moulds:** made from silicone and used to make boiled sugar spheres.

Large items of equipment for sugar work (see list)

The second photo shows some of the larger items of equipment used in sugar work:

1. **Copper pan:** used for sugar boiling because copper is a good conductor of heat.
2. **Induction hob:** ideal for sugar boiling because it cooks quickly and with no naked flames.
3. **Electric sugar boiler:** heats large quantities of sugar rapidly and with no naked flames.
4. **Croquembouche mould:** designed for the assembly of a croquembouche (see Chapter 4, Recipe 38).
5. **Sugar-pulling cabinet:** used to keep boiled sugar warm and malleable while it is handled.
6. **Crème brulée burner:** a piece of electrical equipment designed to caramelise the sugar on the surface of a crème brulée.
7. **Blowtorch:** used to soften or melt boiled sugar preparations (pulled sugar, poured sugar, etc.) so that pieces can be joined together; available in various sizes.
8. **Dipping irons:** used to create decorations from boiled sugar solution taken to 160°C.

Equipment for chocolate work

Hotmix Pro

The Hotmix Pro is a machine that cooks and blends. It is ideal for making ganache, giving a good emulsification.

The photo below shows some of the smaller equipment used in chocolate work:

1. **Wheel cutters:** extending wheels used to cut even shapes through tempered couverture that has been spread onto acetate or transfer sheets.
2. **Figure moulds:** originally made from metal, now made from polycarbonate. Used for moulding figures in chocolate coverture.
3. **Praline moulds:** used to make filled pralines.
4. **Scraper:** specifically made for chocolate work. Uses include table-top tempering, making chocolate copeaux and levelling the surface on praline moulds.
5. **Marble slab:** used for table-top tempering of melted couverture.
6. **Magnetic moulds:** specially designed to hold transfer sheets on the base of praline and lollipop moulds.
7. **Dipping forks:** used to dip cut and layered pralines.
8. **Trigger funnel:** Used to dispense liquid fillings such as pâte des fruits.
9. **Spatulas:** used to mix ganaches and temper chocolate.
10. **Tempering spatula:** used in tempering couverture. This spatula has a temperature probe and gives a temperature reading when seeding the couverture.
11. **Textured sheets:** available in various designs. Used for casting tempered couverture into sheets and poles (see Chapter 10, Recipe 6).

Small equipment for chocolate work (see list)

12. **Dipping irons:** used to create decorations from tempered chocolate and boiled sugar. Frozen before use.
13. **Hoop rings:** tempered chocolate is cast into the rings to make chocolate bases for centerpieces.
14. **Combs:** used to make striped patterns, e.g. on chocolate cigarettes.

15. **Praline frame:** used to make layered and cut pralines.
16. **Silicone bands:** can be shaped into various designs and filled with tempered chocolate or boiled sugar. Used in the construction of centrepieces.
17. **Bubble wrap:** tempered chocolate may be cast onto bubble wrap, giving a textured finish.

Mid-sized equipment for chocolate work (see list)

The photo above shows some of the mid-sized equipment used in chocolate work:

1. **Food processor:** used in chocolate work to break down couverture callets into a paste suitable for modelling.
2. **Stem blender:** used in chocolate work to emulsify boiled cream and chocolate together when making a ganache.
3. **Hot air gun:** used to raise the temperature of tempered chocolate during processing.
4. **Airbrush:** used to spray coloured cocoa butter into praline moulds or to embellish a chocolate centrepiece.
5. **Compressor** and **spray gun:** used to spray a solution of equal parts of melted cocoa butter and couverture onto chocolate desserts and centrepieces.

Other specialist items used in chocolate work include:

- **Melting/holding tank:** an electrically heated tank used for melting chocolate couverture before seeding it with couverture callets
- **Tempering machine:** specially made for tempering and holding chocolate at the required temperature
- **Chocolate guitar:** used to cut pralines, pâte des fruits and set mousses
- **Vibrating table:** used to vibrate filled chocolate moulds, forcing any air bubbles in the chocolate to rise to the surface.

A holding tank

A vibrating table

A chocolate guitar

A wheel tempering tank

Equipment for ice cream production

The photo to the right shows some of the smaller equipment used in the production of ice cream and other frozen desserts:

1. **Refractometer:** measures the amount of sugar in a syrup. It is calibrated using the Brix scale, which gives a percentage reading. Used for water ices, sorbets, savarins and poaching syrups.
2. **Saccharometer:** measures the amount of sugar in a syrup. It is calibrated using the Baumé scale of the density of sugar solutions (Pesse syrup). Like the refractometer, it is used for water ices, sorbets, savarins and poaching syrups.
3. **Ice cream scoop:** available in various sizes. Used to dispense scoops of ice cream and sorbets.
4. **Metal rings:** used to set parfait glacé and mould ice cream.

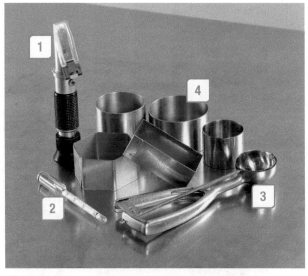

Small equipment for ice cream production (see list)

Other specialist items used in the production of ice cream include:

- **Bombe mould:** made from copper with a screw base. Used to make bombe glacé (see Chapter 7, Recipe 35).
- **Table-top ice cream machine:** used to make traditional ice cream and water ices.
- **Upright commercial ice cream machine:** designed to make traditional ice cream and water ices in large quantities.
- **Pacojet:** designed to make ice cream and water ices. The base for the ice cream or water ice is frozen first in special containers. The machine 'shaves' the surface of the frozen mixture, creating a smooth iced confection.

An upright ice cream machine

A Pacojet

Other specialist equipment

A selection of specialist equipment used in patisserie work (see list)

The photo above shows a selection of specialist equipment and items that have particular uses in patisserie work:

1. **Silicone moulds:** used to set desserts in moulded shapes and for baking.
2. **Rings:** used to make charlottes, mousses, parfaits and truffe au chocolat.
3. **Specialist moulds:** used to create shaped mousse-based desserts.
4. **Thermoformed PVC moulds:** used to mould buttercream products, bavarois and mousses. These moulds leave a design on the surface of the dessert.
5. **Tartlet and barquette moulds:** used to produce shaped pastry cases.
6. **Curved drainpipe:** used to shape tuiles and for chocolate work (see Chapter 10 Recipe 16).
7. **Dariole mould:** used for crème caramel and mousses.
8. **Stepped palette knife:** used to evenly spread out sponges such as roulades, jaconde sponge and dacquoise sponge.

9. **Pastry cutters:** used to cut out pastry or sponge, and in some chocolate work.
10. **Digital scales:** electronic scales used for weighing pastry products.
11. **Metal ruler:** used as a straight edge for cutting sponges and for measuring the portion size of cut-out desserts when using a mousse frame.
12. **Pastry brushes:** used to apply egg wash to pastry products; to clean the sides of the pan when boiling sugar; and to apply glazes to fruit tartlets.
13. **Wood grainer:** used to apply a wood grain effect when using tempered couverture or chocolate cigarette paste.
14. **Roller:** used to apply a thin layer of chocolate to the base of cut pralines or to roll tempered chocolate onto the flat surface of a centrepiece.
15. **Sprayer:** used to spray coloured cocoa butter, or chocolate spray made from equal parts of melted cocoa butter and couverture, when embellishing chocolate centrepieces.
16. **Nitrogen flask:** used to produce and apply e'spumas and foams.

Other items include:

- **Flan ring:** used to produce baked pastry cases for flans.
- **Conical strainer** and **passoir:** used to strain fruit coulis and sauces.
- **Mousse frames:** used in large-scale catering to set mousses, bavarois, buttercream products, gateau opera and other products.
- **Marzipan roller:** used to give a pattern on the surface of marzipan and pastillage.

Larger items used in patisserie work include:

- **Convection oven:** a multi-function oven that can be used for steaming, convection or a combination of both.
- **Microwave oven:** used for tempering chocolate, softening fondant and heating liquids for pastry dishes (e.g. heating fruit coulis before adding gelatine).
- **Sous vide water bath:** used for cooking fruit and making fruit syrups.
- **Pastry break:** used to roll out pastry evenly.
- **Pastry deck oven:** a sectioned deck oven designed for patisserie work. It allows the chef to control top and bottom heat within the baking chamber, which gives a more uniform baked product.
- **Dehydrator cabinet:** used to dry pastry products such as meringues and fruit slices.
- **Prover:** designed to prove bread prior to baking. It provides the right temperature for the yeast to activate and the correct humidity to prevent the surface of the bread from skinning.
- **Blast freezer:** used to quickly chill or freeze pastry products.
- **Spiral mixer:** specially designed for making fermented dough products.
- **Mixing machine:** used for numerous pastry functions such as mixing, whipping, creaming and beating. It has three attachments: whisk, dough hook and paddle beater.

A sous vide water bath

A pastry break

A deck oven

A dehydrator cabinet

A prover

Index of recipes

Subject index